CAMBRIDGE LIBRARY COLLECTION

Books of enduring scholarly value

Technology

The focus of this series is engineering, broadly construed. It covers technological innovation from a range of periods and cultures, but centres on the technological achievements of the industrial era in the West, particularly in the nineteenth century, as understood by their contemporaries. Infrastructure is one major focus, covering the building of railways and canals, bridges and tunnels, land drainage, the laying of submarine cables, and the construction of docks and lighthouses. Other key topics include developments in industrial and manufacturing fields such as mining technology, the production of iron and steel, the use of steam power, and chemical processes such as photography and textile dyes.

Samuel F.B. Morse

The American inventor Samuel Morse (1791–1872) spent decades fighting to be recognised for his key role in devising the electromagnetic telegraph. While he will always be remembered in the history of telecommunications, and for co-developing the code which bears his name, Morse started out as a painter and also involved himself in matters of politics over the course of his career. Published in 1914, this two-volume collection of personal papers was edited by his son, who provides helpful commentary throughout, illuminating the struggles and successes of a remarkable life. Volume 2 begins with Morse's return voyage to the United States; following a conversation with a fellow passenger regarding electromagnetism, Morse began to develop the concept of the single-wire telegraph. The rest of the volume gives much personal background to the development of the invention and particularly to Morse's efforts to gain the recognition he believed he deserved.

Cambridge University Press has long been a pioneer in the reissuing of out-of-print titles from its own backlist, producing digital reprints of books that are still sought after by scholars and students but could not be reprinted economically using traditional technology. The Cambridge Library Collection extends this activity to a wider range of books which are still of importance to researchers and professionals, either for the source material they contain, or as landmarks in the history of their academic discipline.

Drawing from the world-renowned collections in the Cambridge University Library and other partner libraries, and guided by the advice of experts in each subject area, Cambridge University Press is using state-of-the-art scanning machines in its own Printing House to capture the content of each book selected for inclusion. The files are processed to give a consistently clear, crisp image, and the books finished to the high quality standard for which the Press is recognised around the world. The latest print-on-demand technology ensures that the books will remain available indefinitely, and that orders for single or multiple copies can quickly be supplied.

The Cambridge Library Collection brings back to life books of enduring scholarly value (including out-of-copyright works originally issued by other publishers) across a wide range of disciplines in the humanities and social sciences and in science and technology.

Samuel F.B. Morse

His Letters and Journals

Volume 2

Samuel Finley Breese Morse
Edited by Edward Lind Morse

CAMBRIDGE
UNIVERSITY PRESS

University Printing House, Cambridge, CB2 8BS, United Kingdom

Cambridge University Press is part of the University of Cambridge.
It furthers the University's mission by disseminating knowledge in the pursuit of education, learning and research at the highest international levels of excellence.

www.cambridge.org
Information on this title: www.cambridge.org/9781108074391

This edition first published 1914
This digitally printed version 2014

ISBN 978-1-108-07439-1 Paperback

SAMUEL F. B. MORSE

HIS LETTERS AND JOURNALS

IN TWO VOLUMES

VOLUME II

Saml. F. B. Morse.

SAMUEL F. B. MORSE

HIS LETTERS AND JOURNALS

EDITED AND SUPPLEMENTED

BY HIS SON

EDWARD LIND MORSE

ILLUSTRATED
WITH REPRODUCTIONS OF HIS PAINTINGS
AND WITH NOTES AND DIAGRAMS
BEARING ON THE

INVENTION OF THE TELEGRAPH

VOLUME II

BOSTON AND NEW YORK
HOUGHTON MIFFLIN COMPANY
The Riverside Press Cambridge
1914

Published November 1914

"Th' invention all admir'd, and each how he
To be th' inventor miss'd, so easy it seem'd
Once found, which yet unfound most would have thought
Impossible."

MILTON.

CONTENTS

CHAPTER XXVIII

JUNE 20, 1840 — AUGUST 12, 1842

CHAPTER XXIX

JULY 16, 1842 — MARCH 26, 1843

CHAPTER XXX

MARCH 15, 1843 — JUNE 13, 1844

CHAPTER XXXI

JUNE 23, 1844 — OCTOBER 9, 1845

CHAPTER XXXII

DECEMBER 20, 1845 — APRIL 19, 1848

CHAPTER XXXIII

JANUARY 9, 1848 — DECEMBER 19, 1849

CHAPTER XXXIX

NOVEMBER 28, 1867 — JUNE 10, 1871

CHAPTER XL

JUNE 14, 1871 — APRIL 16, 1872

ILLUSTRATIONS

SAMUEL F. B. MORSE

HIS LETTERS AND JOURNALS

VOLUME II

SAMUEL F. B. MORSE
HIS LETTERS AND JOURNALS

CHAPTER XXI

OCTOBER 1, 1832 — FEBRUARY 28, 1833

Packet-ship Sully. — Dinner-table conversation. — Dr. Charles T. Jackson. — First conception of telegraph. — Sketch-book. — Idea of 1832 basic principle of telegraph of to-day. — Thoughts on priority. — Testimony of passengers and Captain Pell. — Difference between "discovery" and "invention." — Professor E. N. Horsford's paper. — Arrival in New York. — Testimony of his brothers. — First steps toward perfection of the invention. — Letters to Fenimore Cooper.

THE history of every great invention is a record of struggle, sometimes heartbreaking, on the part of the inventor to secure and maintain his rights. No sooner has the new step in progress proved itself to be an upward one than claimants arise on every side; some honestly believing themselves to have solved the problem first; others striving by dishonest means to appropriate to themselves the honor and the rewards, and these sometimes succeeding; and still others, indifferent to fame, thinking only of their own pecuniary gain and dishonorable in their methods. The electric telegraph was no exception to this rule; on the contrary, its history perhaps leads all the rest as a chronicle of "envy, hatred, malice, and all uncharitableness." On the other hand, it brings out in strong relief the opposing virtues of steadfastness, perseverance, integrity, and loyalty.

Many were the wordy battles waged in the scientific world over the questions of priority, exclusive discovery or invention, indebtedness to others, and conscious or unconscious plagiarism. Some of these questions are, in many minds, not yet settled. Acrimonious were the legal struggles fought over infringements and rights of way, and, in the first years of the building of the lines to all parts of this country, real warfare was waged by the workers of competing companies.

It is not my purpose to treat exhaustively of any of these battles, scientific, legal, or physical. All this has already been written down by abler pens than mine, and has now become history. My aim in following the career of Morse the Inventor is to shed a light (to some a new light) on his personality, self-revealed by his correspondence, tried first by hardships, poverty, and deep discouragement, and then by success, calumny, and fame. Like other men who have achieved greatness, he was made the target for all manner of abuse, accused of misappropriating the ideas of others, of lying, deceit, and treachery, and of unbounded conceit and vaingloriousness. But a careful study of his notes and correspondence, and the testimony of others, proves him to have been a pure-hearted Christian gentleman, earnestly desirous of giving to every one his just due, but jealous of his own good name and fame, and fighting valiantly, when needs must be, to maintain his rights; guilty sometimes of mistakes and errors of judgment; occasionally quick-tempered and testy under the stress of discouragement and the pressure of poverty, but frank to acknowledge his error and to make amends when convinced of his fault; and the calm verdict of posterity has awarded him the crown of greatness.

Morse was now forty-one years old; he had spent three delightful years in France and Italy; had matured his art by the intelligent study of the best of the old masters; had made new friends and cemented more strongly the ties that bound him to old ones; and he was returning to his dearly loved native land and to his family with high hopes of gaining for himself and his three motherless children at least a competence, and of continuing his efforts in behalf of the fine arts.

From Mr. Cooper's and Mr. Habersham's reminiscences we must conclude that, in the background of his mind, there existed a plan, unformed as yet, for utilizing electricity to convey intelligence. He was familiar with much that had been discovered with regard to that mysterious force, through his studies under Professors Day and Silliman at Yale, and through the lectures and conversation of Professors Dana and Renwick in New York, so that the charge which was brought against him that he knew absolutely nothing of the subject, can be dismissed as simply proving the ignorance of his critics.

Thus prepared, unconsciously to himself, to receive the inspiration which was to come to him like a flash of the subtle fluid which afterwards became his servant, he went on board the good ship Sully, Captain Pell commanding, on the 1st of October, 1832. Among the other passengers were the Honorable William C. Rives, of Virginia, our Minister to France, with his family; Mr. J. F. Fisher, of Philadelphia; Dr. Charles T. Jackson, of Boston, who was destined to play a malign rôle in the subsequent history of the telegraph, and others. The following letter was written to his friend Fenimore Cooper from Havre, on the 2d of October: —

"I have but a moment to write you one line, as in a few hours I shall be under way for dear America. I arrived from England by way of Southampton a day or two since, and have had every moment till now occupied in preparations for embarking. I received yours from Vevay yesterday and thank you for it. Yes, Mr. Rives and family, Mr. Fisher, Mr. Rogers, Mr. Palmer and family, and a full cabin beside accompany me. What shall I do with such an *antistatistical* set? I wish you were of the party to shut their mouths on some points. I shall have good opportunity to talk with Mr. Rives, whom I like notwithstanding. I think he has good American feeling in the main and means well, although I cannot account for his permitting you to suffer in the chambers (of the General). I will find out *that* if I can.

"My journey to England, change of scene and air, have restored me wonderfully. I knew they would. I like John's country; it is a garden beautifully in contrast with France, and John's people have excellent qualities, and he has many good people; but I hate his aristocratic system, and am more confirmed in my views than ever of its oppressive and unjust character. I saw a great deal of Leslie; he is the same good fellow that he always was. Be tender of him, my dear sir; I could mention some things which would soften your judgment of his political feelings. One thing only I can now say, — remember he has married an English wife, whom he loves, and who has never known America. He keeps entirely aloof from politics and is wholly absorbed in his art. Newton is married to a Miss Sullivan, daughter of General Sullivan, of Boston, an accomplished woman and a belle. He is expected in England soon.

"I found almost everybody out of town in London. I called and left a card at Rogers's, but he was in the country, so were most of the artists of my acquaintance. The fine engraver who has executed so many of Leslie's works, Danforth, is a stanch American; he would be a man after your heart; he admires you for that very quality. — I must close in great haste."

The transatlantic traveller did not depart on schedule time in 1832, as we find from another letter written to Mr. Cooper on October 5: —

"Here I am yet, wind-bound, with a tremendous southwester directly in our teeth. Yesterday the Formosa arrived and brought papers, etc., to the 10th September. I have been looking them over. Matters look serious at the South; they are mad there; great decision and prudence will be required to restore them to reason again, but they are so hot-headed, and are so far committed, I know not what will be the issue. Yet I think our institutions are equal to any crisis. . . .

"*October 6, 7 o'clock.* We are getting under way. Good-bye."

It is greatly to be regretted that Morse did not, on this voyage as on previous ones, keep a careful diary. Had he done so, many points relating to the first conception of his invention would, from the beginning, have been made much clearer. As it is, however, from his own accounts at a later date, and from the depositions of the captain of the ship and some of the passengers, the story can be told.

The voyage was, on the whole, I believe, a pleasant one and the company in the cabin congenial. One night at the dinner-table the conversation chanced upon the subject of electro-magnetism, and Dr. Jackson described some of the more recent discoveries of European scientists — the length of wire in the coil of a magnet, the fact that electricity passed instantaneously through any known length of wire, and that its presence could be observed at any part of the line by breaking the circuit. Morse was, naturally, much interested and it was then that the inspiration, which had lain dormant in his brain for many years, suddenly came to him, and he said: "If the presence of electricity can be made visible in any part of the circuit, I see no reason why intelligence may not be transmitted instantaneously by electricity."

The company was not startled by this remark; they soon turned to other subjects and thought no more of it. Little did they realize that this exclamation of Morse's was to mark an epoch in civilization; that it was the germ of one of the greatest inventions of any age, an invention which not only revolutionized the methods by which intelligence was conveyed from place to place, but paved the way for the subjugation, to the uses of man in many other ways, of that mysterious fluid, electricity, which up to this time had remained but a plaything of the laboratory. In short, it ushered in the Age of Electricity. Least of all, perhaps, did that Dr. Jackson, who afterwards claimed to have given Morse all his ideas, apprehend the tremendous importance of that chance remark. The fixed idea had, however, taken root in Morse's brain and obsessed him. He withdrew from the cabin and paced the deck, revolving in his mind the various means by which the object sought could be attained. Soon his ideas were so far focused that he sought to give them expression on paper, and he drew from his pocket one of the little sketch-books which he

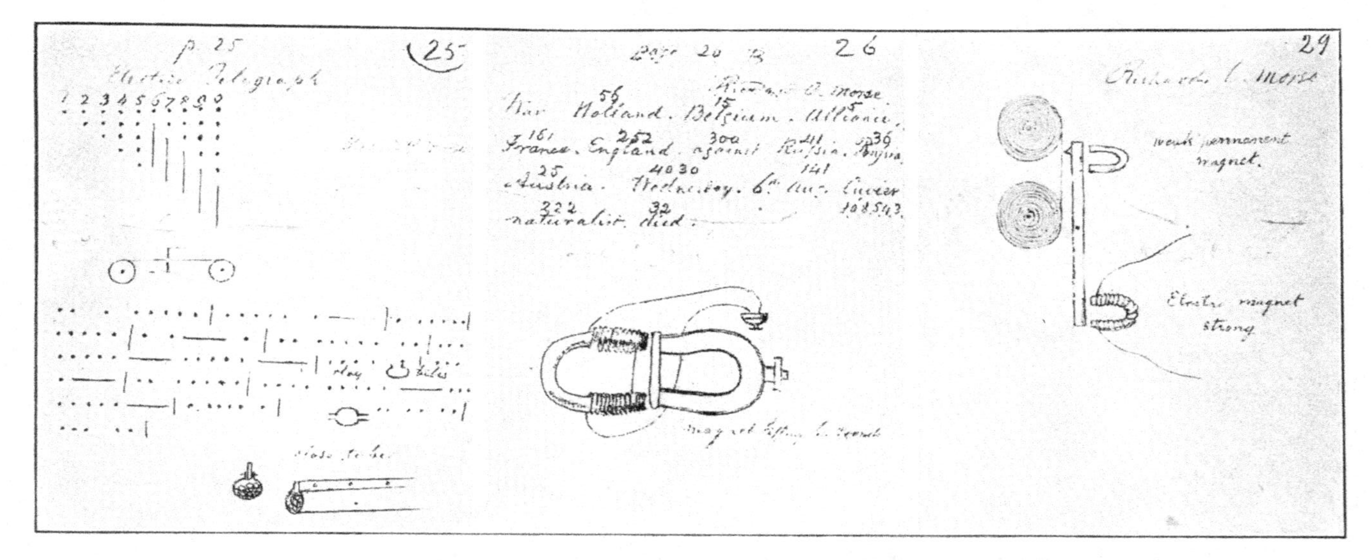

DRAWINGS FROM 1832 SKETCH-BOOK, SHOWING FIRST CONCEPTION OF TELEGRAPH

always carried with him, and rapidly jotted down in sketches and words the ideas as they rushed from his brain. This original sketch-book was burned in a mysterious fire which, some years later, during one of the many telegraph suits, destroyed many valuable papers. Fortunately, however, a certified copy had wisely been made, and this certified copy is now in the National Museum in Washington, and the reproduction here given of some of its pages will show that Morse's first conception of a Recording Electric Magnetic Telegraph is practically the telegraph in universal use today.

His first thought was evidently of some system of signs which could be used to transmit intelligence, and he at once realized that nothing could be simpler than a point or a dot, a line or dash, and a space, and a combination of the three. Thus the first sketch shows the embryo of the dot-and-dash alphabet, applied only to numbers at first, but afterwards elaborated by Morse to represent all the letters of the alphabet.

Next he suggests a method by which these signs may be recorded permanently, evidently by chemical decomposition on a strip of paper passed along over two rollers. He then shows a message which could be sent by this means, interspersed with ideas for insulating the wires in tubes or pipes. And here I want to call attention to a point which has never, to my knowledge, been noticed before. In the message, which, in pursuance of his first idea, adhered to by him for several years, was to be sent by means of numbers, every word is numbered conventionally except the proper name "Cuvier," and for this he put a number for each letter. How this was to be indicated was not made clear, but it is evident that he saw at once that all proper names could not be numbered; that some other means must be employed to indicate them; in other words that each letter of the alphabet must have its own sign. Whether at that early period he had actually devised any form of alphabet does not appear, although some of the depositions of his fellow passengers would indicate that he had. He himself put its invention at a date a few years after this, and it has been bitterly contested that he did not invent it at all. I shall prove, in the proper place, that he did, but I think it is proved that it must have been thought of even at the early date of 1832, and, at all events, the dot-and-dash as the basis of a conventional code were original with Morse and were quite different from any other form of code devised by others.

The next drawing of a magnet lifting sixty pounds shows that Morse was familiar with the discoveries of Arago, Davy, and Sturgeon in electro-magnetism, but what application of them was to be made is not explained.

The last sketch is to me the most important of all, for it embodies the principle of the receiving magnet which is universally used at the present day. The weak permanent magnet has been replaced by a spring, but the electro-magnet still attracts the lever and produces the dots and dashes of the alphabet; and this, simple as it seems to us "once found," was original with Morse, was absolutely different from any other form of telegraph devised by others, and, improved and elaborated by him through years of struggle, is now recognized throughout the world as the Telegraph.

It was not yet in a shape to prove to a skeptical

world its practical utility; much had still to be done to bring it to perfection; new discoveries had still to be made by Morse and by others which were essential to its success; the skill, the means, and the faith of others had to be enlisted in its behalf, but the actual invention was there and Morse was the inventor.

How simple it all seems to us now, and yet its very simplicity is its sublimest feature, for it was this which compelled the admiration of scientists and practical men of affairs alike, and which gradually forced into desuetude all other systems of telegraphy until to-day the Morse telegraph still stands unrivalled.

That many other minds had been occupied with the same problem was a fact unknown to the inventor at the time, although a few years later he was rudely awakened. A fugitive note, written many years later, in his handwriting, although speaking of himself in the third person, bears witness to this. It is entitled "Good thought":—

"A circumstance which tends to confuse, in fairly ascertaining priority of invention, is that a subsequent state of knowledge is confounded in the general mind with the state of knowledge when the invention is first announced as successful. This is certainly very unfair. When Morse announced his invention, what was the general state of knowledge in regard to the telegraph? It should be borne in mind that a knowledge of the futile attempts at electric telegraphs previous to his successful one has been brought out from the lumber garret of science by the research of eighteen years. Nothing was known of such telegraphs to many scientific men of the highest attainments in the centres of civilization.

Professor Morse says himself (and certainly he has not given in any single instance a statement which has been falsified) that, at the time he devised his system, he supposed himself to be the first person that ever put the words 'electric telegraph' together. He supposed himself at the time the originator of the phrase as well as the thing. But, aside from his positive assertion, the truth of this statement is not only possible but very probable. The comparatively few (very few as compared with the mass who now are learned in the facts) who were in the habit of reading the scientific journals may have read of the thought of an electric telegraph about the year 1832, and even of Ronald's, and Betancourt's, and Salva's, and Lomond's inpracticable schemes previously, and have forgotten them again, with thousands of other dreams, as the ingenious ideas of visionary men; ideas so visionary as to be considered palpably impracticable, declared to be so, indeed, by Barlow, a scientific man of high standing and character; yet the mass of the scientific as well as the general public were ignorant even of the attempts that had been made. The fact of any of them having been published in some magazine at the time, whose circulation may be two or three thousand, and which was soon virtually lost amid the shelves of immense libraries, does not militate against the assertion that the world was ignorant of the fact. We can show conclusively the existence of this ignorance respecting telegraphs at the time of the invention of Morse's telegraph."

The rest of this note (evidently written for publication) is missing, but enough remains to prove the point.

Thus we have seen that the idea of his telegraph came

to Morse as a sudden inspiration and that he was quite ignorant of the fact that others had thought of using electricity to convey intelligence to a distance. Mr. Prime in his biography says: "Of all the great inventions that have made their authors immortal and conferred enduring benefit upon mankind, no one was so completely grasped at its inception as this."

One of his fellow passengers, J. Francis Fisher, Esq., counsellor-at-law of Philadelphia, gave the following testimony at Morse's request:—

"In the fall of the year 1832 I returned from Europe as a passenger with Mr. Morse in the ship Sully, Captain Pell master. During the voyage the subject of an electric telegraph was one of frequent conversation. Mr. Morse was most constant in pursuing it, and *alone* the one who seemed disposed to reduce it to a practical test, and I recollect that, for this purpose, he devised *a system of signs for letters* to be indicated and marked by a quick succession of strokes or shocks of the galvanic current, and I am sure of the fact that it was deemed by Mr. Morse perfectly competent to effect the result stated. I did not suppose that any other person on board the ship claimed any merit in the invention, or was, in fact, interested to pursue it to maturity as Mr. Morse then seemed to be, nor have I been able since that time to recall any fact or circumstance to justify the claim of any person other than Mr. Morse to the invention."

This clear statement of Mr. Fisher's was cheerfully given in answer to a request for his recollections of the circumstances, in order to combat the claim of Dr. Charles T. Jackson that he had given Morse all the ideas of the telegraph, and that he should be considered at least its joint inventor. This was the first of the many claims which the inventor was forced to meet. It resulted in a lawsuit which settled conclusively that Morse was the sole inventor, and that Jackson was the victim of a mania which impelled him to claim the discoveries and achievements of others as his own. I shall have occasion to refer to this matter again.

It is to be noted that Mr. Fisher refers to "signs for letters." Whether Morse actually had devised or spoken of a conventional alphabet at that time cannot be proved conclusively, but that it must have been in his mind the "Cuvier" referred to before indicates.

Others of his fellow-passengers gave testimony to the same effect, and Captain Pell stated under oath that, when he saw the completed instrument in 1837, he recognized it as embodying the principles which Morse had explained to him on the Sully; and he added: "Before the vessel was in port, Mr. Morse addressed me in these words: 'Well, Captain, should you hear of the telegraph one of these days as the wonder of the world, remember the discovery was made on board the good ship Sully.'"

Morse always clung tenaciously to the date of 1832 as that of his invention, and, I claim, with perfect justice. While it required much thought and elaboration to bring it to perfection; while he used the published discoveries of others in order to make it operate over long distances; while others labored with him in order to produce a practical working apparatus, and to force its recognition on a skeptical world, the basic idea on which everything else depended was his; it was original with him, and he pursued it to a successful issue, himself making certain new and essential discoveries and inventions. While, as I have said, he made use of the discoveries of others, these men in turn were dependent on the earlier investigations of scientists who preceded them, and so the chain lengthens out.

There will always be a difference of opinion as to the comparative value of a new discovery and a new invention, and the difference between these terms should be clearly apprehended. While they are to a certain extent interchangeable, the word "discovery" in science is usually applied to the first enunciation of some property of nature till then unrecognized; "invention," on the other hand, is the application of this property to the uses of mankind. Sometimes discovery and invention are combined in the same individual, but often the discoverer is satisfied with the fame arising from having called attention to something new, and leaves to others the practical application of his discovery. Scientists will always claim that a new discovery, which marks an advance in knowledge in their chosen field, is of paramount importance; while the world at large is more grateful to the man who, by combining the discoveries of others and adding the culminating link, confers a tangible blessing upon humanity.

Morse was completely possessed by this new idea. He worked over it that day and far into the night. His vivid imagination leaped into the future, brushing aside all obstacles, and he realized that here in his hands was an instrument capable of working inconceivable good. He recalled the days and weeks of anxiety when he was hungry for news of his loved ones; he foresaw that in affairs of state and of commerce rapid communication might mean the avoidance of war or the saving of a fortune; that, in affairs nearer to the heart of the people, it might bring a husband to the bedside of a dying wife, or save the life of a beloved child; apprehend the fleeing criminal, or commute the sentence of an innocent man. His great ambition had always been to work some good for his fellow-men, and here was a means of bestowing upon them an inestimable boon.

After several days of intense application he disclosed his plan to Mr. Rives and to others. Objections were raised, but he was ready with a solution. While the idea appeared to his fellow-passengers as chimerical, yet, as we have seen, his earnestness made so deep an impression that when, several years afterwards, he exhibited to some of them a completed model, they, like Captain Pell, instantly recognized it as embodying the principles explained to them on the ship.

Without going deeply into the scientific history of the successive steps which led up to the invention of the telegraph, I shall quote a few sentences from a long paper written by the late Professor E. N. Horsford, of Cambridge, Massachusetts, and included in Mr. Prime's biography: —

"What was needed to the *original conception* of the Morse recording telegraph?

"1. A knowledge that soft wire, bent in the form of a horseshoe, could be magnetized by sending a galvanic current through a coil wound round the iron, and that it would lose its magnetism when the current was suspended.

"2. A knowledge that such a magnet had been made to lift and drop masses of iron of considerable weight.

"3. A knowledge, or a belief, that the galvanic current could be transmitted through wires of great length.

"These were all. Now comes the conception of devices for employing an agent which could produce reciprocal motion to effect registration, and the invention of an alphabet. In order to this invention it must be seen how up and down — reciprocal — motion could be produced by the opening and closing of the circuit. Into this simple band of vertical tracery of paths in space must be thrown the shuttle of time and a ribbon of paper. It must be seen how a lever-pen, alternately dropping upon and rising at defined intervals from a fillet of paper moved by independent clock-work, would produce the fabric of the alphabet and writing and printing.

"Was there anything required to produce these results which was not known to Morse? . . .

"He knew, for he had witnessed it years before, that, by means of a battery and an electro-magnet, reciprocal motion could be produced. He knew that the force which produced it could be transmitted along a wire. He *believed* that the battery current could be made, through an electro-magnet, to produce physical results at a *distance*. He saw in his mind's eye the existence of an agent and a medium by which reciprocal motion could be not only produced but controlled at a distance. The question that addressed itself to him at the outset was, naturally, this: 'How can I make use of the simple up-and-down motion of opening and closing a circuit to write an intelligible message at one end of a wire, and at the same time print it at the other?' . . . Like many a kindred work of genius it was in nothing more wonderful than in its simplicity. . . . Not one of the brilliant scientific men who have attached their names to the history of electro-magnetism had brought the means to produce the practical registering telegraph. Some of them had ascended the tower that looked out on the field of conquest. Some of them brought keener vision than others. Some of them stood higher than others. But the genius of invention had not recognized them. There was needed an inventor. Now what sort of a want is this?

"There was required a rare combination of qualities and conditions. There must be ingenuity in the adaptation of available means to desired ends; there must be the genius to see through non-essentials to the fundamental principle on which success depends; there must be a kind of skill in manipulation; great patience and pertinacity; a certain measure of culture, and the inventor of a recording telegraph must be capable of being inspired by the grandeur of the thought of writing, figuratively speaking, with a pen a thousand miles long — with the thought of a postal system without the element of time. Moreover the person who is to be the inventor must be free from the exactions of well-compensated, everyday, absorbing duties — perhaps he must have had the final baptism of poverty.

"Now the inventor of the registering telegraph did not rise from the perusal of any brilliant paper; he happened to be at leisure on shipboard, ready to contribute and share in the after-dinner conversation of a ship's cabin, when the occasion arose. Morse's electro-magnetic telegraph was mainly an invention employing powers and agencies through mechanical devices to produce a given end. It involved the combination of the results of the labors of others with a succession of special contrivances and some discoveries of the inventor himself. There was an ideal whole almost at the outset, but involving great thought, and labor, and patience, and invention to produce an art harmonious in its organization and action."

After a voyage of over a month Morse reached home and landed at the foot of Rector Street on November 15, 1832. His two brothers, Sidney and Richard, met him on his arrival, and were told at once of his invention. His brother Richard thus described their meeting: —

"Hardly had the usual greetings passed between us three brothers, and while on our way to my house, before he informed us that he had made, during his voyage, an important invention, which had occupied almost all his attention on shipboard — one that would astonish the world and of the success of which he was perfectly sanguine; that this invention was a means of communicating intelligence by electricity, so that a message could be written down in a permanent manner by characters at a distance from the writer. He took from his pocket and showed from his sketch-book, in which he had drawn them, the kind of characters he proposed to use. These characters were dots and spaces representing the ten digits or numerals, and in the book were sketched other parts of his electro-magnetic machinery and apparatus, actually drawn out in his sketch-book."

The other brother, Sidney, also bore testimony: —

"He was full of the subject of the telegraph during the walk from the ship, and for some days afterwards could scarcely speak about anything else. He expressed himself anxious to make apparatus and try experiments for which he had no materials or facilities on shipboard. In the course of a few days after his arrival he made a kind of cogged or saw-toothed type, the object of which I understood was to regulate the interruptions of the electric current, so as to enable him to make dots, and regulate the length of marks or spaces on the paper upon which the information transmitted by his telegraph was to be recorded.

"He proposed at that time a single circuit of wire, and only a single circuit, and letters, words, and phrases were to be indicated by numerals, and these numerals were to be indicated by dots and other marks and spaces on paper. It seemed to me that, as wire was cheap, it would be better to have twenty-four wires, each wire representing a letter of the alphabet, but my brother always insisted upon the superior advantages of his single circuit."

Thus we see that Morse, from the very beginning, and from intuition, or inspiration, or whatever you please, was insistent on one of the points which differentiated his invention from all others in the same field, namely, its simplicity, and it was this feature which eventually won for it a universal adoption. But, simple as it was, it still required much elaboration in order to bring it to perfection, for as yet it was but an idea roughly sketched on paper; the appliances to put this idea to a practical test had yet to be devised and made, and Morse now entered upon the most trying period of his career. His three years in Europe, while they had been enjoyed to the full and had enabled him to perfect himself in his art, had not yielded him large financial returns; he had not expected that they would, but based his hopes on increased patronage after his return. He was entirely dependent on his brush for the support of himself and his three motherless children, and now this new inspiration had come as a disturbing element. He was on the horns of a dilemma. If he devoted himself to his art, as he must in order to keep the wolf from the door, he would not have the leisure to perfect his invention, and others might grasp the prize before him. If he allowed thoughts of electric currents, and magnets, and batteries to monopolize his attention, he could not give to his art, notoriously a jealous mistress, that worship which alone leads to success.

An added bar to the rapid development of his invention was the total lack (hard to realize at the present day) of the simplest essentials. There were no manufacturers of electrical appliances; everything, even to the winding of the wires around the magnets, had to be done laboriously by hand. Even had they existed Morse had but scant means with which to purchase them.

This was his situation when he returned from Europe in the fall of 1832, and it is small wonder that twelve years elapsed before he could prove to the world that his revolutionizing invention was a success, and the wonder is great that he succeeded at all, that he did not sink under the manifold discouragements and hardships, and let fame and fortune elude him. Unknown to him many men in different lands were working over the same problem, some of them of assured scientific position and with good financial backing; is it then remarkable that Morse in later years held himself to be but an instrument in the hands of God to carry out His will? He never ceased to marvel at the amazing fact that he, poor, scoffed at or pitied, surrounded by difficulties of every sort, should have been chosen to wrest the palm from the hands of trained scientists of two continents. To us the wonder is not so great, for we, if we have read his character aright as revealed by his correspondence, can see that in him, more than in any other man of his time, were combined the qualities necessary to a great inventor as specified by Professor Horsford earlier in this chapter.

In following Morse's career at this critical period it will be necessary to record his experiences both as painter and inventor, for there was no thought of abandoning his profession in his mind at first; on the contrary, he still had hopes of ultimate success, and it was his sole means of livelihood. It is true that he at times gave way to fits of depression. In a letter to his brother Richard before leaving Europe he had thus given expression to his fears: —

"I have frequently felt melancholy in thinking of my prospects for encouragement when I return, and your letter found me in one of those moments. You cannot, therefore, conceive with what feelings I read your offer of a room in your new house. Give me a resting-place and I will yet move the country in favor of the arts. I return with some hopes but many fears. Will my country employ me on works which may do it honor? I want a commission from Government to execute two pictures from the life of Columbus, and I want eight thousand dollars for each, and on these two I will stake my reputation as an artist."

It was in his brother Richard's house that he took the

first step towards the construction of the apparatus which was to put his invention to a practical test. This was the manufacture of the saw-toothed type by which he proposed to open and close the circuit and produce his conventional signs. He did not choose the most appropriate place for this operation, for his sister-in-law rather pathetically remarked: "He melted the lead which he used over the fire in the grate of my front parlor, and, in his operation of casting the type, he spilled some of the heated metal upon the drugget, or loose carpeting, before the fireplace, and upon a flag-bottomed chair upon which his mould was placed."

He was also handicapped by illness just after his return, as we learn from the following letter to his friend Fenimore Cooper. In this letter he also makes some interesting comments on New York and American affairs, but, curiously enough, he says nothing of his invention:

"*February 21, 1833.* Don't scold at me. I don't deserve a scolding if you knew all, and I do if you don't know all, for I have not written to you since I landed in November. What with severe illness for several weeks after my arrival, and the accumulation of cares consequent on so long an absence from home, I have been overwhelmed and distracted by calls upon my time for a thousand things that pressed upon me for immediate attention; and so I have put off and put off what I have been longing (I am ashamed to say for weeks if not months) to do, I mean to write to you.

"The truth is, my dear sir, I have so much to say that I know not where to commence. I throw myself on your indulgence, and, believing you will forgive me, I commence without further apology.

"First, as to things at home. New York is *improved*, as the word goes, wonderfully. You will return to a strange city; you will not recognize many of your acquaintances among the old buildings; brand-new buildings, stores, and houses are taking the place of the good, staid, modest houses of the early settlers. *Improvement* is all the rage, and houses and churchyards must be overthrown and upturned whenever the Corporation plough is set to work for the widening of a narrow, or the making of a new, street.

"I believe you sometimes have a fit of the blues. It is singular if you do not with your temperament. I confess to many fits of this disagreeable disorder, and I know nothing so likely to induce one as the finding, after an absence of some years from home, the great hour-hand of life sensibly advanced on all your former friends. What will be your sensations after six or seven years if mine are acute after three years' absence?

"I have not been much in society as yet. I have many visitations, but, until I clear off the accumulated rubbish of three years which lies upon my table, I must decline seeing much of my friends. I have seen twice your sisters the Misses Delancy, and was prevented from being at their house last Friday evening by the severest snow-storm we have had this season. Our friends the Jays I have met several times, and have had much conversation with them about you and your delightful family. Mr. P. A. Jay is a member of the club, so I see him every Friday evening. Chancellor Kent also is a member, and both warm friends of yours. . . .

"My time for ten or twelve days past has been occupied in answering a pamphlet of Colonel Trumbull, who came out for the purpose of justifying his opposition to measures which had been devised for uniting the two Academies. I send you the first copy hot from the press. There is a great deal to dishearten in the state of feeling, or rather state of no feeling, on the arts in this city. The only way I can keep up my spirits is by resolutely resisting all disposition to repine, and by fighting perseveringly against all the obstacles that hinder the progress of art.

"I have been told several times since my return that I was born one hundred years too soon for the arts in our country. I have replied that, if that be the case, I will try and make it but fifty. I am more and more persuaded that I have quite as much to do with the pen for the arts as the pencil, and if I can in my day so enlighten the public mind as to make the way easier for those that come after me, I don't know that I shall not have served the cause of the fine arts as effectively as by painting pictures which might be appreciated one hundred years after I am gone. If I am to be the Pioneer and am fitted for it, why should I not glory as much in felling trees and clearing away the rubbish as in showing the decorations suited to a more advanced state of cultivation? . . .

"You will certainly have the blues when you first arrive, but the longer you stay abroad the more severe will be the disease. Excuse my predictions. . . . The Georgia affair is settled after a fashion; not so the nullifiers; they are infatuated. Disagreeable as it will be, they will be put down with disgrace to them."

In another letter to Mr. Cooper, dated February 28, 1833, he writes in the same vein: —

"The South Carolina business is probably settled by this time by Mr. Clay's compromise bill, so that the legitimates of Europe may stop blowing their two-penny trumpets in triumph at our *disunion*. The same clashing of interests in Europe would have caused twenty years of war and torrents of bloodshed; with us it has caused three or four years of wordy war and some hundreds of gallons of ink; but no necks are broken, nor heads; all will be in *statu ante bello* in a few days. . . .

"My dear sir, you are wanted at home. I want you to encourage me by your presence. I find the pioneer business has less of romance in the reality than in the description, and I find some tough stumps to pry up and heavy stones to roll out of the way, and I get exhausted and desponding, and I should like a little of your sinew to come to my aid at such times, as it was wont to come at the Louvre. . . .

"There is nothing new in New York; everybody is driving after money, as usual, and there is an alarm of fire every half-hour, as usual, and the pigs have the freedom of the city, as usual; so that, in these respects at least, you will find New York as you left it, except that they are not the same people that are driving after money, nor the same houses burnt, nor the same pigs at large in the street. . . . You will all be welcomed home, but come prepared to find many, very many things in taste and manners different from your own good taste and manners. Good taste and good manners would not be conspicuous if all around possessed the same manners."

CHAPTER XXII

1833 — 1836

Still painting. — Thoughts on art. — Picture of the Louvre. — Rejection as painter of one of the pictures in the Capitol. — John Quincy Adams. — James Fenimore Cooper's article. — Death blow to his artistic ambition. — Washington Allston's letter. — Commission by fellow artists. — Definite abandonment of art. — Repayment of money advanced. — Death of Lafayette. — Religious controversies. — Appointed Professor in University of City of New York. — Description of first telegraphic instrument. — Successful experiments. — Relay. — Address in 1853.

It was impossible for the inventor during the next few years to devote himself entirely to the construction of a machine to test his theories, impatient though he must have been to put his ideas into practical form. His two brothers came nobly to his assistance, and did what lay in their power and according to their means to help him; but it was always repugnant to him to be under pecuniary obligations to any one, and, while gratefully accepting his brothers' help, he strained every nerve to earn the money to pay them back. We, therefore, find little or no reference in the letters of those years to his invention, and it was not until the year 1835 that he was able to make any appreciable progress towards the perfection of his telegraphic apparatus. The intervening years were spent in efforts to rouse an interest in the fine arts in this country; in hard work in behalf of the still young Academy of Design; and in trying to earn a living by the practice of his profession.

"During this time," he says, "I never lost faith in the practicability of the invention, nor abandoned the intention of testing it as soon as I could command the means." But in order to command the means, he was obliged to devote himself to his art, and in this he did not meet with the encouragement which he had expected and which he deserved. His ideals were always high, perhaps too high for the materialistic age in which he found himself. The following fugitive note will illustrate the trend of his thoughts, and is not inapplicable to conditions at the present day: —

"Are not the refining influences of the fine arts needed, doubly needed, in our country? Is there not a tendency in the democracy of our country to low and vulgar pleasures and pursuits? Does not the contact of those more cultivated in mind and elevated in purpose with those who are less so, and to whom the former look for political favor and power, necessarily debase that cultivated mind and that elevation of purpose? When those are exalted to office who best can flatter the low appetites of the vulgar; when boorishness and ill manners are preferred to polish and refinement, and when, indeed, the latter, if not avowedly, are in reality made an objection, is there not danger that those who would otherwise encourage refinement will fear to show their favorable inclination lest those to whom they look for favor shall be displeased; and will not habit fix it, and another generation bear it as its own inherent, native character?"

That he was naturally optimistic is shown by a footnote which he added to this thought, dated October, 1833: —

"These were once my fears. There is doubtless danger, but I believe in the possibility, by the diffusion of the highest moral and intellectual cultivation through every class, of raising the lower classes in refinement."

But while in his leisure moments he could indulge in such hopeful dreams, his chief care at that time, as stated at the beginning of this chapter, was to earn money by the exercise of his profession. His important painting of the Louvre, from which he had hoped so much, was placed on exhibition, and, while it received high praise from the artists, its exhibition barely paid expenses, and it was finally sold to Mr. George Clarke, of Hyde Hall, on Otsego Lake, for thirteen hundred dollars, although the artist had expected to get at least twenty-five hundred dollars for it. In a letter to Mr. Clarke, of June 30, 1834, he says: —

"The picture of the Louvre was intended originally for an exhibition picture, and I painted it in the expectation of disposing of it to some person for that purpose who could amply remunerate himself from the receipts of a well-managed exhibition. The time occupied upon this picture was fourteen months, and at much expense and inconvenience, so that that sum [$2500] for it, if sold under such circumstances, would not be more than a fair compensation.

"I was aware that but few, if any, gentlemen in our country would be willing to expend so large a sum on a single picture, although in fact they would, in this case, purchase seven-and-thirty in one.

"I have lately changed my plans in relation to this picture and to my art generally, and consequently I am able to dispose of it at a much less price. I have need of funds to prosecute my new plans, and, if this picture could now realize the sum of twelve hundred dollars it would at this moment be to me equivalent in value to the sum first set upon it."

The change of plans no doubt referred to his desire to pursue his electrical experiments, and for this ready money was most necessary, and so he gladly, and even gratefully, accepted Mr. Clarke's offer of twelve hundred dollars for the painting and one hundred dollars for the frame. Even this was not cash, but was in the form of a note payable in a year! His enthusiasm for his art seems at this period to have been gradually waning, although he still strove to command success; but it needed a decisive stroke to wean him entirely from his first love, and Fate did not long delay the blow.

His great ambition had always been to paint historical pictures which should commemorate the glorious events in the history of his beloved country. In the early part of the year 1834 his great opportunity had, apparently, come, and he was ready and eager to grasp it. There were four huge panels in the rotunda of the Capitol at Washington, which were still to be filled by historical paintings, and a committee in Congress was appointed to select the artists to execute them.

Morse, president of the National Academy of Design, and enthusiastically supported by the best artists in the country, had every reason to suppose that he would be chosen to execute at least one of these paintings. Confident that he had but to make his wishes known to secure the commission, he addressed the following circular letter to various members of Congress, among whom were such famous men as Daniel Webster, John C. Calhoun, Henry Clay, and John Quincy Adams, all personally known to him: —

March 7, 1834.

My dear Sir, — I perceive that the Library Committee have before them the consideration of a resolution on the expediency of employing four artists to paint the remaining four pictures in the Rotunda of the Capitol. If Congress should pass a resolution in favor of the measure, I should esteem it a great honor to be selected as one of the artists.

I have devoted twenty years of my life, of which seven were passed in England, France, and Italy, studying with special reference to the execution of works of the kind proposed, and I must refer to my professional life and character in proof of my ability to do honor to the commission and to the country.

May I take the liberty to ask for myself your favorable recommendation to those in Congress who have the disposal of the commissions?

With great respect, Sir,

Your most obedient servant,

S. F. B. Morse.

While this letter was written in 1834, the final decision of the committee was not made until 1837, but I shall anticipate a little and give the result which had such a momentous effect on Morse's career. There was every reason to believe that his request would be granted, and he and his friends, many of whom endorsed by letter his candidacy, had no fear as to the result; but here again Fate intervened and ordered differently.

Among the committee men in Congress to whom this matter was referred was John Quincy Adams, ex-President of the United States. In discussing the subject, Mr. Adams submitted a resolution opening the competition to foreign artists as well as to American, giving it as his opinion that there were no artists in this country of sufficient talent properly to execute such monumental works. The artists and their friends were, naturally, greatly incensed at this slur cast upon them, and an indignant and remarkably able reply appeared anonymously in the New York "Evening Post." The authorship of this article was at once saddled on Morse, who was known to wield a facile and fearless pen. Mr. Adams took great offense, and, as a result, Morse's name was rejected and his great opportunity passed him by. There can be no reasonable doubt that, had he received this commission, he would have deferred the perfecting of his telegraphic device until others had so far distanced him in the race that he could never have overtaken them.

Instead of his having been the author of the "Evening Post" article, it transpired that he had not even heard of Mr. Adams's resolution until his friend Fenimore Cooper, the real author of the answer, told him of both attack and reply.

This was the second great tragedy of Morse's life; the first was the untimely death of his young wife, and this other marked the death of his hopes and ambitions as an artist. He was stunned. The blow was as unexpected as it was overwhelming, and what added to its bitterness was that it had been innocently dealt by the hand of one of his dearest friends, who had sought to render him a favor. The truth came out too late to influence the decision of the committee; the die was cast, and his whole future was changed in the twinkling of an eye; for what had been to him a joy and an inspiration, he now turned from in despair. He could not, of course, realize at the time that Fate, in dealing him this cruel blow, was dedicating him to a higher destiny. It is doubtful if he ever fully realized this, for in after years he could never speak of it unmoved. In a letter to this same friend, Fenimore Cooper, written on November 20, 1849, he thus laments:—

"Alas! My dear sir, the very name of *pictures* produces a sadness of heart I cannot describe. Painting has been a smiling mistress to many, but she has been a cruel jilt to me. I did not abandon her, she abandoned me. I have taken scarcely any interest in painting for many years. Will you believe it? When last in Paris, in 1845, I did not go into the Louvre, nor did I visit a single picture gallery.

"I sometimes indulge a vague dream that I may paint again. It is rather the memory of past pleasures, when hope was enticing me onward only to deceive me at last. Except some family portraits, valuable to me from their likenesses only, I could wish that every picture I ever painted was destroyed. I have no wish to be remembered as a painter, for I never was a painter. My ideal of that profession was, perhaps, too exalted — I may say is too exalted. I leave it to others more worthy to fill the niches of art."

Of course his self-condemnation was too severe, for we have seen that present-day critics assign him an honorable place in the annals of art, and while, at the time of writing that letter, he had definitely abandoned the brush, he continued to paint for some years after his rejection by the committee of Congress. He had to, for it was his only means of earning a livelihood, but the old enthusiasm was gone never to return. Fortunately for himself and for the world, however, he transferred it to the perfecting of his invention, and devoted all the time he could steal from the daily routine of his duties to that end.

His friends sympathized with him most heartily and were indignant at his rejection. Washington Allston wrote to him: —

I have learned the disposition of the pictures. I had hoped to find your name among the commissioned artists, but I was grieved to find that all my efforts in your behalf have proved fruitless. I know what your disappointment must have been at this result, and most sincerely do I sympathize with you. That my efforts were both sincere and conscientious I hope will be some consolation to you.

But let not this disappointment cast you down, my friend. You have it still in your power to let the world know what you can do. Dismiss it, then, from your mind, and determine to paint all the better for it. God bless you. Your affectionate friend

WASHINGTON ALLSTON.

The following sentences from a letter written on March 14, 1837, by Thomas Cole, one of the most celebrated of the early American painters, will show in what estimation Morse was held by his brother artists: —

"I have learned with mortification and disappointment that your name was not among the *chosen*, and I have feared that you would carry into effect your resolution of abandoning the art and resigning the presidency of our Academy. I sincerely hope you will have reason to cast aside that resolution. To you our Academy owes its existence and present prosperity, and if, in after times, it should become a great institution, your name will always be coupled with its greatness. But, if you leave us, I very much fear that the fabric will crumble to pieces. You are the keystone of the arch; if you remain with us time may furnish the Academy with another block for the place. I hope my fears may be vain, and that circumstances will conspire to induce you to remain our president."

Other friends were equally sympathetic and Morse did retain the presidency of the Academy until 1845.

To emphasize further their regard for him, a number of artists, headed by Thomas S. Cummings, unknown to Morse, raised by subscription three thousand dollars, to be given to him for the painting of some historical subject. General Cummings, in his "Annals of the Academy," thus describes the receipt of the news by the discouraged artist: —

"The effect was electrical; it roused him from his depression and he exclaimed that never had he read or known of such an act of professional generosity, and that he was fully determined to paint the picture — his favorite subject, 'The Signing of the First Compact on board the Mayflower,' — not of small size, as requested, but of the size of the panels in the Rotunda. That was immediately assented to by the committee, thinking it possible that one or the other of the pictures so ordered might fail in execution, in which case it would afford favorable inducements to its substitution, and, of course, much to Mr. Morse's profit; as the artists from the first never contemplated taking possession of the picture so executed. It was to remain with Mr. Morse, and for his use and benefit."

The enthusiasm thus roused was but a flash in the pan, however; the wound he had received was too deep to be thus healed. Some of the money was raised and paid to him, and he made studies and sketches for the painting, but his mind was now on his invention, and the painting of the picture was deferred from year to year and finally abandoned. It was characteristic of him that, when he did finally decide to give up the execution of this work, he paid back the sums which had been advanced to him, with interest.

Another grief which came to him in the summer of 1834 (to return to that year) was the death of his illustrious friend General Lafayette. The last letter received from him was written by his amanuensis and unsigned, and simply said: —

"General Lafayette, being detained by sickness, has sent to the reporter of the committee the following note, which the said reporter has read to the House."

The note referred to is, unfortunately, missing. This letter was written on April 29 and the General died on May 20. Morse sent a letter of sympathy to the son, George Washington Lafayette, a member of the Chamber of Deputies, in which the following sentiments occur: —

"In common with this whole country, now clad in mourning, with the lovers of true liberty and of exalted philanthropy throughout the world, I bemoan the departure from earth of your immortal parent. Yet I may be permitted to indulge in additional feelings of more private sorrow at the loss of one who honored me with his friendship, and had not ceased, till within a few days of his death, to send to me occasional marks of his affectionate remembrance. Be assured, my dear Sir, that the memory of your father will be especially endeared to me and mine."

Morse's admiration of Lafayette was most sincere, and he was greatly influenced in his political feelings by his intercourse with that famous man. Among other opinions which he shared with Lafayette and other thoughtful men, was the fear of a Roman Catholic plot to gain control of the Government of the United States. He defended his views fearlessly and vigorously in the public press and by means of pamphlets, and later entered into a heated controversy with Bishop Spaulding of Kentucky.

I shall not attempt to treat exhaustively of these controversies, but think it only right to refer to them from time to time, not only that the clearest possible light may be shed upon Morse's character and convictions, but to show the extraordinary activity of his brain, which, while he was struggling against obstacles of all kinds, not only to make his invention a success, but for the very means of existence, could yet busy itself with the championing of what he conceived to be the right.

To illustrate his point of view I shall quote a few extracts from a letter to R. S. Willington, Esq., who was the editor of a journal which is referred to as the "Courier." This letter was written on May 20, 1835, when Morse's mind, we should think, would have been wholly absorbed in the details of the infant telegraph: —

"With regard to the more important matter of the Conspiracy, I perceive with regret that the evidence which has been convincing to so many minds of the first order, and which continues daily to spread conviction of the truth of the charge I have made, is still viewed by the editors of the 'Courier' as inconclusive. My situation in regard to those who dissent from me is somewhat singular. I have brought against the absolute Governments of Europe a charge of conspiracy against the liberties of the United States. I support the charge by facts, and by reasonings from those facts, which produce conviction on most of those who examine the matter. ... But those that dissent simply say, 'I don't think there is a conspiracy'; yet give no reasons for dissent. The Catholic journals very artfully make no defense themselves, but adroitly make use of the Protestant defense kindly prepared for them....

"No Catholic journal has attempted any refutation of the charge. It cannot be refuted, for it is true. And be assured, my dear sir, it is no extravagant prediction when I say that the question of Popery and Protestantism, or Absolutism and Republicanism, which in these two opposite categories are convertible terms, is fast becoming and will shortly be the *great absorbing question*, not only of this country but of the whole civilized world. I speak not at random; I speak from long and diligent observation in Europe, and from comparison of the state of affairs in this country with the state of public opinion in Europe.

"We are asleep, sir, when every freeman should be awake and look to his arms. ... Surely, if the danger is groundless, there can be no harm in endeavoring to ascertain its groundlessness. If you were told your house was on fire you would hardly think of calling the man a maniac for informing you of it, even if he should use a tone of voice and gestures somewhat earnest and impassioned. The course of some of our journals on the subject of Popery has led to the belief that they are covertly under the control of the Jesuits. And let me say, sir, that the modes of control in the resources of this insidious society, notorious for its political arts and intrigues, are more numerous, more powerful, and more various than an unsuspicious people are at all conscious of. . . .

"Mr. Y. falls into the common error and deprecates what he calls a *religious* controversy, as if the subject of Popery was altogether religious. History, it appears to me, must have been read to very little purpose by any one who can entertain such an error in regard to the cunningest political despotism that ever cursed mankind. I must refer you to the preface of the second edition, which I send you, for my reasonings on that point. If they are not conclusive, I should be glad to be shown wherein they are defective. If they are conclusive, is it not time for every patriot to open his eyes to the truth of the fact that we are politically attacked under guise of a religious system, and is it not a serious question whether our political press should advocate the cause of foreign enemies to our government, or help to expose and repel them?"

It was in the year 1835 that Morse was appointed Professor of the Literature of the Arts of Design in the University of the City of New York, and here again we can mark the guiding hand of Fate. A few years earlier he had been tentatively offered the position of instructor of drawing at the United States Military Academy at West Point, but this offer he had promptly but courteously declined. Had he accepted it he would have missed the opportunity of meeting certain men who gave him valuable assistance. As an instructor in the University he not only received a small salary which relieved him, in a measure, from the grinding necessity of painting pot-boilers, but he had assigned to him spacious rooms in the building on Washington Square, which he could utilize not only as studio and living apartments, but as a workshop. For these rooms, however, he paid a rent, at first of $325 a year, afterwards of $400.

Three years had elapsed since his first conception of the invention, and, although burning to devote himself to its perfecting, he had been compelled to hold himself in check and to devote all his time to painting. Now, however, an opportunity came to him, for he moved into the University building before it was entirely finished, and the stairways were in such an embryonic state that he could not expect sitters to attempt their perilous ascent. This enforced leisure gave him the chance he had long desired and he threw himself heart and soul into his electrical experiments. Writing of this period in later years he thus records his struggles: —

"There I immediately commenced, with very limited means, to experiment upon my invention. My first instrument was made up of an old picture or canvas frame fastened to a table; the wheels of an old wooden clock moved by a weight to carry the paper forward;

FIRST TELEGRAPH INSTRUMENT, 1837

Now in the National Museum, Washington

three wooden drums, upon one of which the paper was wound and passed over the other two; a wooden pendulum, suspended to the top piece of the picture or stretching-frame, and vibrating across the paper as it passes over the centre wooden drum; a pencil at the lower end of the pendulum in contact with the paper; an electro-magnet fastened to a shelf across the picture or stretching frame, opposite to an armature made fast to the pendulum; a type rule and type, for breaking the circuit, resting on an endless band composed of carpet-binding, which passed over two wooden rollers, moved by a wooden crank, and carried forward by points projecting from the bottom of the rule downward into the carpet-binding; a lever, with a small weight on the upper side, and a tooth projecting downward at one end, operated on by the type, and a metallic fork, also projecting downward, over two mercury cups; and a short circuit of wire embracing the helices of the electro-magnet connected with the positive and negative poles of the battery and terminating in the mercury cups."

This first rude instrument was carefully preserved by the inventor, and is now in the Morse case in the National Museum at Washington. A reproduction of it is here given.

I shall omit certain technical details in the inventor's account of this first instrument, but I wish to call attention to his ingenuity in adapting the means at his disposal to the end desired. Much capital has been made, by those who opposed his claims, out of the fact that this primitive apparatus could only produce a V-shaped mark, thus — —and not a dot and a dash, which they insist was of later introduction and by another hand. But a reference to the sketches made on board the Sully will show that the original system of signs consisted of dots and lines, and that the first conception of the means to produce these signs was by an up-and-down motion of a lever controlled by an electro-magnet. It is easy to befog an issue by misstating facts, but the facts are here to speak for themselves, and that Morse temporarily abandoned his first idea, because he had not the means at his disposal to embody it in workable form and had recourse to another method for producing practically the same result, only shows wonderful ingenuity on his part. It can easily be seen that the waving line traced by the first instrument — thus, — can be translated by reading the lower part into a ·— i ·· u ··— of the final Morse alphabet.

The beginnings of every great invention have been clumsy and uncouth compared with the results attained by years of study and elaboration participated in by many clever brains. Contrast the Clermont of Fulton with the floating palaces of the present day, the Rocket of Stephenson with the powerful locomotives of our mile-a-minute fliers, and the hand-press of Gutenberg with the marvellous and intricate Hoe presses of modern times. And yet the names of those who first conceived and wrought these primitive contrivances stand highest in the roll of fame; and with justice, for it is infinitely easier to improve on the suggestion of another than to originate a practical advance in human endeavor.

Returning again to Morse's own account of his early experiments I shall quote the following sentences: —

"With this apparatus, rude as it was, and completed before the first of the year 1836, I was enabled to and did mark down telegraphic, intelligible signs, and to make and did make distinguishable sounds for telegraphing; and, having arrived at that point, I exhibited it to some of my friends early in that year, and among others to Professor Leonard D. Gale, who was a college professor in the University. I also experimented with the chemical power of the electric current in 1836, and succeeded in marking my telegraphic signs upon paper dipped in turmeric and solution of the sulphate of soda (as well as other salts) by passing the current through it. I was soon satisfied, however, that the electro-*magnetic* power was more available for telegraphic purposes and possessed many advantages over any other, and I turned my thoughts in that direction.

"Early in 1836 I procured forty feet of wire, and, putting it in the circuit, I found that my battery of one cup was not sufficient to work my instrument. This result suggested to me the probability that the magnetism to be obtained from the electric current would diminish in proportion as the circuit was lengthened, so as to be insufficient for any practical purposes at great distances; and, to remove that probable obstacle to my success, I conceived the idea of combining two or more circuits together in the manner described in my first patent, each with an independent battery, making use of the magnetism of the current on the first to close and break the second; the second the third; and so on."

Thus modestly does he refer to what was, in fact, a wonderful discovery, the more wonderful because of its simplicity. Professor Horsford thus comments on it: —

"In 1835 Morse made the discovery of the *relay*, the most brilliant of all the achievements to which his name must be forever attached. It was a discovery of a means by which the current, which through distance from its source had become feeble, could be reënforced or renewed. This discovery, according to the different objects for which it is employed, is variously known as the registering magnet, the local circuit, the marginal circuit, the repeater, etc."

Professor Horsford places the date of this discovery in the year 1835, but Morse himself, in the statement quoted above, assigned it to the early part of 1836.

It is only fair to note that the discovery of the principle of the relay was made independently by other scientists, notably by Davy, Wheatstone, and Henry, but Morse apparently antedated them by a year or two, and could not possibly have been indebted to any of them for the idea. This point has given rise to much discussion among scientists which it will not be necessary to enter into here, for all authorities agree in according to Morse independent invention of the relay.

"Up to the autumn of 1837," again to quote Morse's own words, "my telegraphic apparatus existed in so rude a form that I felt a reluctance to have it seen. My means were very limited — so limited as to preclude the possibility of constructing an apparatus of such mechanical finish as to warrant my success in venturing upon its public exhibition. I had no wish to expose to ridicule the representative of so many hours of laborious thought.

"Prior to the summer of 1837, at which time Mr. Alfred Vail's attention became attracted to my telegraph, I depended upon my pencil for subsistence. Indeed, so straitened were my circumstances that, in order to save time to carry out my invention and to economize my scanty means, I had for months lodged and eaten in my studio, procuring my food in small quantities from some grocery, and preparing it myself. To conceal from my friends the stinted manner in which I lived, I was in the habit of bringing my food to my room in the evenings, and this was my mode of life for many years."

Nearly twenty years later, in 1853, Morse referred to this trying period in his career at a meeting of the Association of the Alumni of the University: —

"Yesternight, on once more entering your chapel, I saw the same marble staircase and marble floors I once so often trod, and so often with a heart and head overburdened with almost crushing anxieties. Separated from the chapel by but a thin partition was that room I occupied, now your Philomathean Hall, whose walls — had thoughts and mental struggles, with the alternations of joys and sorrows, the power of being daguerreotyped upon them — would show a thickly studded gallery of evidence that there the Briarean infant was born who has stretched forth his arms with the intent to encircle the world. Yes, that room of the University was the birthplace of the Recording Telegraph. Attempts, indeed, have been made to assign to it other parentage, and to its birthplace other localities. Personally I have very little anxiety on this point, except that the truth should not suffer; for I have a consciousness, which neither sophistry nor ignorance can shake, that that room is the place of its birth, and a confidence, too, that its cradle is in hands that will sustain its rightful claim."

The old building of the University of the City of New York on Washington Square has been torn down to be replaced by a mercantile structure; the University has moved to more spacious quarters in the upper part of the great city; but one of its notable buildings is the Hall of Fame, and among the first names to be immortalized in bronze in the stately colonnade was that of Samuel F. B. Morse.

CHAPTER XXIII

1835 — 1837

First exhibitions of the Telegraph. — Testimony of Robert G. Rankin and Rev. Henry B. Tappan. — Cooke and Wheatstone. — Joseph Henry, Leonard D. Gale, and Alfred Vail. — Professor Gale's testimony. — Professor Henry's discoveries. — Regrettable controversy of later years. — Professor Charles T. Jackson's claims. — Alfred Vail. — Contract of September 23, 1837. — Work at Morristown, New Jersey. — The "Morse Alphabet." — Reading by sound. — First and second forms of alphabet.

In after years the question of the time when the telegraph was first exhibited to others was a disputed one; it will, therefore, be well to give the testimony of a few men of undoubted integrity who personally witnessed the first experiments.

Robert G. Rankin, Esq., gave his reminiscences to Mr. Prime, from which I shall select the following passages: —

"Professor Morse was one of the purest and noblest men of any age. I believe I was among the earliest, outside of his family circle, to whom he communicated his design to encircle the globe with wire. . . .

"Some time in the fall of 1835 I was passing along the easterly walk of Washington Parade-Ground, leading from Waverly Place to Fourth Street, when I heard my name called. On turning round I saw, over the picket-fence, an outstretched arm from a person standing in the middle or main entrance door of the unfinished University building of New York, and immediately recognized the professor, who beckoned me toward him. On meeting and exchanging salutations, — and you know how genial his were, — he took me by the arm and said:

'I wish you to go up in my sanctum and examine a piece of mechanism, which, if you may not believe in, *you*, at least, will not laugh at, as I fear some others will. I want you to give me your frank opinion as a friend, for I know your interest in and love of the applied sciences.'"

Here follow a description of what he saw and Morse's explanation, and, then he continues: —

"A long silence on the part of each ensued, which was at length broken by my exclamation: 'Well, professor, you have a pretty play! — theoretically true but practically useful only as a mantel ornament, or for a mistress in the parlor to direct the maid in the cellar! But, professor, *cui bono?* In imagination one can make a new earth and improve all the land communications of our old one, but my unfortunate practicality stands in the way of my comprehension as yet.'

"We then had a long conversation on the subject of magnetism and its modifications, and if I do not recollect the very words which clothed his thoughts, they were substantially as follows.

"He had been long impressed with the belief that God had created the great forces of nature, not only as manifestations of his own infinite power, but as expressions of good-will to man, to do him good, and that every one of God's great forces could yet be utilized for man's welfare; that modern science was constantly evolving from the hitherto hidden secrets of nature some new development promotive of human welfare; and that, at no distant day, magnetism would do more for the advancement of human sociology than any of the material forces yet known; that he would scarcely dare to compare spiritual with material forces, yet that, analogically, magnetism would do in the advancement of human welfare what the Spirit of God would do in the moral renovation of man's nature; that it would educate and enlarge the forces of the world. . . . He said he had felt as if he was doing a great work for God's glory as well as for man's welfare; that such had been his long cherished thought. His whole soul and heart appeared filled with a glow of love and good-will, and his sensitive and impassioned nature seemed almost to transform him in my eyes into a prophet."

It required, indeed, the inspirational vision of a prophet to foresee, in those narrow, skeptical days, the tremendous part which electricity was to play in the civilization of a future age, and I wish again to lay stress on the fact that it was the telegraph which first harnessed this mysterious force, and opened the eyes of the world to the availability of a power which had lain dormant through all the ages, but which was now, for the first time, to be brought under the control of man, and which was destined to rival, and eventually to displace, in many ways, its elder brother steam. Was not Morse's ambition to confer a lasting good on his fellowmen more fully realized than even he himself at that time comprehended?

The Reverend Henry B. Tappan, who in 1835 was a colleague of Morse's in the New York University and afterwards President of the University of Michigan, gave his testimony in reply to a request from Morse, and, among other things, he said: —

"In 1835 you had advanced so far that you were prepared to give, on a small scale, a practical demonstration of the possibility of transmitting and recording words through distance by means of an electro-magnetic arrangement. I was one of the limited circle whom you invited to witness the first experiments. In a long room of the University you had wires extended from end to end, where the magnetic apparatus was arranged.

"It is not necessary for me to describe particulars which have now become familiar to every one. The fact which I recall with the liveliest interest, and which I mentioned in conversation at Mr. Bancroft's as one of the choicest recollections of my life, was that of the first transmission and recording of a telegraphic dispatch.

"I suppose, of course, that you had already made these experiments before the company arrived whom you had invited. But I claim to have witnessed *the first transmission and recording of words* by lightning ever made public. . . .The arrangement which you exhibited on the above mentioned occasion, as well as the mode of receiving the dispatches, were substantially the same as those you now employ. I feel certain that you had then already grasped the whole invention, however you may have since perfected the details."

Others bore testimony in similar words, so that we may regard it as proved that, both in 1835 and 1836, demonstrations were made which, uncouth though they were, compared to present-day perfection, proved that the electric telegraph was about to emerge from the realms of fruitless experiment. Among these witnesses were Daniel Huntington, Hon. Hamilton Fish, and Commodore Shubrick; and several of these gentlemen asserted that, at that early period, Morse confidently predicted that Europe and America would eventually be united by an electric wire.

The letters written by Morse during these critical years have become hopelessly dispersed, and but few have come into my possession. His brothers were both in New York, so that there was no necessity of writing to them, and the letters written to others cannot, at this late day, be traced. As he also, unfortunately, did not keep a journal, I must depend on the testimony of others, and on his own recollections in later years for a chronicle of his struggles. The pencil copy of a letter written to a friend in Albany, on August 27, 1837, has, however, survived, and the following sentences will, I think, be found interesting: —

"Thanks to you, my dear C——, for the concern you express in regard to my health. It has been perfectly good and is now, with the exception of a little anxiety in relation to the telegraph and to my great pictorial undertaking, which wears the furrows of my face a little deeper. My Telegraph, in all its essential points, is tested to my own satisfaction and that of the scientific gentlemen who have seen it; but the machinery (all which, from its peculiar character, I have been compelled to make myself) is imperfect, and before it can be perfected I have reason to fear that other nations will take the hint and rob me both of the credit and the profit. There are indications of this in the foreign journals lately received. I have a defender in the 'Journal of Commerce' (which I send you that you may know what is the progress of the matter), and doubtless other journals of our country will not allow foreign nations to take the credit of an invention of such vast importance as they assign to it, when they learn that it certainly belongs to America.

"There is not a thought in any one of the foreign journals relative to the Telegraph which I had not expressed nearly five years ago, on my passage from France, to scientific friends; and when it is considered how quick a hint flies from mind to mind and is soon past all tracing back to the original suggester of the hint, it is certainly by no means improbable that the excitement on the subject in England has its origin from my giving the details of the plan of my Telegraph to some of the Englishmen or other fellow-passengers on board the ship, or to some of the many I have since made acquainted with it during the five years past."

In this he was mistaken, for the English telegraph of Cooke and Wheatstone was quite different in principle, using the deflection, by a current of electricity, of a delicately adjusted needle to point to the letters of the alphabet. While this was in use in England for a number of years, it was gradually superseded by the Morse telegraph which proved its decided superiority. It is also worthy of note that in this letter, and in all future letters and articles, he, with pardonable pride, uses a capital T in speaking of his Telegraph.

One of the most difficult of the problems which confront the historian who sincerely wishes to deal dispassionately with his subject is justly to apportion the credit which must be given to different workers in the same field of endeavor, and especially in that of invention; for every invention is but an improvement on something which has gone before. The sail-boat was an advance on the rude dugout propelled by paddles. The first clumsy steamboat seemed a marvel to those who had known no other propulsive power than that of the wind or the oar. The horse-drawn vehicle succeeded the litter and the palanquin, to be in turn followed by the locomotive; and so the telegraph, as a means of rapidly communicating intelligence between distant points, was the logical successor of the signal fire and the semaphore.

In all of these improvements by man upon what man had before accomplished, the pioneer was not only dependent upon what his predecessors had achieved, but, in almost every case, was compelled to call to his assistance other workers to whom could be confided some of the minutiæ which were essential to the successful launching of the new enterprise.

I have shown conclusively that the idea of transmitting intelligence by electricity was original with Morse in that he was unaware, until some years after his first conception, that anyone else had ever thought of it. I have also shown that he, unaided by others, invented and made with his own hands a machine, rude though it may have been, which actually did transmit and record intelligence by means of the electric current, and in a manner entirely different from the method employed by others. But he had now come to a point where knowledge of what others had accomplished along the same line would greatly facilitate his labors, and when the assistance of one more skilled in mechanical construction was a great desideratum, and both of these essentials were at hand. It is quite possible that he might have succeeded in working out the problem absolutely unaided, just as a man might become a great painter without instruction, without a knowledge of the accumulated wisdom of those who preceded him, and without the assistance of the color-maker and the manufacturer of brushes and canvas. But the artist is none the less a genius because he listens to the counsels of his master, profits by the experience of others, and purchases his supplies instead of grinding his own colors and laboriously manufacturing his own canvas and brushes.

The three men to whom Morse was most indebted for material assistance in his labors at this critical period were Professor Joseph Henry, Professor Leonard D. Gale, and Alfred Vail, and it is my earnest desire to do full justice to all of them. Unfortunately after the telegraph had become an assured success, and even down to the present day, the claims of Morse have been bitterly assailed, both by well-meaning persons and by the unscrupulous who sought to break down his patent rights; and the names of these three men were freely used in the effort to prove that to one or all of them more credit was due than to Morse.

Now, after the lapse of nearly three quarters of a century, the verdict has been given in favor of Morse, his name alone is accepted as that of the Inventor of the Telegraph, and in this work it is my aim to prove that the judgment of posterity has not erred, but also to give full credit to those who aided him when he was most in need of assistance. My task in some instances will be a delicate one; I shall have to prick some bubbles, for the friends of some of these men have claimed too much for them, and, on that account, have been bitter in their accusations against Morse. I shall also have to acknowledge some errors of judgment on the part of Morse, for the malice of others fomented a dispute between him and one of these three men, which caused a permanent estrangement and was greatly to be regretted.

The first of the three to enter into the history of the telegraph was Leonard D. Gale, who, in 1836, was a professor in the University of the City of New York, and he has given his recollections of those early days. Avoiding a repetition of facts already recorded I shall quote some sentences from Professor Gale's statement. After describing the first instrument, which he saw in January of 1836, he continues: —

"During the years 1836 and beginning of 1837 the studies of Professor Morse on his telegraph I found much interrupted by his attention to his professional duties. I understood that want of pecuniary means prevented him from procuring to be made such mechanical improvements, and such substantial workmanship, as would make the operation of his invention more exact.

"In the months of March and April, 1837, the announcement of an extraordinary telegraph on the visual plan (as it afterwards proved to be), the invention of two French gentlemen of the names of Gonon and Servell, was going the rounds of the papers. The thought occurred to me, as well as to Professor Morse and some others of his friends, that the invention of his electro-magnetic telegraph had somehow become known, and was the origin of the new telegraph thus conspicuously announced. This announcement at once aroused Professor Morse to renewed exertions to bring the new invention creditably before the public, and to consent to a public announcement of the existence of his invention. From April to September, 1837, Professor Morse and myself were engaged together in the work of preparing magnets, winding wire, constructing batteries, etc., in the University for an experiment on a larger, but still very limited scale, in the little leisure that each had to spare, and being at the same time much cramped for funds. . . .

"The latter part of August, 1837, the operation of the instruments was shown to numerous visitors at the University. . . .

"On Saturday, the 2d of September, 1837, Professor Daubeny, of the English Oxford University, being on a visit to this country, was invited with a few friends to see the operation of the telegraph, in its then rude form, in the cabinet of the New York University, where it had then been put up with a circuit of seventeen hundred feet of copper wire stretched back and forth in that long room. Professor Daubeny, Professor Torrey, and Mr. Alfred Vail were present among others. This exhibition of the telegraph, although of very rude and imperfectly constructed machinery, demonstrated to all present the practicability of the invention, and it resulted in enlisting the means, the skill, and the zeal of Mr. Alfred Vail, who, early the next week, called at the rooms and had a more perfect explanation from Professor Morse of the character of the invention."

It was Professor Gale who first called Morse's attention to the discoveries of Professor Joseph Henry, especially to that of the intensity magnet, and he thus describes the interesting event: —

"Morse's machine was complete in all its parts and operated perfectly through a circuit of some forty feet, but there was not sufficient force to send messages to a distance. At this time I was a lecturer on chemistry, and from necessity was acquainted with all kinds of galvanic batteries, and knew that a battery of one or a few cups generates a large quantity of electricity capable of producing heat, etc., but not of projecting electricity to a great distance, and that, to accomplish this, a battery of many cups is necessary. It was, therefore, evident to me that the one large cup-battery of Morse should be made into ten or fifteen smaller ones to make it a battery of intensity so as to project the electric fluid. . . . Accordingly I substituted the battery of many cups for the battery of one cup. The remaining defect in the Morse machine, as first seen by me, was that the coil of wire around the poles of the electro-magnet consisted of but a few turns only, while, to give the greatest projectile power, the number of turns should be increased from tens to hundreds, as shown by Professor Henry in his paper published in the 'American Journal of Science,' 1831. . . . After substituting the battery of twenty cups for that of a single cup, we added some hundred or more turns to the coil of wire around the poles of the magnet and sent a message through two hundred feet of conductors, then through one thousand feet, and then through ten miles of wire arranged on reels in my own lecture-room in the New York University in the presence of friends."

This was a most important step in hastening the reduction of the invention to a practical, workable basis and I wish here to bear testimony to the great services of Professor Henry in making this possible. His valuable discoveries were freely given to the world with no attempt on his part to patent them, which is, perhaps, to be regretted, but much more is it to be deplored that, in the litigation which ensued a few years later, Morse and Henry were drawn into a controversy, fostered and fomented by others for their own pecuniary benefit, which involved the honor and veracity of both of these distinguished men. Both were men of the greatest sensitiveness, proud and jealous of their own integrity, and the breach once made was never healed. Of the rights and wrongs of this controversy I may have occasion later on to treat more in detail, although I should much prefer to dismiss it with the acknowledgment that there was much to deplore in what was said and written by Morse, although he sincerely believed himself to be in the right, and much to regret in some of the statements and actions of Henry.

At this late day, when the mists which enveloped the questions have rolled away, it seems but simple justice to admit that the wonderful discoveries of Henry were essential to the successful working over long distances of Morse's discoveries and inventions; just as the discoveries and inventions of earlier and contemporary scientists were essential to Henry's improvements. But it is also just to place emphasis on the fact that Henry's experiments were purely scientific. He never attempted to put them in concrete form for the use of mankind in general; they led up to the telegraph; they were not a practical telegraph in themselves. It was Morse who added the final link in the long chain, and, by combining the discoveries of others with those which he had himself made, gave to the world this wonderful new agent.

A recent writer in the "Scientific American" gave utterance to the following sentiment, which, it seems to me, most aptly describes this difference: "We need physical discoveries and revere those who seek truth for its own sake. But mankind with keen instinct saves its warmest acclaim for those who also make discoveries of some avail in adding to the length of life, its joys, its possibilities, its conveniences."

We must also remember that, while the baby telegraph had, in 1837, been recognized as a promising infant by a very few scientists and personal friends of the inventor, it was still regarded with suspicion, if not with scorn, by the general public and even by many men of scholarly attainments, and a long and heart-breaking struggle for existence was ahead of it before it should reach maturity and develop into the lusty giant of the present day. Here again Morse proved that he was the one man of his generation most eminently fitted to fight for the child of his brain, to endure and to persevere until the victor's crown was grasped.

It is always idle to speculate on what might have happened if certain events had not taken place; if certain men had not met certain other men. A telegraph would undoubtedly have been invented if Morse had never been born; or he might have perfected his invention without the aid and advice of others, or with the assistance of different men from those who appeared at the psychological moment. But we are dealing with facts and not with suppositions, and the facts are that through Professor Gale he was made acquainted with the discoveries of Joseph Henry, which had been published to the world several years before, and could have been used by others if they had had the wit or genius to grasp their significance and hit upon the right means to make them of practical utility.

Morse was ever ready cheerfully to acknowledge the assistance which had been given to him by others, but, at the same time, he always took the firm stand that this did not give them a claim to an equal share with himself in the honor of the invention. In a long letter to Professor Charles T. Jackson, written on September 18, 1837, he vigorously but courteously repudiates the claim of the latter to have been a co-inventor on board the Sully, and he proves his point, for Jackson not only knew nothing of the plan adopted by Morse, and carried by him to a successful issue, but had never suggested anything of a practical nature. At the same time Morse freely acknowledges that the conversation between them on the ship suggested to him the train of thought which culminated in the invention, for he adds: —

"You say, 'I trust you will take care that the proper share of credit shall be given to me when you make public your doings.' This I always have done and with pleasure. I have always given you credit for great genius and acquirements, and have always said, in giving any account of my Telegraph, that it was during a scientific conversation with you on board the ship that I first conceived the thought of an electric Telegraph. Is there really any more that you will claim or that I could in truth and justice give?

"I have acknowledgments of a similar kind to make to Professor Silliman and to Professor Gale; to the former of whom I am under precisely similar obligations with yourself for several useful hints; and to the latter I am most of all indebted for substantial and effective aid in many of my experiments. If any one has a claim to be considered as a mutual inventor on the score of aid by hints, it is Professor Gale, but he prefers no claim of the kind."

And he never did prefer such a claim (although it was made for him by others), but remained always loyal to Morse. Jackson, on the other hand, insisted on pressing his demand, although it was an absurd one, and he was a thorn in the flesh to Morse for many years. It will not be necessary to go into the matter in detail, as Jackson was, through his wild claims to other inventions and discoveries, thoroughly discredited, and his views have now no weight in the scientific world.

The third person who came to the assistance of Morse at this critical period was Alfred Vail, son of Judge Stephen Vail, of Morristown, New Jersey. In 1837 he was a young man of thirty and had graduated from the University of the City of New York in 1836. He was present at the exhibition of Morse's invention on the 2d of September, 1837, and he at once grasped its great possibilities. After becoming satisfied that Morse's device of the relay would permit of operation over great distances, he expressed a desire to become associated with the inventor in the perfecting and exploitation of the invention. His father was the proprietor of the Speedwell Iron Works in Morristown, and young Vail had had some experience in the manufacture of mechanical appliances in the factory, although he had taken the theological course at the University with the intention of entering the Presbyterian ministry. He had abandoned the idea of becoming a clergyman, however, on account of ill-health, and was, for a time, uncertain as to his future career, when the interest aroused by the sight of Morse's machine settled the matter, and, after consulting with his father and brother, he entered into an agreement with Morse on the 23d day of September, 1837.

In the contract drawn up between them Vail bound himself to construct, at his own expense, a complete set of instruments; to defray the costs of securing patents in this country and abroad; and to devote his time to both these purposes. It was also agreed that each should at once communicate to the other any improvement or new invention bearing on the simplification or perfecting of the telegraph, and that such improvements or inventions should be held to be the property of each in the proportion in which they were to share in any pecuniary benefits which might accrue.

As the only way in which Morse could, at that time, pay Vail for his services and for money advanced, he gave him a one-fourth interest in the invention in this country, and one half in what might be obtained from Europe. This was, in the following March, changed to three sixteenths in the United States and one fourth in Europe.

Morse had now secured two essentials most necessary to the rapid perfection of his invention, the means to purchase materials and an assistant more skilled than he in mechanical construction, and who was imbued with faith in the ultimate success of the enterprise. Now began the serious work of putting the invention into such a form that it could demonstrate to the skeptical its capability of performing what was then considered a miracle. It is hard for us at the present time, when new marvels of science and invention are of everyday occurrence, to realize the hidebound incredulousness which prevailed during the first half of the nineteenth century. Men tapped their foreheads and shook their heads in speaking of Morse and his visionary schemes, and deeply regretted that here was the case of a brilliant man and excellent artist evidently gone wrong. But he was not to be turned from his great purpose by the jeers of the ignorant and the anxious solicitations of his friends, and he was greatly heartened by the encouragement of such men as Gale and Vail. They all three worked over the problems yet to be solved, Morse going backwards and forwards between New York and Morristown. That both Gale and Vail suggested improvements which were adopted by Morse, can be taken for granted, but, as I have said before, to modify or elaborate something originated by another is a comparatively easy matter, and the basic idea, first conceived by Morse on the Sully, was retained throughout.

All the details of these experiments have not been recorded, but I believe that at first an attempt was made to put into a more finished form the principle of the machine made by Morse, with its swinging pendulum tracing a waving line, but this was soon abandoned in favor of an instrument using the up-and-down motion of a lever, as drawn in the 1832 sketch-book. In other words, it was a return to first principles as thought out by Morse, and not, as some would have us believe, something entirely new suggested and invented independently by Vail.

It was rather unfortunate and curious, in view of Morse's love of simplicity, that he at first insisted on using the dots and dashes to indicate numbers only, the numbers to correspond to words in a specially prepared dictionary. His arguments in favor of this plan were specious, but the event has proved that his reasoning was faulty. His first idea was that the telegraph should belong to the Government; that intelligence sent should be secret by means of a kind of cipher; that it would take less time to send a number than each letter of each word, especially in the case of the longer words; and, finally, that although the labor in preparing a dictionary of all the most important words in the language and giving to each its number would be great, once done it would be done for all time.

I say that this was unfortunate because the fact that the telegraphic alphabet of dots and dashes was not used until after his association with Vail has lent strength to the claims on the part of Vail's family and friends that he was the inventor of it and not Morse. This claim has been so insistently, and even bitterly, made, especially after Morse's death, that it gained wide credence and has even been incorporated in some encyclopedias and histories. Fortunately it can be easily disproved, and I am desirous of finally settling this vexed question because I consider the conception of this simplest of all conventional alphabets one of the grandest of Morse's inventions, and one which has conferred great good upon mankind. It is used to convey intelligence not only by electricity, but in many other ways. Its cabalistic characters can be read by the eye, the ear, and the touch.

Just as the names of Ampère, Volta, and Watt have been used to designate certain properties or things discovered by them, so the name of Morse is immortalized in the alphabet invented by him. The telegraph operators all over the world send "Morse" when they tick off the dots and dashes of the alphabet, and happily I can prove that this is not an honor filched from another.

It is a matter of record that Vail himself never claimed in any of his letters or diaries (and these are voluminous) that he had anything to do with the devising of this conventional alphabet, even with the modification of the first form. On the other hand, in several letters to Morse he refers to it as being Morse's. For instance, in a letter of April 20, 1848, he uses the words "your system of marking, *lines* and *dots*, which you have patented." All the evidence brought forward by the advocates of Vail is purely hearsay; he is said to have said that he invented the alphabet.

Morse, however, always, in every one of his many written references to the matter, speaks of it as "my conventional alphabet." In an article which I contributed to the "Century Magazine" of March, 1912, I treated this question at length and proved by documentary evidence that Morse alone devised the dot-and-dash alphabet. It will not be necessary for me to repeat all this evidence here; I shall simply give enough to prove conclusively that the Morse Alphabet has not been misnamed.

The following is a fugitive note which was reproduced photographically in the "Century" article: —

"Mr. Vail, in his work on the Telegraph, at p. 32, intimates that the saw-teeth type for letters, as he has described them in the diagram (9), were devised by me as early as the year 1832. Two of the elements of these letters, indeed, were then devised, the dot and space, and used in constructing the type for numerals, but, so far as my recollection now serves me, it was not until I experimented with the first instrument in 1835 that I added the — dash, which supplied me with the three elements for combinations for letters. It was on noticing the fact that, when the circuit was closed a longer time than was necessary to make a dot, there was produced a line or dash, that, if I rightly remember, the broken parts of a continuous line as the means of imprinting at a distance were suggested to me; since the inequalities of long and short lines, separated by long and short spaces, gave me all the variations or combinations of long and short lines necessary to form the alphabet. The date of the code complete must, therefore, be put at 1835, and not 1832, although at the date of 1832 the principle of the code was *evolved.*"

In addition to this being a definite claim in writing on the part of Morse that he had devised an alphabetic code in 1835, two years before Vail had ever heard of the telegraph, it is well to note his scrupulous insistence on historical accuracy.

In a letter to Professor Gale, referring to reading by sound as well as by sight, occur the following sentences. (Let me remark, by the way, that it is interesting to note that Morse thus early recognized the possibility of reading by sound, an honor which has been claimed for many others.)

"Exactly at what time I recognized the adaptation of the difference in the intervals in reading the *letters* as well as the numerals, I have now no means of fixing except in a general manner. It was, however, almost immediately on the construction of the letters by dots and lines, and this was some little time previous to your seeing the instrument.

"Soon after the first operation of the instrument in 1835, in which the type for writing numbers were used, I not only conceived the letter type, but made them from some leads used in the printing-office. I have still quite a quantity of these type. They were used in Washington as well as the type for numerals in the winter of 1837–38.

"In the earlier period of the invention it was a matter which experience alone could determine whether the *numerical* system, by means of a numbered dictionary, or the alphabetic mode, by spelling of the words, was the better. While I perceived some advantages in the alphabetic system, especially in the writing of proper names, I at that time leaned rather towards the *numerical* mode under the impression that it would, on the whole, be the more rapid. A very short experience, however, showed the superiority of the alphabetic mode, and the big leaves of the numbered dictionary, which cost me a world of labor, and which you, perhaps, remember, were discarded and the alphabetic installed in its stead."

Perhaps the most conclusive evidence that Vail did not invent this alphabet is contained in his own book on the "American Electro-Magnetic Telegraph," published in 1845, in which he lays claim to certain improvements. After describing the dot-and-dash alphabet, he says:—

"This conventional alphabet was originated on board the packet Sully by Professor Morse, the very first elements of the invention, and arose from the necessity of the case; the motion produced by the magnet being limited to a single action. During the period of the thirteen years *many plans have been devised by the inventor* to bring the telegraphic alphabet to its simplest form."

The italics are mine, for the advocates of Vail have always quoted the first sentence only, and have said that the word "originated" implies that, while Vail admitted that the embryo of the alphabet — the dots and dashes to represent numbers only — was conceived on the Sully, he did not admit that the alphabetical code was Morse's. But when we read the second sentence with the words "devised by the inventor," the meaning is so plain that it is astonishing that any one at all familiar with the facts could have been misled.

The first form of the alphabet which was attached to Morse's caveat of October 3, 1837, is shown in the drawing of the type in the accompanying figure.

It has been stated by some historians that the system of signs for letters was not attached to the caveat, but a careful reading of the text, in which reference is made to the drawing, will prove conclusively that it was. Moreover, in this caveat under section 5, "The Dictionary or Vocabulary," the very first sentence reads: "The dictionary is a complete vocabulary of words alphabetically arranged and regularly numbered, *beginning with the letters of the alphabet.*" The italics are mine. The mistake arose because the drawing was detached from the caveat and affixed to the various patents which were issued, even after the first form of the alphabet had been superseded by a better one, the principle, however, remaining the same, so that it was not necessary to patent the new form.

The following is the form of the type, and the code as drawn in the caveat of 1837.

The changes from this original arrangement of the dots, spaces and lines, are seen on comparison.

ROUGH DRAWING OF ALPHABET BY MORSE

Showing the first form of the alphabet and the changes to the present form

As soon as it was proved that it would be simpler to use the letters of the alphabet in sending intelligence, the first form of the alphabet was changed in the manner shown in the preceding figure. Exactly when this was done has not been recorded, but it was after Vail's association with Morse, and it is quite possible that they worked over the problem together, but there is no written proof of this, whereas the accompanying reproduction of calculations in Morse's handwriting will prove that he gave himself seriously to its consideration.

The large numbers represent the quantities of type found in the type-cases of a printing-office; for, after puzzling over the question of the relative frequency of the occurrence of the different letters in the written language, a visit to the printing-office easily settled the matter.

This dispute, concerning the paternity of the alphabet, lasting for many years after the death of both principals, and regrettably creating much bad feeling, is typical of many which arose in the case of the telegraph, as well as in that of every other great invention, and it may not be amiss at this point to introduce the following fugitive note of Morse's, which, though evidently written many years later, is applicable to this as well as to other cases:—

"It is quite common to misapprehend the nature and extent of an improvement without a thorough knowledge of an original invention. A casual observer is apt to confound the new and the old, and, in noting a new arrangement, is often led to consider the whole as new. It is, therefore, necessary to exercise a proper discrimination lest injustice be done to the various laborers in the same field of invention. I trust it will not be deemed egotistical on my part if, while conscious of the unfeigned desire to concede to all who are attempting improvements in the art of telegraphy that which belongs to them, I should now and then recognize the familiar features of my own offspring and claim their paternity."

Quantities of the Letters.

Letter	Quantity
a	8000
b	1600
c	3000
d	4400
e	12000
f	2500
g	1700
h	6400
i	8000
j	400
k	800
l	4000
m	3000
n	8000
o	8000
p	1700
q	500
r	6200
s	8000
t	9000
u	3400
v	1200
w	2000
x	400
y	2000
z	200

QUANTITIES OF THE TYPE FOUND IN A PRINTING-OFFICE

Calculation made by Morse to aid him in simplifying alphabet

CHAPTER XXIV

OCTOBER 3, 1837 — MAY 16, 1838

The Caveat. — Work at Morristown. — Judge Vail. — First success. — Resolution in Congress regarding telegraphs. — Morse's reply. — Illness. — Heaviness of first instruments. — Successful exhibition in Morristown. — Exhibition in New York University. — First use of Morse alphabet. — Change from first form of alphabet to present form. — Trials of an inventor. — Dr. Jackson. — Slight friction between Morse and Vail. — Exhibition at Franklin Institute, Philadelphia. — Exhibitions in Washington. — Skepticism of public. — F. O. J. Smith. — F. L. Pope's estimate of Smith. — Proposal for government telegraph. — Smith's report. — Departure for Europe.

I HAVE incidentally mentioned the caveat in the preceding chapter, but a more detailed account of this important step in bringing the invention into the light of day should, perhaps, be given. The reports in the newspapers of the activities of others, especially of scientists in Europe, led Morse to decide that he must at once take steps legally to protect himself if he did not wish to be distanced in the race. He accordingly wrote to the Commissioner of Patents, Henry L. Ellsworth, who had been a classmate of his at Yale, for information as to the form to be used in applying for a caveat, and, after receiving a cordial reply enclosing the required form, he immediately set to work to prepare his caveat. This was in the early part of September, 1837, before he had met Vail. The rough draft, which is still among his papers, was completed on September 28, and the finished copy was sent to Washington on October 3, and the receipt acknowledged by Commissioner Ellsworth on October 6. The drawing containing the signs for both numbers and letters was attached to this caveat.

Having now safeguarded himself, he was able to give

his whole mind to the perfecting of the mechanical parts of his invention, and in this he was ably assisted by his new partner, Alfred Vail, and by Professor Gale.

The next few months were trying ones to both Morse and Vail. It must not be supposed that the work went along smoothly without a hitch. Many were the discouragements, and many experiments were tried and then discarded. To add to the difficulties, Judge Vail, who, of course, was supplying the cash, piqued by the sneers of his neighbors and noting the feverish anxiety of his son and of Morse, lost faith, and would have willingly abandoned the whole enterprise. The two enthusiasts worked steadily on, however, avoiding the Judge as much as possible, and finally, on the 6th of January, 1838, they proudly invited him to come to the workshop and witness the telegraph in operation.

His hopes renewed by their confident demeanor, he hastened down from his house. After a few words of explanation he handed a slip of paper to his son on which he had written the words — "A patient waiter is no loser." He knew that Morse could not possibly know what he had written, and he said: "If you can send this and Mr. Morse can read it at the other end, I shall be convinced."

Slowly the message was ticked off, and when Morse handed him the duplicate of his message, his enthusiasm knew no bounds, and he proposed to go at once to Washington and urge upon Congress the establishment of a government line. But the instrument was not yet in a shape to be seen of all men, and many years were yet to elapse before the legislators of the country awoke to their opportunity.

Morse and Vail were, of course, greatly encouraged by this first triumph, and worked on with increased enthusiasm.

Many years after their early struggles, when the telegraph was an established success and Morse had been honored both at home and abroad, he thus spoke of his friend: —

"Alfred Vail, then a student in the university, and a young man of great ingenuity, having heard of my invention, came to my rooms and I explained it to him, and from that moment he has taken the deepest interest in the Telegraph. Finding that I was unable to command the means to bring my invention properly before the public, and believing that he could command those means through his father and brother, he expressed the belief to me, and I at once made such an arrangement with him as to procure the pecuniary means and the skill of these gentlemen. It is to their joint liberality, but especially to the attention, and skill, and faith in the final success of the enterprise maintained by Alfred Vail, that is due the success of my endeavors to bring the Telegraph at that time creditably before the public."

The idea of telegraphs seems to have been in the air in the year 1837, for the House of Representatives had passed a resolution on the 3d of February, 1837, requesting the Secretary of the Treasury, Hon. Levi Woodbury, to report to the House upon the propriety of establishing a system of telegraphs for the United States. The term "telegraph" in those days included semaphores and other visual appliances, and, in fact, anything by which intelligence could be transmitted to a distance.

The Secretary issued a circular to "Collectors of

Customs, Commanders of Revenue Cutters, and other Persons," requesting information. Morse received one of these circulars, and in reply sent a long account of his invention. But so hard to convince were the good people of that day, and so skeptical and even flippant were most of the members of Congress that six long years were to elapse, years filled with struggles, discouragements, and heart-breaking disappointments, before the victory was won.

Morse had still to contend with occasional fits of illness, for he writes to his brother Sidney from Morristown on November 8, 1837: —

"You will perhaps be surprised to learn that I came out here to be sick. I caught a severe cold the day I left New York from the sudden change of temperature, and was taken down the next morning with one of my bilious attacks, which, under other treatment and circumstances, might have resulted seriously. But, through a kind Providence, I have been thrown among most attentive, and kind, and skilful friends, who have treated me more like one of their own children than like a stranger. Mrs. Vail has been a perfect mother to me; our good Nancy Shepard can alone compare with her. Through her nursing and constant attention I am now able to leave my room and have been downstairs to-day, and hope to be out in a few days. This sickness will, of course, detain me a while longer than I intended, for I must finish the portraits before I return."

This refers to portraits of various members of the Vail family which he had undertaken to execute while he was in Morristown. Farther on in the letter he says: —

"The machinery for the Telegraph goes forward daily; slowly but well and thorough. You will be surprised at the strength and quantity of machinery, greater, doubtless, than will eventually be necessary, yet it gives the main points, certainty and accuracy."

It may be well to note here that Morse evidently foresaw that the machinery constructed by Alfred Vail was too heavy and cumbersome; that more delicate workmanship would later be called for, and this proved to be the case. The iron works at Morristown were only adapted to the manufacture of heavy machinery for ships, etc., and Alfred Vail had had experience in that class of work only, so that he naturally made the telegraphic instruments much heavier and more unwieldy than was necessary. While these answered the purpose for the time being, they were soon superseded by instruments of greater delicacy and infinitely smaller bulk made by more skilful hands.

The future looked bright to the sanguine inventor in the early days of the year 1838, as we learn from the following letter to his brother Sidney, written on the 13th of January: —

"Mr. Alfred Vail is just going in to New York and will return on Monday morning. The machinery is at length completed and we have shown it to the Morristown people with great *éclat*. It is the talk of all the people round, and the principal inhabitants of Newark made a special excursion on Friday to see it. The success is complete. We have tried the experiment of sending a pretty full letter, which I set up from the numbers given me, transmitting through two miles of wire and deciphered with but a single unimportant error.

"I am staying out to perfect a modification of my portrule and hope to see you on Tuesday, or, at the farthest, on Wednesday, when I shall tell you all about it. The matter looks well now, and I desire to feel grateful to Him who gives success, and be always prepared for any disappointment which He in infinite wisdom may have in store."

We see from this letter, and from an account which appeared in the Morristown "Journal," that in these exhibitions the messages were sent by numbers with the aid of the cumbersome dictionary which Morse had been at such pains to compile. Very soon after this, however, as will appear from what follows, the dictionary was discarded forever, and the Morse alphabet came into practical use.

The following invitation was sent from the New York University on January 22, 1838: —

"Professor Morse requests the honor of Thomas S. Cummings, Esq., and family's company in the Geological Cabinet of the University, Washington Square, to witness the operation of the Electro-Magnetic Telegraph at a private exhibition of it to a few friends, previous to its leaving the city for Washington.

"The apparatus will be prepared at precisely twelve o'clock on Wednesday, 24th instant. The time being limited punctuality is specially requested."

Similar invitations were sent to other prominent persons and a very select company gathered at the appointed hour. That the exhibition was a success we learn from the following account in the "Journal of Commerce" of January 29, 1838: —

"THE TELEGRAPH. — We did not witness the operation of Professor Morse's Electro-Magnetic Telegraph on Wednesday last, but we learn that the numerous company of scientific persons who were present pronounced it entirely successful. Intelligence was instantaneously transmitted through a circuit of TEN MILES, and legibly written on a cylinder at the extremity of the circuit. The great advantages which must result to the public from this invention will warrant an outlay on the part of the Government sufficient to test its practicability as a general means of transmitting intelligence.

"Professor Morse has recently improved on his mode of marking by which he can dispense altogether with the telegraphic dictionary, using *letters* instead of *numbers*, and he can transmit ten words per minute, which is more than double the number which can be transmitted by means of the dictionary."

A charming and rather dramatic incident occurred at this exhibition which was never forgotten by those who witnessed it. General Cummings had just been appointed to a military command, and one of his friends, with this fact evidently in mind, wrote a message on a piece of paper and, without showing it to any one else, handed it to Morse. The assembled company was silent and only the monotonous clicking of the strange instrument was heard as the message was ticked off in the dots and dashes, and then from the other end of the ten miles of wire was read out this sentence pregnant with meaning: —

"Attention, the Universe, by kingdoms right wheel."

The name of the man who indited that message seems

not to have been preserved, but, whoever he was, he must have been gifted with prophetic vision, and he must have realized that he was assisting at an occasion which was destined to mark the beginning of a new era in civilization. The attention of the universe was, indeed, before long attracted to this child of Morse's brain, and kingdom after kingdom wheeled into line, vying with each other in admiration and acceptance.

The message was recorded fourfold by means of a newly invented fountain pen, and was given to General Cummings and preserved by him. It is here reproduced.

It will be noticed that the signs for the letters are those, not of the first form of the alphabet as embodied in the drawing attached to the caveat, but of the finally adopted code. This has led some historians, notably Mr. Franklin Leonard Pope, to infer that some mistake has been made in giving out this as a facsimile of this early message; that the letters should have been those of the earlier alphabet. I think, however, that this is but an added proof that Morse devised the first form of the code long before he met Vail, and that the changes to the final form, a description of which I have given, were made by Morse in 1837, or early in 1838, as soon as he became convinced of the superiority of the alphabetic mode, in plenty of time to have been used in this exhibition.

The month of January, 1838, was a busy one at Morristown, for Morse and Vail were bending all their energies toward the perfecting and completion of the instruments, so that a demonstration of the telegraph could be given in Washington at as early a date as possible. Morse refers feelingly to the trials and anxa t t ė n t i o n t he

ė u n i v ė r s ė b

iy k iy n g d o m s

r j g h t w h ė ė l

ieties of an inventor in a letter to a friend, dated January 22, 1838: —

"I have just returned from nearly six weeks' absence at Morristown, New Jersey, where I have been engaged in the superintendence of the making of my Telegraph for Washington.

Be thankful, C——, that you are not an inventor. Invention may seem an easy way to *fame*, or, what is the same thing to many, *notoriety*, different as are in reality the two objects. But it is far otherwise. I, indeed, desire the first, for true fame implies well-deserving, but I have no wish for the latter, which yet seems inseparable from it.

"The condition of an inventor is, indeed, not enviable. I know of but one condition that renders it in any degree tolerable, and that is the reflection that his fellow-men may be benefited by his discoveries. In the outset, if he has really made a *discovery*, which very word implies that it was before unknown to the world, he encounters the incredulity, the opposition, and even the sneers of many, who look upon him with a kind of pity, as a little beside himself if not quite mad. And, while maturing his invention, he has the comfort of reflection, in all the various discouragements he meets with from petty failures, that, should he by any means fail in the grand result, he subjects himself rather to the ridicule than the sympathy of his acquaintances, who will not be slow in attributing his failure to a want of that common sense in which, by implication, they so much abound, and which preserves them from the consequences of any such delusions.

"But you will, perhaps, think that there is an offset

in the honors and emoluments that await the successful inventor, one who has really demonstrated that he has made an important discovery. This is not so. Trials of another kind are ready for him after the appropriate difficulties of his task are over. Many stand ready to snatch the prize, or at least to claim a share, so soon as the success of an invention seems certain, and honor and profit alone remain to be obtained.

"This long prelude, C——, brings me at the same time to the point of my argument and to my excuse for my long silence. My argument goes to prove that, unless there is a benevolent consideration in our discoveries, one which enables us to rejoice that others are benefited even though we should suffer loss, our happiness from any honor awarded to a successful invention is exposed to constant danger from the designs of the unprincipled. My excuse is that, ever since the receipt of your most welcome letter, I have been engaged in preparing to repel a threatened invasion of my rights to the invention of the Telegraph by a fellow-passenger from France, one from whom I least expected any such insidious design. The attempt startled me and put me on my guard, and set me to the preparation for any attack. I have been compelled for some weeks to use my pen only for this purpose, and have written much in the hope of preventing the public exposure of my antagonist; but I fear my labor will be vain on this point, from what I hear and the tone in which he writes. I have no fear for myself, being now amply prepared with evidence to repel any attempt which may be made to sustain any claim he may prefer to a share with me in the invention of the Telegraph."

I have already shown that this claim of Dr. Jackson's was proved to be but the hallucination of a disordered brain, and it will not be necessary to go into the details of the controversy.

These were anxious and nerve-racking days for both Morse and Vail, and it is small wonder that there should have been some slight friction. Vail in his private correspondence makes some mention of this. For instance, in a letter to his brother George, of January 22, 1838, he says: —

"We received the machine on Thursday morning, and in an hour we made the first trial, which did not succeed, nor did it with perfect success until Saturday — all which time Professor M. was rather *unwell.* To-morrow we shall make our first exhibition, and continue it until Wednesday, when we must again box up. Professor M. has received a letter from Mr. Patterson inviting us to exhibit at Philadelphia, and has answered it, but has said nothing to *me* about his intentions. He is altogether inclined to operate in his own name, so much so that he has had printed five hundred blank invitations in his own name at your expense."

On the other hand, this same George Vail, writing to Morse on January 25, 1838, asks him to "bear with A., which I have no doubt you will. He is easily vexed. Trusting to your universal coolness, however, there is nothing to fear. Keep him from running ahead too fast."

Again writing to his brother George from Washington, on February 20, 1838, Alfred says: "In regard to Professor M. calling me his '*assistant,*' this is also settled, and he has said as much as to apologize for using the term."

Why Vail should have objected to being called Morse's assistant, I cannot quite understand, for he was so designated in the contract later made with the Government; but Morse was evidently willing to humor him in this.

I have thought it best to refer to these little incidents partly in the interest of absolute candor, partly to emphasize the nervous tension under which both were working at that time. That there was no lasting resentment in the mind of Vail is amply proved by the following extract from a long letter written by him on March 19, 1838:—

"The great expectations I had on my return home of going into partnership with George, founded, or semi-founded, on the promises made by my father, have burst. I am again on vague promises for three months, and they resting upon the success of the printing machine.

"I feel, Professor Morse, that, if I am ever worth anything, it will be wholly attributable to your kindness. I now should have no *earthly* prospect of happiness and domestic bliss had it not been for what you have done. For which I shall ever remember [you] with the liveliest emotions of gratitude, whether it is eventually successful or not."

Aside from the slight friction to which I have referred, and which was most excusable under the circumstances, the joint work on the telegraph proceeded harmoniously. The invitation from Mr. Patterson, to exhibit the instrument before the Committee of Science and Arts of the Franklin Institute of Philadelphia, was accepted. The exhibition took place on February 8, and was a pronounced success, and the committee, in expressing their gratification, voiced the hope that the Government would provide the funds for an experiment on an adequate scale.

From Philadelphia Morse proceeded to Washington accompanied by Vail, confidently believing that it would only be necessary to demonstrate the practicability of his invention to the country's legislators assembled in Congress, in order to obtain a generous appropriation to enable him properly to test it. But he had not taken into account that trait of human nature which I shall dignify by calling it "conservatism," in order not to give it a harder name.

The room of the Committee on Commerce was placed at his disposal, and there he hopefully strung his ten miles of wire and connected them with his instruments. Outwardly calm but inwardly nervous and excited, as he realized that he was facing a supreme moment in his career, he patiently explained to all who came, Congressmen, men of science, representatives of foreign governments, and hard-headed men of business, the workings of the instrument and proved its feasibility. The majority saw and wondered, but went away unconvinced. On February 21, President Martin Van Buren and his entire Cabinet, at their own special request, visited the room and saw the telegraph in operation. But no action was taken by Congress; the time was not yet ripe for the general acceptance of such a revolutionary departure from the slow-going methods of that early period. While individuals here and there grasped the full significance of what the mysterious ticking of that curious instrument foretold, they were vastly in the minority. The world, through its representatives in the capital city of the United States, remained incredulous.

Among those who at once recognized the possibilities of the invention was Francis O. J. Smith, member of Congress from Portland, Maine, and chairman of the Committee on Commerce. He was a lawyer of much shrewdness and a man of great energy, and he very soon offered to become pecuniarily interested in the invention. Morse was, unfortunately, not a keen judge of men. Scrupulously honest and honorable himself, he had an almost childlike faith in the integrity of others, and all through his life he fell an easy victim to the schemes of self-seekers. In this case a man of more acute intuition would have hesitated, and would have made some enquiries before allying himself with one whose ideas of honor proved eventually to be so at variance with his own. Smith did so much in later years to injure Morse, and to besmirch his fame and good name, that I think it only just to give the following estimate of his character, made by the late Franklin Leonard Pope in an article contributed to the "Electrical World" in 1895:—

"A sense of justice compels me to say that the uncorroborated statements of F. O. J. Smith, in any matter affecting the credit or honor due to Professor Morse, should be allowed but little weight. . . . For no better reason than that Morse in 1843–1844 courteously but firmly refused to be a party to a questionable scheme devised by Smith for the irregular diversion into his own pocket of a portion of the governmental appropriation of $30,000 for the construction of the experimental line, he ever after cherished toward the inventor the bitterest animosity; a feeling which he took no pains to conceal. Many of his letters to him at that time, and for many years afterward, were couched in studiously insulting language, which must have been in the highest degree irritating to a sensitive artistic temperament like that of Morse.

"It probably by no means tended to mollify the disposition of such a man as Smith to find that Morse, in reply to these covert sneers and open insinuations, never once lost his self-control, nor permitted himself to depart from the dignified tone of rejoinder which becomes a gentleman in his dealings with one who, in his inmost nature, was essentially a blackguard."

However, it is an old saying that we must "give the devil his due," and the cloven foot did not appear at first. On the other hand, a man of business acumen and legal knowledge was greatly needed at this stage of the enterprise, and Smith possessed them both. Morse was so grateful to find any one with faith enough to be willing to invest money in the invention, and to devote his time and energy to its furtherance, that he at once accepted Smith's offer, and he was made a partner and given a one-fourth interest, Morse retaining nine sixteenths, Vail two sixteenths, and Professor Gale, also admitted as a partner, being allotted one sixteenth. It was characteristic of Morse that he insisted, before signing the contract, that Smith should obtain leave of absence from Congress for the remainder of the term, and should not stand for reëlection. It was agreed that Smith should accompany Morse to Europe as soon as possible and endeavor to secure patents in foreign countries, and, if successful, the profits were to be divided differently, Morse receiving eight sixteenths, Smith five, Vail two, and Gale one.

In spite of the incredulity of the many, Morse could not help feeling encouraged, and in a long letter to Smith, written on February 15, 1838, proposing an experiment of one hundred miles, he thus forecasts the future and proposes an intelligent plan of government control: —

"If no insurmountable obstacles present themselves in a distance of one hundred miles, none may be expected in one thousand or in ten thousand miles; and then will be presented for the consideration of the Government the propriety of completely organizing this *new telegraphic system as a part of the Government*, attaching it to some department already existing, or creating a new one which may be called for by the accumulating duties of the present departments.

"It is obvious, at the slightest glance, that this mode of instantaneous communication must inevitably become an instrument of immense power, to be wielded for good or for evil, as it shall be properly or improperly directed. In the hands of a company of speculators, who should monopolize it for themselves, it might be the means of enriching the corporation at the expense of the bankruptcy of thousands; and even in the hands of Government alone it might become the means of working vast mischief to the Republic.

"In considering these prospective evils, I would respectfully suggest a remedy which offers itself to my mind. Let the sole right of using the Telegraph belong, in the first place, to the Government, who should grant, for a specified sum or bonus, to any individual or company of individuals who may apply for it, and under such restrictions and regulations as the Government may think proper, the right to lay down a communication between any two points for the purpose of transmitting intelligence, and thus would be promoted a general competition. The Government would have a Telegraph of its own, and have its modes of communicating with its own officers and agents, independent of private permission or interference with and interruption to the ordinary transmissions on the private telegraphs. Thus there would be a system of checks and preventives of abuse operating to restrain the action of this otherwise dangerous power within those bounds which will permit only the good and neutralize the evil. Should the Government thus take the Telegraph solely under its own control, the revenue derived from the bonuses alone, it must be plain, will be of vast amount.

"From the enterprising character of our countrymen, shown in the manner in which they carry forward any new project which promises private or public advantage, it is not visionary to suppose that it would not be long ere the whole surface of this country would be channelled for those *nerves* which are to diffuse, with the speed of thought, a knowledge of all that is occurring throughout the land, making, in fact, *one neighborhood* of the whole country.

"If the Government is disposed to test this mode of telegraphic communication by enabling me to give it a fair trial for one hundred miles, I will engage to enter into no arrangement to dispose of my rights, as the inventor and patentee for the United States, to any individual or company of individuals, previous to offering it to the Government for such a just and reasonable compensation as shall be mutually agreed upon."

We have seen that Morse was said to be a hundred years ahead of his time as an artist. From the sentences above quoted it would appear that he was far in advance of his contemporaries in some questions of national policy, for the plan outlined by him for the proper governmental control of a great public utility, like the telegraph, it seems to me, should appeal to those who, at the present time, are agitating for that very thing. Had the legislators and the people of 1838 been as wise and clear-sighted as the poor artist-inventor, a great step forward in enlightened statecraft could have been taken at a cost inconceivably less than would now be the case. Competent authorities estimate that to purchase the present telegraph lines in this country at their market valuation would cost the Government in the neighborhood of $500,000,000; to parallel them would cost some $25,000,000. The enormous difference in these two sums represents what was foretold by Morse would happen if the telegraph should become a monopoly in the hands of speculators. The history of the telegraph monopoly is too well known to be more than alluded to here, but it is only fair to Morse to state that he had sold all his telegraph stock, and had retired from active participation in the management of the different companies, long before the system of stock-watering began which has been carried on to the present day.

And for what sum could the Government have kept this great invention under its own control? It is on record that Morse offered, in 1844, after the experimental line between Washington and Baltimore had demonstrated that the telegraph was a success, to sell all the rights in his invention to the Government for $100,000, and would have considered himself amply remunerated.

But the legislators and the people of 1838, and even those of 1844, were not wise and far-sighted; they failed utterly to realize what a magnificent opportunity had been offered to them for a mere song; and this in spite of the fact that the few who did glimpse the great future of the telegraph painted it in glowing terms.

It is true that the House of Representatives had passed the resolution referred to earlier in this chapter, but that is as far as they went for several years. On the 6th of April, 1838, Mr. F. O. J. Smith made a long report on the petition of Morse asking for an appropriation sufficient to enable him to test his invention adequately. In the course of this report Mr. Smith indulged in the following eulogistic words: —

"It is obvious, however, that the influence of this invention over the political, commercial, and social relations of the people of this widely extended country, looking to nothing beyond, will, in the event of success, of itself amount to a revolution unsurpassed in moral grandeur by any discovery that has been made in the arts and sciences, from the most distant period to which anthentic history extends to the present day. With the means of almost instantaneous communication of intelligence between the most distant points of the country, and simultaneously between any given number of intermediate points which this invention contemplates, space will be, to all practical purposes of information, completely annihilated between the States of the Union, as also between the individual citizens thereof.

The citizen will be invested with, and reduce to daily and familiar use, an approach to the HIGH ATTRIBUTE OF UBIQUITY in a degree that the human mind, until recently, has hardly dared to comtemplate seriously as belonging to human agency, from an instinctive feeling of religious reverence and reserve on a power of such awful grandeur."

In the face of these enthusiastic, if somewhat stilted, periods the majority of his colleagues remained cold, and no appropriation was voted. Morse, however, was prepared to meet with discouragements, for he wrote to Vail on March 15: —

"Everything looks encouraging, but I need not say to you that in this world a continued course of prosperity is not a rational expectation. We shall, doubtless, find troubles and difficulties in store for us, and it is the part of true wisdom to be prepared for whatever may await us. If our hearts are right we shall not be taken by surprise. I see nothing now but an unclouded prospect, for which let us pay to Him who shows it to us the homage of grateful and obedient hearts, with most earnest prayers for grace to use prosperity aright."

This was written while there was still hope that Congress might take some action at that session, and Morse was optimistic. On March 31, he thus reports progress to Vail: —

"I write you a hasty line to say, in the first place, that I have overcome all difficulties in regard to a portrule, and have invented one which will be perfect. It is very simple, and will not take much time or expense to make it. Mr. S. has incorporated it into the specification for the patent. Please, therefore, not to proceed with the type or portrule as now constructed. I will see you on my return and explain it in season for you to get one ready for us.

"I find it a most arduous and tedious process to adjust the specification. I have been engaged steadily for three days with Mr. S., and have not yet got half through, but there is one consolation, when done it will be well done. The drawings, I find on enquiry, would cost you from forty to fifty dollars if procured from the draughtsman about the Patent Office. I have, therefore, determined to do them myself and save you that sum."

The portrule, referred to above, was a device for sending automatically messages which were recorded permanently on the tape at the other end of the line. It worked well enough, but it was soon superseded by the key manipulated by hand, as this was much simpler and the dots and dashes could be sent more rapidly. It is curious to note, however, that down to the present day inventors have been busy in an effort to devise some mechanism by which messages could be sent automatically, and consequently more rapidly than by hand, which was Morse's original idea, but, to the best of my knowledge, no satisfactory solution of the problem has yet been found.

Morse was now preparing to go to Europe with Smith to endeavor to secure patents abroad, and, while he had put in his application for a patent in this country, he requested that the issuing of it should be held back until his return, so that a publication on this side should not injure his chances abroad.

All the partners were working under high pressure along their several lines to get everything in readiness for a successful exhibition of the telegraph in Europe. Vail sent a long letter to Morse on April 18, detailing some of the difficulties which he was encountering, and Morse answered on the 24th: —

"I write in greatest haste, just to say that the boxes have safely arrived, and we shall proceed immediately to examine into the difficulties which have troubled you, but about which we apprehend no serious issue. . . .

"If you can possibly get the circular portrule completed before we go it will be a great convenience, not to say an indispensable matter, for I have just learned so much of Wheatstone's Telegraph as to be pretty well persuaded that my superiority over him will be made evident more by the rapidity with which I can make the portrule work than in almost any other particular."

At last every detail had been attended to, and in a postscript to a letter of April 28 he says: "We sail on the 16th of May for Liverpool in the ship Europe, so I think you will have time to complete circular portrule. Try, won't you?"

CHAPTER XXV

JUNE, 1838 — JANUARY 21, 1839

Arrival in England. — Application for letters patent. — Cooke and Wheatstone's telegraph. — Patent refused. — Departure for Paris. — Patent secured in France. — Earl of Elgin. — Earl of Lincoln. — Baron de Meyendorff. — Russian contract. — Return to London. — Exhibition at the Earl of Lincoln's. — Letter from secretary of Lord Campbell, Attorney-General. — Coronation of Queen Victoria. — Letters to daughter. — Birth of the Count of Paris. — Exhibition before the Institute of France. — Arago; Baron Humboldt. — Negotiations with the Government and Saint-Germain Railway. — Reminiscences of Dr. Kirk. — Letter of the Honorable H. L. Ellsworth. — Letter to F. O. J. Smith. — Dilatoriness of the French.

It seems almost incredible to us, who have come to look upon marvel after marvel of science and invention as a matter of course, that it should have taken so many years to convince the world that the telegraph was a possibility and not an iridescent dream. While men of science and a few far-sighted laymen saw that the time was ripe for this much-needed advance in the means of conveying intelligence, governments and capitalists had held shyly aloof, and, even now, weighed carefully the advantages of different systems before deciding which, if any, was the best. For there were at this time several different systems in the field, and Morse soon found that he would have to compete with the trained scientists of the Old World, backed, at last, by their respective governments, in his effort to prove that his invention was the simplest and the best of them all. That he should have persisted in spite of discouragement after discouragement, struggling to overcome obstacles which to the faint-hearted would have seemed insuperable, constitutes one of his greatest claims to undying fame. He left on record an account of his experiences in Europe on this voyage, memorable in more ways than one, and extracts from this, and from letters written to his daughter and brothers, will best tell the story: —

"On May 16, 1838, I left the United States and arrived in London in June, for the purpose of obtaining letters patent for my Electro-Magnetic Telegraph System. I learned before I left the United States that Professor Wheatstone and Mr. Cooke, of London, had obtained letters patent in England for a '*Magnetic-Needle Telegraph*,' based, as the name implies, on the *deflection of the magnetic needle.* Their telegraph, at that time, required *six conductors* between the two points of intercommunication *for a single instrument* at each of the two termini. Their mode of indicating signs for communicating intelligence was by deflecting *five magnetic needles* in various directions, in such a way as to point to the required letters upon a diamond-shaped dial-plate. It was necessary that the signal should be *observed at the instant*, or it was lost and vanished forever.

"I applied for letters patent for my system of communicating intelligence at a distance by electricity, differing in all respects from Messrs. Wheatstone and Cooke's system, invented five years before theirs, and having nothing in common in the whole system but the use of *electricity* on *metallic conductors*, for which use no one could obtain an exclusive privilege, since this much had been used for nearly one hundred years. My system is peculiar in the employment of *electro-magnetism*, or the *motive* power of electricity, *to imprint permanent signs at a distance.*

"I made no use of the deflections of the magnetic needle as *signs.* I required but *one conductor* between the two termini, or any number of intermediate points of intercommunication. I used *paper moved by clock-work* upon which I caused a *lever* moved by *magnetism* to *imprint the letters* and *words* of any required dispatch, having also invented and adapted to telegraph writing a *new and peculiar alphabetic character* for that purpose, a *conventional alphabet,* easily acquired and easily made and used by the operator. It is obvious at once, from a simple statement of these facts, that the system of Messrs. Wheatstone and Cooke and my system were wholly unlike each other. As I have just observed, there was *nothing in common in the two systems* but the use of electricity upon metallic conductors, for which no one could obtain an exclusive privilege.

"The various steps required by the English law were taken by me to procure a patent for my mode, and the fees were paid at the Clerk's office, June 22, and at the Home Department, June 25, 1838; also, June 26, caveats were entered at the Attorney and Solicitor-General's, and I had reached that part of the process which required the sanction of the Attorney-General. At this point I met the opposition of Messrs. Wheatstone and Cooke, and also of Mr. Davy, and a hearing was ordered before the Attorney-General, Sir John Campbell, on July 12, 1838. I attended at the Attorney-General's residence on the morning of that day, carrying with me my telegraphic apparatus for the purpose of explaining to him the total dissimilarity between my system and those of my opponents. But, contrary to my expectation, the similarity or dissimilarity of my mode from that of my opponents was not considered by the Attorney-General. He neither examined my instrument, which I had brought for that purpose, nor did he ask any questions bearing upon its resemblance to my opponents' system. I was met by the single declaration that my '*invention had been published*,' and in proof a copy of the London 'Mechanics' Magazine,' No. 757, for February 10, 1838, was produced, and I was told that 'in consequence of said publication I could not proceed.'

"At this summary decision I was certainly surprised, being conscious that there had been no such publication of my method as the law required to invalidate a patent; and, even if there had been, I ventured to hint to the Attorney-General that, if I was rightly informed in regard to the British law, it was the province of a court and jury, and not of the Attorney-General, to try, and to decide that point."

The publication to which the Attorney-General referred had merely stated results, with no description whatever of the means by which these results were to be obtained and it was manifestly unfair to Morse on the part of this official to have refused his sanction; but he remained obdurate. Morse then wrote him a long letter, after consultation with Mr. Smith, setting forth all these points and begging for another interview.

"In consequence of my request in this letter I was allowed a second hearing. I attended accordingly, but, to my chagrin, the Attorney-General remarked that he had not had time to examine the letter. He carelessly took it up and turned over the leaves without reading it, and then asked me if I had not taken measures for a patent in my own country. And, upon my reply in the affirmative, he remarked that: 'America was a large country and I ought to be satisfied with a patent there.' I replied that, with all due deference, I did not consider that as a point submitted for the Attorney-General's decision; that the question submitted was whether there was any legal obstacle in the way of my obtaining letters patent for my Telegraph in England. He observed that he considered my invention as having been *published*, and that he must *therefore* forbid me to proceed.

"Thus forbidden to proceed by an authority from which there was no appeal, as I afterward learned, but to Parliament, and this at great cost of time and money, I immediately left England for France, where I found no difficulty in securing a patent. My invention there not only attracted the regards of the distinguished savants of Paris, but, in a marked degree, the admiration of many of the English nobility and gentry at that time in the French capital. To several of these, while explaining the operation of my telegraphic system, I related the history of my treatment by the English Attorney-General. The celebrated Earl of Elgin took a deep interest in the matter and was intent on my obtaining a special Act of Parliament to secure to me my just rights as the inventor of the Electro-Magnetic Telegraph. He repeatedly visited me, bringing with him many of his distinguished friends, and on one occasion the noble Earl of Lincoln, since one of Her Majesty's Privy Council. The Honorable Henry Drummond also interested himself for me, and through his kindness and Lord Elgin's I received letters of introduction to Lord Brougham and to the Marquis of Northampton, the

President of the Royal Society, and several other distinguished persons in England. The Earl of Lincoln showed me special kindness. In taking leave of me in Paris he gave me his card, and, requesting me to bring my telegraphic instruments with me to London, pressed me to give him the earliest notice of my arrival in London.

"I must here say that for weeks in Paris I had been engaged in negotiation with the Russian Counselor of State, the Baron Alexander de Meyendorff, arranging measures for putting the telegraph in operation in Russia. The terms of a contract had been mutually agreed upon, and all was concluded but the signature of the Emperor to legalize it. In order to take advantage of the ensuing summer season for my operations in Russia, I determined to proceed immediately to the United States to make some necessary preparations for the enterprise, without waiting for the formal completion of the contract papers, being led to believe that the signature of the Emperor was sure, a matter of mere form.

"Under these circumstances I left Paris on the 13th of March, 1839, and arrived in London on the 15th of the same month. The next day I sent my card to the Earl of Lincoln and my letter and card to the Marquis of Northampton, and in two or three days received a visit from both. By Earl Lincoln I was at once invited to send my Telegraph to his house in Park Lane, and on the 19th of March I exhibited its operation to members of both Houses of Parliament, of the Royal Society, and the Lords of the Admiralty, invited to meet me by the Earl of Lincoln. From the circumstances mentioned my time in London was necessarily short, my passage having been secured in the Great Western to sail on the 23d of March. Although solicited to remain a while in London, both by the Earl of Lincoln and the Honorable Henry Drummond, with a view to obtaining a special Act of Parliament for a patent, I was compelled by the circumstances of the case to defer till some more favorable opportunity, on my expected return to England, any attempt of the kind. The Emperor of Russia, however, refused to ratify the contract made with me by the Counselor of State, and my design of returning to Europe was frustrated, and I have not to this hour [April 2, 1847] had the means to prosecute this enterprise to a result in England. All my exertions were needed to establish my telegraphic system in my own country.

"Time has shown conclusively the essential difference of my telegraphic system from those of my opponents; time has also shown that my system *was not published* in England, as alleged by the Attorney-General, for, to this day, no work in England has published anything that does not show that, as yet, it is perfectly misunderstood. . . .

"The refusal to grant me a patent was, at that period, very disastrous. It was especially discouraging to have made a long voyage across the Atlantic in vain, incurring great expenditure and loss of time, which in their consequences also produced years of delay in the prosecution of my enterprise in the United States."

The long statement, from which I have taken the above extracts, was written, as I have noted, on April 2, 1847, but the following interesting addition was made to it on December 11, 1848: —

"At the time of preparing this statement I lacked one

item of evidence, which it was desirable to have aside from my own assertion, viz., evidence that the refusal of the Attorney-General was on the ground '*that a publication of the invention had been made.*' I deemed it advisable rather to suffer from the delay and endure the taunts, which my unscrupulous opponents have not been slow to lavish upon me in consequence, if I could but obtain this evidence in proper shape. I accordingly wrote to my brother, then in London, to procure, if possible, from Lord Campbell or his secretary an acknowledgment of the ground on which he refused my application for a patent in 1838, since no public report or record in such cases is made.

"My brother, in connection with Mr. Carpmael, one of the most distinguished patent agents in England, addressed a note to Mr. H. Cooper, the Attorney-General's secretary at the time, and the only official person besides Lord Campbell connected with the matter. The following is Mr. Cooper's reply: —

"'Wilmington Square, May 23d, 1848.

"'Gentlemen, — In answer to yours of the 20th inst., I beg to state that I have a distinct recollection of Professor Morse's application for a patent, strengthened by the fact of his not having paid the fees for the hearing, etc., and these being now owing. I understood at the time that the patent was stopped on the ground that a publication of the invention had been made, but I cannot procure Lord Campbell's certificate of that fact.

"'I am, gentlemen

"'Your obedient servant

"'H. Cooper.'

"I thus have obtained the evidence I desired in the most authentic form, but accompanied with as gross an insult as could well be conceived. On the receipt of this letter I immediately wrote to F. O. J. Smith, Esq., at Portland, who accompanied me to England, and at whose sole expense, according to agreement, all proceedings in taking out patents in Europe were to be borne, to know if this charge of the Attorney-General's secretary could possibly be true; not knowing but through some inadvertence on his (Mr. Smith's) part, this bill might have been overlooked.

"Mr. Smith writes me in answer, sending me a copy *verbatim* of the following receipt, which he holds and which speaks for itself: —

"'Mr. Morse to the Attorney-General, Dr.

	£	s.	d.
Hearing on a patent	3	10	0
Giving notice on the same	1	1	0
	4	11	0

Settled the 13th of August, 1838.

"'(Signed) H. COOPER.'

"This receipt is signed, as will be perceived, by the same individual, H. Cooper, who, nearly ten years after his acknowledgment of the money, has the impudence to charge me with leaving my fees unpaid. I now leave the public to make their own comments both on the character of the whole transaction in England, and on the character and motives of those in this country who have espoused Lord Campbell's course, making it an occasion to charge me with having *invented nothing.*

"SAMUEL F. B. MORSE."

I have, in these extracts from an account of his European experiences, written by Morse at a later date, given but a brief summary of certain events; it will now be necessary to record more in detail some of the happenings on that memorable trip.

Attention has been called before to the fact that it was Morse's good fortune to have been an eye-witness of many events of historic interest. Still another was now to be added to the list, for, while he was in London striving unsuccessfully to secure a patent for his invention, he was privileged to witness the coronation of Queen Victoria; our Minister, the Honorable Andrew Stevenson, having procured for him a ticket of admission to Westminster Abbey.

Writing to his daughter Susan on June 19, 1838, before he had met with his rebuff from the Attorney-General, he comments briefly on the festivities incident to the occasion: —

"London is filling fast with crowds of all characters, from ambassadors and princes to pickpockets and beggars, all brought together by the coronation of the queen, which takes place in a few days (the 28th of June). Everything in London now is colored by the coming pageant. In the shop windows are the robes of the nobility, the crimson and ermine dresses, coronets, etc. Preparations for illuminations are making all over the city.

"I have scarcely entered upon the business of the Telegraph, but have examined (tell Dr. Gale) the specification of Wheatstone at the Patent Office, and except the alarum part, he has nothing which interferes with mine. His invention is ingenious and beautiful, but very complicated, and he must use twelve wires where I use but four. I have also seen a telegraph exhibiting at Exeter Hall invented by Davy, something like Wheatstone's but still complicated. I find mine is yet the simplest and hope to accomplish something, but always keep myself prepared for disappointment."

At a later date he recounted the following pretty incident, showing the kindly character of the young queen, which may not be generally known: —

"I was in London in 1838, and was present with my excellent friend, the late Charles R. Leslie, R.A., at the imposing ceremonies of the coronation of the queen in Westminster Abbey. He then related to me the following incident which, I think, may truly be said to have been the first act of Her Majesty's reign.

"When her predecessor, William IV, died, a messenger was immediately dispatched by his queen (then become by his death queen dowager) to Victoria, apprising her of the event. She immediately called for paper and indited a letter of condolence to the widow. Folding it, she directed it 'To the Queen of England.' Her maid of honor in attendance, noting the inscription, said: 'Your Majesty, you are Queen of England.' 'Yes,' she replied, 'but the widowed queen is not to be reminded of that fact first by me.'"

Writing to his daughter from Havre, on July 26, 1838, while on his way to Paris, after telling her of the unjust decision of the Attorney-General, he adds: —

"Professor Wheatstone and Mr. Davy were my opponents. They have each very ingenious inventions of their own, particularly the former, who is a man of genius and one with whom I was personally much pleased. He has invented his, I believe, without knowing that I was engaged in an invention to produce a similar result; for, although he dates back into 1832, yet, as no publication of our thoughts was made by either, we are eivdently independent of each other. My time has not been lost, however, for I have ascertained with certainty that the *Telegraph of a single circuit* and a *recording apparatus* is mine. . . .

"I found also that both Mr. Wheatstone and Mr. Davy were endeavoring to simplify theirs by adding a recording apparatus and reducing theirs to a single circuit. The latter showed to the Attorney-General a drawing, which I obtained sight of, of a method by which he proposed a bungling imitation of my first characters, those that were printed in our journals, and one, however plausible on paper, and sufficiently so to deceive the Attorney-General, was perfectly impracticable. Partiality, from national or other motives, aside from the justice of the case, I am persuaded, influenced the decision against me.

"We are now on our way to Paris to try what we can do with the French Government. I confess I am not sanguine as to any favorable pecuniary result in Europe, but we shall try, and, at any rate, we have seen enough to know that the matter is viewed with great interest here, and the plan of such telegraphs will be adopted, and, of course, the United States is secured to us, and I do hope something from that.

"Be economical, my dear child, and keep your wants within bounds, for I am preparing myself for an unsuccessful result here, yet every proper effort will be made. I am in excellent health and spirits and leave to-morrow morning for Paris."

"*Paris, August 29, 1838.* I have obtained a patent here and it is exciting some attention. The prospects of future benefit from the invention are good, but I shall not probably realize much, or even anything, immediately.

"I saw by the papers, before I got your letter, that Congress had not passed the appropriation bill for the Telegraph. On some accounts I regret it, but it is only delayed, and it will probably be passed early in the winter."

Little did he think, in his cheerful optimism, that nearly five long years must elapse before Congress should awaken to its great opportunity.

"You will be glad to learn, my dear daughter, that your father's health was never so good, and probably before this reaches you he will be on the ocean on his return. I think of leaving Paris in a very few days. I am only waiting to show the Telegraph to the King, from whom I expect a message hourly. The birth of a prince occupies the whole attention just now of the royal family and the court. He was born on the 24th inst., the son of the Duke and Duchess of Orleans. My rooms are as delightfully situated, perhaps, as any in Paris; they are close to the palace of the Tuileries and overlook the gardens, and are within half a stone's throw of the rooms of the Duke and Duchess of Orleans. From my balcony I look directly into their rooms. I saw the company that was there assembled on the birthday of the little prince, and saw him in his nurse's arms at the window the next day after his birth. He looked very much like any other baby, and not half so handsome as little Hugh Peters.

"I received from the Minister of War, General Bernard, who has been very polite to me, a ticket to be present at the *Te Deum* performed yesterday in the great cathedral of Paris, Notre Dame, on account of the birth of the prince. The king and all the royal family and the court, with all the officers of state, were present. The cathedral was crowded with all the fashion of Paris. Along the ways and around the church were soldiers without number, almost; a proof that some danger was apprehended to the king, and yet he ought to be popular for he is the best ruler they have had for years. The ceremonies were imposing, appealing to the senses and the imagination, and not at all to the reason or the heart."

The king was Louis Philippe; the little prince, his grandson, was the Count of Paris.

"*Paris, September 29, 1838.* Since my last matters have assumed a totally different aspect. At the request of Monsieur Arago, the most distinguished astronomer of the day, I submitted the Telegraph to the Institute at one of their meetings, at which some of the most celebrated philosophers of France and of Germany and of other countries were present. Its reception was in the highest degree flattering, and the interest which they manifested, by the questions they asked and the exclamations they used, showed to me then that the invention had obtained their favorable regard. The papers of Paris immediately announced the Telegraph in the most favorable terms, and it has literally been the topic of the day ever since. The Baron Humboldt, the celebrated traveller, a member of the Institute and who saw its operation before that body, told Mr. Wheaton, our

Minister to Prussia, that my Telegraph was the best of all the plans that had been devised.

"I received a call from the administrator-in-chief of all the telegraphs of France, Monsieur Alphonse Foy. I explained it to him; he was highly delighted with it, and told me that the Government was about to try an experiment with the view of testing the practicability of the Electric Telegraph, and that he had been requested to see mine and report upon it; that he should report that '*mine was the best that had been submitted to him*'; and he added that I had better forthwith get an introduction to the Minister of the Interior, Mons. the Count Montalivet. I procured a letter from our Minister, and am now waiting the decision of the Government.

"Everything looks promising thus far, as much so as I could expect, but it involves the possibility, not to say the probability, of my remaining in Paris during the winter.

"If I should be delayed till December it would be prudent to remain until April. If it be possible, without detriment to my affairs, to make such arrangements that I may return this autumn, I shall certainly do it; but, if I should not, you must console yourselves that it is in consequence of meeting with success that I am detained, and that I shall be more likely to return with advantage to you all on account of the delay.

"I ought to say that the directors of the Saint-Germain Railroad have seen my Telegraph, and that there is some talk (as yet vague) of establishing a line of my Telegraph upon that road. I mention these, my dear child, to show you that I cannot at this moment leave Paris without detriment to my principal object."

"*Paris, October 10, 1838.* You are at an age when a parent's care, and particularly a mother's care, is most needed. You cannot know the depth of the wound that was inflicted when I was deprived of your dear mother, nor in how many ways that wound was kept open. Yet I know it is all well; I look to God to take care of you; it is his will that you should be almost truly an orphan, for, with all my efforts to have a home for you and to be near you, I have met hitherto only with disappointment. But there are now indications of a change, and, while I prepare for disappointment and wish you to prepare for disappointment, we ought to acknowledge the kind hand of our Heavenly Father in so far prospering me as to put me in the honorable light before the world which is now my lot. With the eminence is connected the prospect of pecuniary prosperity, yet this is not consummated, but only in prospect; it may be a long time before anything is realized. Study, therefore, prudence and economy in all things; make your wants as few as possible, for the habit thus acquired will be of advantage to you whether you have much or little."

Thus did hope alternate with despondency as the days and weeks wore away and nothing tangible was accomplished. All who saw the working of the telegraph were loud in their expressions of wonder and admiration, but, for reasons which shall presently be explained, nothing else was gained by the inventor at that time.

An old friend of Morse's, the Reverend Dr. Kirk, was then living in Paris, and the two friends not only roomed together but Dr. Kirk, speaking French fluently, which Morse did not, acted as interpreter in the many exhibitions given. Writing of this in later years, Dr. Kirk says: —

"I remember rallying my friend frequently about the experience of great inventors, who are generally permitted to starve while living and are canonized after death.

"When the model telegraph had been set up in our rooms, Mr. Morse desired to exhibit it to the savants of Paris, but, as he had less of the talking propensity than myself, I was made the grand exhibitor.

"Our levee-day was Tuesday, and for weeks we received the visits of distinguished citizens and strangers, to whom I explained the principles and operation of the Telegraph. The visitors would agree upon a word among themselves which I was not to hear; then the Professor would receive it at the writing end of the wires, while it devolved upon me to interpret the characters which recorded it at the other end. As I explained the hieroglyphics the announcement of the word, which they saw could have come to me only through the wire, would often create a deep sensation of delighted wonder; and much do I now regret that I did not take notes of these interviews, for it would be an interesting record of distinguished names and of valuable remarks."

On the 10th of September, 1838, Morse enjoyed the greatest triumph of all, for it was on that day that, by invitation of M. Arago, the exhibition of his invention before the Institute of France, casually mentioned in one of his letters to his daughter, took place. Writing of the occasion to Alfred Vail, he says: —

"I exhibited the Telegraph to the Institute and the sensation produced was as striking as at Washington.

It was evident that hitherto the assembled science of Europe had considered the plan of an Electric Telegraph as ingenious but visionary, and, like aëronautic navigation, practicable in little more than theory and destined to be useless.

"I cannot describe to you the scene at the Institute when your box with the registering-machine, just as it left Speedwell, was placed upon the table and surrounded by the most distinguished men of all Europe, celebrated in the various arts and sciences — Arago, Baron Humboldt, Gay-Lussac, and a host of others whose names are stars that shine in both hemispheres. Arago described it to them, and I showed its action. A buzz of admiration and approbation filled the whole hall and the exclamations '*Extraordinaire!*' '*Très bien!*' '*Très admirable!*' I heard on all sides. The sentiment was universal."

Another American at that time in Paris, the Honorable H. L. Ellsworth, also wrote home about the impression which was produced by the exhibition of this new wonder: —

"I am sure you will be glad to learn that our American friend, Professor Morse, is producing a very great sensation among the learned men of this kingdom by his ingenious and wonderful Magnetic Telegraph. He submitted it to the examination of the Academy of Sciences of the Royal Institute of France, at their sitting on Monday last, and the deepest interest was excited among the members of that learned body on the subject. Its novelty, beauty, simplicity, and power were highly commended. . . .

"Other projects for the establishment of a magnetic telegraph have been broached here, especially from Professor Wheatstone, of London, and Professor Steinheil, of Munich. It is said, however, to be very manifest that our Yankee Professor is ahead of them all in the essential requisitions of such an invention, and that he is in the way to bear off the palm. In simplicity of design, cheapness of construction and efficiency, Professor Morse's Telegraph transcends all yet made known. In each of these qualities it is admitted, by those who have inspected it closely, there seems to be little else to desire. It is certain, moreover, that in priority of discovery he antedates all others."

Encouraged by the universal praise which was showered upon him, the hopeful inventor redoubled his efforts to secure in some way, either through the Government or through private parties, the means to make a practical test of his invention.

Mr. F. O. J. Smith had, in the mean time, returned to America, and Morse kept him informed by letter of the progress of affairs in Paris. Avoiding, as far as possible, repetitions and irrelevant details, I shall let extracts from these letters tell the story: —

"*September 29, 1838.* On Monday I received a very flattering letter from our excellent Minister, Governor Cass, introducing me to the Count Montalivet, and I accordingly called the next day. I did not see him, but had an interview with his secretary, who told me that the Administrator of the Telegraphs had not yet reported to the Minister, but that he would see him the next day, and that, if I would call on Friday, he would inform me of the result. I called on Friday. The secretary informed me that he had seen M. Foy, and that he had more than confirmed the flattering accounts in the

American Minister's letter respecting the Telegraph, but was not yet prepared with his report to the Minister — he wished to make a detailed account of the *differences in favor of mine over all others that had been presented to him*, or words to that effect; and the secretary assured me that the report would be all I could wish. This is certainly flattering and I am to call on Monday to learn further."

"*October 24.* I can only add, in a few words, that everything here is as encouraging as could be expected. The report of the Administrator of Telegraphs has been made to the Minister of the Interior, and I have been told that I should be notified of the intentions of the Government in a few days. I have also shown the railroad telegraph to the Saint-Germain directors, who are delighted with it, and from them I expect a proposition within a few days."

"*November 22.* I intend sending this letter by the packet of the 24th inst., and am in hopes of sending with it some intelligence from those from whom I have been so long expecting something. Everything moves at a snail's pace here. I find delay in all things; at least, so it appears to me, who have too strong a development of the American organ of 'go-ahead-ativeness' to feel easy under its tantalizing effects. A Frenchman ought to have as many lives as a cat to bring to pass, on his dilatory plan of procedure, the same results that a Yankee would accomplish in his single life."

"*Afternoon, November 22.* Called on the Ministre de l'Intérieur; no one at home; left card and will call again to-morrow, and hope to be in time yet for the packet."

"*November 23.* I have again called, but do not find

at home the chief secretary, M. Merlin. . . . I shall miss the packet of the 24th, but I am told she is a slow ship and that I shall probably find the letters reach home quite as soon by the next. I will leave this open to add if anything occurs between this and next packet day."

"*November 30.* I have been called off from this letter until the last moment by stirring about and endeavoring to expedite matters with the Government. I have been to see General Cass since my last date. I talked over matters with him. He complains much of their dilatoriness, but sees no way of quickening them. . . . I called again this morning at the Minister's and, as usual, the secretary was absent; at the palace they said. If I could once get them to look at it I should be sure of them, for I have never shown it to any one who did not seem in raptures. I showed it a few days ago to M. Fremel, the Director of Light-Houses, who came with Mr. Vail and Captain Perry. He was cautious at first, but afterwards became as enthusiastic as any.

"The railroad directors are as dilatory as the Government, but I know they are discussing the matter seriously at their meetings, and I was told that the most influential man among them said they 'must have it.' There is nothing in the least discouraging that has occurred, but, on the contrary, everything to confirm the practicability of the plan, both on the score of science and expense."

"*January 21, 1839.* I learn that the Telegraph is much talked of in all society, and I learn that the *Théâtre des Variétés*, which is a sort of mirror of the popular topics, has a piece in which persons are made to converse by means of this Telegraph some hundreds of miles off.

This is a straw which shows the way of the wind, and although matters move too slow for my impatient spirit, yet the Telegraph is evidently gaining on the popular notice, and in time will demand the attention of Governments.

"I have the promise of a visit from the Count Boudy, Chief of the Household of the King, and who, I understand, has great influence with the king and can induce him to adopt the Telegraph between some of his palaces.

"Hopes, you perceive, continue bright, but they are somewhat unsubstantial to an empty purse. I look for the first fruits in America. My confidence increases every day in the certainty of the eventual adoption of this means of communication throughout the civilized world. Its practicability, hitherto doubted by savants here, is completely established, and they do not hesitate to give me the credit of having established it. I rejoice quite as much for my country's sake as for my own that both priority and superiority are awarded to my invention."

CHAPTER XXVI

JANUARY 6, 1839 — MARCH 9, 1839

Despondent letter to his brother Sidney. — Longing for a home. — Letter to Smith. — More delays. — Change of ministry. — Proposal to form private company. — Impossible under the laws of France. — Telegraphs a government monopoly. — Refusal of Czar to sign Russian contract. — Dr. Jackson. — M. Amyot. — Failure to gain audience of king. — Lord Elgin. — Earl of Lincoln. — Robert Walsh prophesies success. — Meeting with Earl of Lincoln in later years. — Daguerre. — Letter to Mrs. Cass on lotteries. — Railway and military telegraphs. — Skepticism of a Marshal of France.

Thus hopefully the inventor kept writing home, always maintaining that soon all obstacles would be overcome, and that he would then have a chance to demonstrate in a really practical way the great usefulness of his invention. But, instead of melting away, new obstacles kept arising at every turn. The dilatoriness of the French Government seems past all belief, and yet, in spite of his faith in the more expeditious methods of his own country, he was fated to encounter the same exasperating slowness at home. It was, therefore, only natural that in spite of the courageous optimism of his nature, he should at times have given way to fits of depression, as is instanced by the following extracts from a letter written to his brother Sidney on January 6, 1839: —

"I know not that I feel right to indulge in the despondency which, in spite of all reason to the contrary, creeps over me when I think of returning. I know the feelings of Tantalus perfectly. All my prospects in regard to the Telegraph are bright and encouraging, and so they have been for months, and they still continue to be so; but the sober *now* is that I am expending and not acquiring; it has, as yet, been all *outgo* and no *income.* At the rate business is done here, the slow, dilatory manner in which the most favorable projects are carried forward, I have no reason to believe that anything will be realized before I must leave France, which will probably be in about six weeks. If so, then I return penniless, and, worse than penniless, I return to find debts and no home; to find homeless children with all hope extinguished of ever seeing them again in a family. Indeed, I may say that, in this latter respect, the last ray is departed; I think no more of it.

"I now feel anxious to see my children educated with the means they have of their own, and in a way of usefulness, and for myself I desire to live secluded, without being burdensome to my friends. I should be glad to exchange my rooms in the university for one or two in your new building. I shall probably resign both Professorship and Presidency on my return. The first has become merely nominal, and the latter is connected with duties which properly confine to the city, and, as I wish to be free to go to other places, I think it will be best to resign.

"If our Government should take the Telegraph, or companies should be formed for that purpose, so that a sum is realized from it when I get home, this will, of course, change the face of things; but I dare not expect it and ought not to build any plans on such a contingency. So far as praise goes I have every reason to be satisfied at the state of things here in regard to the Telegraph. All the savants, committees of learned societies, members of the Chamber of Deputies, and officers of Government have, without exception, been as enthusiastic in its reception as any in the United States. Both the priority and superiority of my invention are established, and thus the credit, be it more or less, is secured to our country. The Prefect of the Seine expressed a desire to see it and called by appointment yesterday. He was perfectly satisfied, and said of his own accord that he should see the king last evening and should mention the Telegraph to him. I shall probably soon be requested, therefore, to show the Telegraph to the king.

"All these are most encouraging prospects; there is, indeed, nothing that has arisen to throw any insurmountable obstacle in the way of its adoption with complete success; and for all this I ought to feel gratitude, and I wish to acknowledge it before Him to whom gratitude is due. Is it right or is it wrong, in view of all this, to feel despondency?

"In spite of all I do feel sad. I am no longer young; I have children, but they are orphans, and orphans they are likely to be. I have a country, but *no home.* It is this *no home* that perpetually haunts me. I feel as if it were duty, duty most urgent, for me to settle in a family state at all hazards on account of these children. I know they suffer in this forming period of their lives for the want of a home, of the care of a father and a mother, and that no care and attention from friends, be they ever so kind, can supply the place of parents. But all efforts, direct and indirect, to bring this about have been frustrated.

"My dear brother, may you never feel, as I have felt, *the loss of a wife.* That wound bleeds afresh daily, as if it were inflicted but yesterday. There is a meaning in all these acute mental trials, and they are at times so severe as almost to deprive me of reason, though few around me would suspect the state of my mind."

These last few lines are eminently characteristic of the man. While called upon to endure much, both mentally and physically, he possessed such remarkable self-control that few, if any, of those around him were aware of his suffering. Only to his intimates did he ever reveal the pain which sometimes gnawed at his heart, and then only occasionally and under great stress. It was this self-control, united to a lofty purpose and a natural repugnance to wearing his heart on his sleeve, which enabled him to accomplish what he did. Endowed also with a saving sense of humor, he made light of his trials to others and was a welcome guest in every social gathering.

The want of a place which he could really call home was an ever-present grief. It is the dominant note in almost all the letters to his brothers and his children, and it is rather quaintly expressed in a letter, of November 14, 1838, to his daughter: —

"Tell Uncle Sidney to take good care of you, and to have a little snug room in the upper corner of his new building, where a bed can be placed, a chair, and a table, and let me have it as my own, that there may be one little particular spot which I can call *home*. I will there make three wooden stools, one for you, one for Charles, and one for Finley, and invite you to your father's house."

In spite of the enthusiasm which the exhibition of his invention aroused among the learned men and others in Paris, he met with obstructions of the most vexatious kind at every turn, in his effort to bring it into practical use. Just as the way seemed clear for its adoption by the French Government, something happened which is thus described in a letter to Mr. Smith, of January 28, 1839:

"I wrote by the Great Western a few days ago. The event then anticipated in regard to the Ministry has occurred. The Ministers have resigned, and it is expected that the new Cabinet will be formed this day with Marshal Soult at its head. Thus you perceive new causes of delay in obtaining any answer from the Government. As soon as I can learn the name of the new Minister of the Interior I will address a note to him, or see him, as I may be advised, and see if I can possibly obtain an answer, or at least a report of the administration of the Telegraphs. Nothing has occurred in other respects but what is agreeable. . . .

"All my leisure (if that may be called leisure which employs nearly all my time) is devoted to perfecting the whole matter. The invention of the correspondent, I think you will say, is a more essential improvement. It has been my winter's labor, and, to avoid expense, I have been compelled to make it entirely with my own hands. I can now give you its exact dimensions — twelve and a half inches long, six and a half wide, and six and a half deep. It dispenses entirely with boxes of type (one set alone being necessary) and dispenses also with the rules, and with all machinery for moving the rules. There is no winding up and it is ready at all times. You touch the letter and the letter is written immediately at the other extremity. . . . In my next I hope to send you reports of my further progress. One thing seems certain, my Telegraph has driven out of the field all the other plans on the magnetic principle. I hear nothing of them in public or private. No society notices them."

"*February 2.* I can compare the state of things here to an April day, at one moment sunshine, at the next cloudy. The Telegraph is evidently growing in favor; testimonials of approbation and compliments multiply, and yesterday I was advised by the secretary of the *Académie Industrielle* to interest moneyed men in the matter if I intended to profit by it; and he observed that now was the precise time to do it in the interval of the Chambers.

"I am at a loss how to act. I am not a business man and fear every movement which suggests itself to me. I am thinking of proposing a company on the same plan you last proposed in your letter from Liverpool, and which you intend to create in case the Government shall choose to do nothing; that is to say, a company taking the right at one thousand francs per mile, paying the proprietors fifty per cent in stocks and fifty per cent in cash, raising about fifty thousand francs for a trial some distance. I shall take advice and let you know the result.

"I wish you were here; I am sure something could be done by an energetic business man like yourself. As for poor me I feel that I am a child in business matters. I can invent and perfect the invention, and demonstrate its uses and practicability, but 'further the deponent saith not.' Perhaps I underrate myself in this case, but that is not a usual fault in human nature."

It was natural that a keen business man like F. O. J. Smith should have leaned rather toward a private corporation, with its possibilities of great pecuniary gain, than toward government ownership. Morse, on the contrary, would have preferred, both at home and abroad, to place the great power which he knew his invention was destined to wield in the hands of a responsible government. However, so eager was he to make a practical test of the telegraph that, governments apparently not appreciating their great opportunity, he was willing to entrust the enterprise to capitalists. Here again he was balked, however, for, writing of his trials later, he says: —

"An unforeseen obstacle was interposed which has rendered my patent in France of no avail to me. By the French patent law at the time one who obtained a patent was obliged to put into operation his invention within two years from the issue of his patent, under the penalty of forfeiture if he does not comply with the law. In pursuance of this requisition of the law I negotiated with the president (Turneysen) of the Saint-Germain Railroad Company to construct a line of my Telegraph on their road from Paris to Saint-Germain, a distance of about seven English miles. The company was favorably disposed toward the project, but, upon application (as was necessary) to the Government for permission to have the Telegraph on their road, they received for answer that telegraphs were a government monopoly, and could not, therefore, be used for private purposes. I thus found myself crushed between the conflicting forces of two opposing laws."

This was, indeed, a crushing blow, and ended all hope of accomplishing anything in France, unless the Government should, in the short time still left to him, decide to take it up. The letters home, during the remainder of his stay in Europe, are voluminous, but as they are, in the main, a repetition of experiences similar to those already recorded, it will not be necessary to give them in full. He tells of the enthusiastic reception accorded to his invention by the savants, the high officials of the Government and the Englishmen of note then stopping in Paris. He tells also of the exasperating delays to which he was subjected, and which finally compelled him to return home without having accomplished anything tangible. He goes at length into his negotiations with the representative of the Czar, Baron Meyendorf, from which he entertained so many hopes, hopes which were destined in the end to be blasted, because the Czar refused to put his signature to the contract, his objection being that "Malevolence can easily interrupt the communication." This was a terrible disappointment to the inventor, for he had made all his plans to return to Europe in the spring of 1839 to carry out the Russian contract, which he was led to believe was perfectly certain, and the Czar's signature simply a matter of form.

While at the time, and probably for all his life, Morse considered his failure in Europe as a cruel stroke of Fate, we cannot but conclude, in the light of future developments, that here again Fate was cruel in order to be kind. The invention, while it had been pronounced a scientific success, and had been awarded the palm over all other systems by the foremost scientists of the world, had yet to undergo the baptism of fire on the field of battle. It had never been tried over long distances in the open air, and many practical modifications had yet to be made, the necessity for which could only be ascertained during the actual construction of a commercial line. Morse's first idea, adhered to by him until found by experience, in the building of the first line between Washington and Baltimore, to be impracticable, had been to bury the wires in a trench in the ground. I say it was found to be impracticable, but that is true only of the conditions at that early date. The inventor was here again ahead of his time, for the underground system is now used in many cities, and may in time become universal. However, we shall see, when the story of the building of that first historic line is told, that in this respect, and in many others, great difficulties were encountered and failure was averted only by the ingenuity, the resourcefulness, and the quick-wittedness of the inventor himself and his able assistants. Is it too much to suppose that, had the Russian, or even the French, contract gone through, and had Morse been compelled to recruit his assistants from the people of an alien land, whose language he could neither speak nor thoroughly understand, the result would have been a dismal failure, calling down only ridicule on the head of the luckless inventor, and perhaps causing him to abandon the whole enterprise, discouraged and disheartened?

Be this as it may, the European trip was considered a failure in a practical sense, while having resulted in a personal triumph in so far as the scientific elements of the invention were concerned. I shall, therefore, give only occasional extracts from the letters, some of them dealing with matters not in any way related to the telegraph.

He writes to Mr. Smith on February 13, 1839: —

"I have been wholly occupied for the last week in copying out the correspondence and other documents to defend myself against the infamous attack of Dr. Jackson, notice of which my brother sent me. . . . I have sent a letter to Dr. Jackson calling on him to save his character by a total disclaimer of his presumptuous claim within one week from the receipt of the letter, and giving him the plea of a 'mistake' and 'misconception of my invention' by which he may retreat. If he fails to do this, I have requested my brother to publish immediately my defense, in which I give a history of the invention, the correspondence between Dr. Jackson and myself, and close with the letters of Hon. Mr. Rives, Mr. Fisher, of Philadelphia, and Captain Pell.

"I cannot conceive of such infatuation as has possessed this man. He can scarcely be deceived. It must be his consummate self-conceit that deceives him, if he is deceived. But this cannot be; he knows he has no title whatever to a single hint of any kind in the matter."

I have already alluded to the claim of Dr. Jackson, and have shown that it was proved to be utterly without foundation, and have only introduced this reference to it as an instance of the attacks which were made upon Morse, attacks which compelled him to consume much valuable time, in the midst of his other labors, in order to repel them, which he always succeeded in doing.

In writing of his negotiations with the Russian Government he mentions M. Amyot, "who has proposed also an Electric Telegraph, but upon seeing mine he could not restrain his gratification, and with his whole soul he is at work to forward it with all who have influence. He is the right-hand man of the Baron Meyendorf, and he is exerting all his power to have the Russian Government adopt my Telegraph. . . . He is really a noble-minded man. The baron told me he had a *large soul,* and I find he has. I have no claim on him and yet he seems to take as much interest in my invention as if it were his own. How different a conduct from Jackson's! . . . Every day is clearing away all the difficulties that prevent its adoption; the only difficulty that remains, it is universally said, is the protection of the wires from malevolent attack, and this can be prevented by proper police and secret and deep interment. I have no doubt of its universal adoption; it may take time but it is certain."

"*Paris, March 2, 1839.* By my last letter I informed you of the more favorable prospects of the telegraphic enterprise. These prospects still continue, and I shall return with the gratifying reflection that, after all my anxieties, and labors, and privations, and your and my other associates' expenditures and risks, we are all in a fair way of reaping the fruits of our toil. The political troubles of France have been a hindrance hitherto to the attention of the Government to the Telegraph, but in the mean time I have gradually pushed forward the invention into the notice of the most influential individuals of France. I had Colonel Lasalle, aide-de-camp to the king, and his lady to see the Telegraph a few days ago. He promised that, without fail, it should be mentioned to the king. You will be surprised to learn, after all the promises hitherto made by the Prefect of the Seine, Count Remberteau, and by various other officers of the Government, and after General Cass's letter to the aide on service, four or five months since, requesting it might be brought to the notice of the king, that the king has not yet heard of it. But so things go here.

Such dereliction would destroy a man with us in a moment, but here there is a different standard (this, of course, *entre nous*). . . . Among the numerous visitors that have thronged to see the Telegraph, there have been a great many of the principal English nobility. Among them the Lord and Lady Aylmer, former Governor of Canada, Lord Elgin and son, the celebrated preserver, not depredator (as he has been most slanderously called) of the Phidian Marbles. Lord Elgin has been twice and expressed a great interest in the invention. He brought with him yesterday the Earl of Lincoln, a young man of unassuming manners; he was delighted and gave me his card with a pressing invitation to call on him when I came to London.

"I have not failed to let the English know how I was treated in regard to my application for a patent in England, and contrasted the conduct of the French in this respect to theirs. I believe they felt it, and I think it was Lord Aylmer, but am not quite sure, who advised that the subject be brought up in Parliament by some member and made the object of special legislation, which he said might be done, the Attorney-General to the contrary notwithstanding. I really believe, if matters were rightly managed in England, something yet might be done there, if not by patent, yet by a parliamentary grant of a proper compensation. It is remarkable that they have not yet made anything like mine in England. It is evident that neither Wheatstone nor Davy comprehended my mode, after all their assertions that mine had been published.

"If matters move slower here than with us, yet they gain surely. I am told every hour that the two great wonders of Paris just now, about which everybody is conversing, are Daguerre's wonderful results in fixing permanently the image of the *camera obscura*, and Morse's Electro-Magnetic Telegraph, and they do not hesitate to add that, beautiful as are the results of Daguerre's experiments, the invention of the Electro-Magnetic Telegraph is that which will surpass, in the greatness of the revolution to be effected, all other inventions. Robert Walsh, Esq., who has just left me, is beyond measure delighted. I was writing a word from one room to another; he came to me and said: —'The next word you may write is IMMORTALITY, for the sublimity of this invention is of surpassing grandeur. *I see now that all physical obstacles, which may for a while hinder, will inevitably be overcome; the problem is solved;* MAN MAY INSTANTLY CONVERSE WITH HIS FELLOW-MEN IN ANY PART OF THE WORLD.'"

This prophecy of the celebrated American author, who was afterwards Consul-General to France for six years, is noteworthy considering the date at which it was made. There were indeed many "physical obstacles which for a while hindered" the practical adoption of the invention, but they were eventually overcome, and the problem was solved. Five years of heart-breaking struggle, discouragement and actual poverty had still to be endured by the brave inventor before the tide should turn in his favor, but Robert Walsh shared with Morse the clear conviction that the victory would finally be won.

Reference having been made to Lord Elgin, the following letter from him will be found interesting: —

PARIS, 12th March, 1839.

DEAR SIR, — I cannot help expressing a very strong desire that, instead of delaying till your return from America your wish to take out a patent in England for your highly scientific and simple mode of communicating intelligence by an Electric Telegraph, you would take measures to that effect at this moment, and for that purpose take your model now with you to London. Your discovery is now much known as well as appreciated, and the ingenuity now afloat is too extensive for one not to apprehend that individuals, even in good faith, may make some addition to qualify them to take out a *first patent* for the principle; whereas, if you brought it at once, now, before the competent authorities, especially under the advantage of an introduction such as Mr. Drummond can give you to Lord Brougham, a short delay in your proceeding to America may secure you this desirable object immediately.

With every sincere good wish for your success and the credit you so richly deserve, I am, dear sir,

Yours faithfully

ELGIN.

While it is futile to speculate on what might have been, it does seem as if Morse made a serious mistake in not taking Lord Elgin's advice, for there is no doubt that, with the influential backing which he had now secured, he could have overcome the churlish objections of the Attorney-General, and have secured a patent in England much to his financial benefit. But with the glamour of the Russian contract in his eyes, he decided to return home at once, and the opportunity was lost.

We must also marvel at the strange fact that the fear expressed by Lord Elgin, that another might easily appropriate to himself the glory which was rightly due to Morse, was not realized. Is it to be wondered at that Morse should have always held that he, and he alone, was the humble instrument chosen by an All-Wise Providence to carry to a successful issue this great enterprise?

Regarding one of his other visitors, the Earl of Lincoln, it is interesting to learn that there was another meeting between the two men under rather dramatic circumstances, in later years. This was on the occasion of the visit of the Prince of Wales, afterward Edward VII, to America, accompanied by a suite which included, among others, the Duke of Newcastle. Morse was invited to address the Prince at a meeting given in his honor at the University of the City of New York, and in the course of his address he said: —

"An allusion in most flattering terms to me, rendered doubly so in such presence, has been made by our respected Chancellor, which seems to call for at least the expression of my thanks. At the same time it suggests the relation of an incident in the early history of the Telegraph which may not be inappropriate to this occasion. The infant Telegraph, born and nursed within these walls, had scarcely attained a feeble existence ere it essayed to make its voice heard on the other side of the Atlantic. I carried it to Paris in 1838. It attracted the warm interest, not only of the continental philosophers, but also of the intelligent and appreciative among the eminent nobles of Britain then on a visit to the French capital. Foremost among these was the late

Marquis of Northampton, then President of the Royal Society, the late distinguished Earl of Elgin, and, in a marked degree, the noble Earl of Lincoln. The last-named nobleman in a special manner gave it his favor. He comprehended its important future, and, in the midst of the skepticism that clouded its cradle, he risked his character for sound judgment in venturing to stand godfather to the friendless child. He took it under his roof in London, invited the statesmen and the philosophers of Britain to see it, and urged forward with kindly words and generous attentions those who had the infant in charge. It is with no ordinary feelings, therefore, that, after the lapse of twenty years, I have the singular honor this morning of greeting with hearty welcome, in such presence, before such an assemblage, and in the cradle of the Telegraph, this noble Earl of Lincoln in the person of the present Duke of Newcastle."

Reference was made by Morse, in the letter to Mr. Smith of March 2, to Daguerre and his wonderful discovery. Having himself experimented along the same lines many years before, he was, naturally, much interested and sought the acquaintance of Daguerre, which was easily brought about. The two inventors became warm friends, and each disclosed to the other the minutiæ of his discoveries. Daguerre invited Morse to his workshop, selecting a Sunday as a day convenient to him, and Morse replied in the following characteristic note:—

"Professor Morse asks the indulgence of M. Daguerre. The *time* M. Daguerre, in his great kindness, has fixed to show his most interesting experiments is, unfortunately, one that will deprive Mr. M. of the pleasure he anticipated, as Mr. M. has an engagement for the entire Sunday of a nature that cannot be broken. Will Monday, or any other day, be agreeable to M. Daguerre?

"Mr. M. again asks pardon for giving M. Daguerre so much trouble."

Having thus satisfied his Puritan conscience, another day was cheerfully appointed by Daguerre, who generously imparted the secret of this new art to the American, by whom it was carried across the ocean and successfully introduced into the United States, as will be shown further on.

Writing of this experience to his brothers on March 9, 1839, he says: —

"You have, perhaps, heard of the Daguerreotype, so called from the discoverer, M. Daguerre. It is one of the most beautiful discoveries of the age. I don't know if you recollect some experiments of mine in New Haven, many years ago, when I had my painting-room next to Professor Silliman's, — experiments to ascertain if it were possible to fix the image of the *camera obscura*. I was able to produce different degrees of shade on paper, dipped into a solution of nitrate of silver, by means of different degrees of light, but finding that light produced dark, and dark light, I presumed the production of a true image to be impracticable, and gave up the attempt. M. Daguerre has realized in the most exquisite manner this idea."

Here follows the account of his visit to Daguerre and an enthusiastic description of the wonders seen in his workshop, and he closes by saying: —

"But I am near the end of my paper, and I have, unhappily, to give a melancholy close to my account of this ingenious discovery. M. Daguerre appointed yesterday at noon to see my Telegraph. He came and passed more than an hour with me, expressing himself highly gratified at its operation. But, while he was thus employed, the great building of the Diorama, with his own house, all his beautiful works, his valuable notes and papers, the labor of years of experiment, were, unknown to him, at that moment the prey of the flames. His secret, indeed, is still safe with him, but the steps of his progress in the discovery and his valuable researches in science, are lost to the scientific world. I learn that his Diorama was insured, but to what extent I know not.

"I am sure all friends of science and improvement will unite in expressing the deepest sympathy in M. Daguerre's loss, and the sincere hope that such a liberal sum will be awarded him by his Government as shall enable him, in some degree at least, to recover from his loss."

It is pleasant to record that the French Government did act most generously toward Daguerre.

The reader may remember that, when Morse was a young man in London, lotteries were considered such legitimate ways of raising money, that not only did he openly purchase tickets in the hope of winning a money prize, but his pious father advised him to dispose of his surplus paintings and sketches in that way. As he grew older, however, his views on this question changed, as will be seen by the following letter addressed to Mrs. Cass, wife of the American Minister, who was trying to raise money to help a worthy couple, suddenly reduced from wealth to poverty: —

January 31, 1839.

I am sure I need make no apology to you, my dear madam, for returning the three lottery tickets enclosed in the interesting note I have just had the honor to receive from you, because I know you can fully appreciate the motive which prompts me. In the measures taken some years since for opposing the lottery system in the State of New York, and which issued in its entire suppression, I took a very prominent part under the conviction that the principle on which the lottery system was founded was wrong. But while, on this account, I cannot, my dear madam, consistently take the tickets, I must beg of you to put the price of them, which I enclose, into such a channel as shall, in your judgment, best promote the benevolent object in which you have interested yourself.

Poverty is a bitter lot, even when the habit of long endurance has reconciled the mind and body to its severities, but how much more bitter must it be when it comes in sudden contrast to a life of affluence and ease.

I thank you for giving me the opportunity of contributing my mite to the relief of such affliction, hoping sincerely that all their earthly wants may lead the sufferers to the inexhaustible fountain of true riches.

With sincere respect and Christian regard I remain, my dear madam

Your most obedient servant

S. F. B. Morse.

Before closing the record of this European trip, so disappointing in many ways and yet so encouraging in others, it may be well to note that, while he was in Paris, Morse in 1838 not only took out a patent on his recording telegraph, but also on a system to be used on railways to report automatically the presence of a train at any point on the line. A reproduction of his own drawing of the apparatus to be used is here given, and the mechanism is so simple that an explanation is hardly necessary. From it can be seen not only that he did, at this early date, realize the possibilities of his invention along various lines, but that it embodies the principle of the police and fire-alarm systems now in general use.

It is not recorded that he ever realized anything financially from this ingenious modification of his main invention. Commenting on it, and on his plans for a military telegraph, he gives this amusing sketch: —

"On September 10, 1838, a telegraph instrument constructed in the United States on the same principles, but slightly modified to make it portable, was exhibited to the Academy of Sciences in Paris, and explained by M. Arago at the session of that date. An account of this exhibition is recorded in the *Comptes Rendus*.

"A week or two after I exhibited at my lodgings, in connection with this instrument, my railroad telegraph, an application of signals by sound, for which I took out letters patent in Paris, and at the same time I communicated to the Minister of War, General Bernard, my plans for a military telegraph with which he was much pleased.

"I dined with him by invitation, and in the evening, repairing with him to his billiard-room, while the rest of the guests were amusing themselves with the game, I gave him a general description of my plan. He listened with deep attention while I advocated its use on the

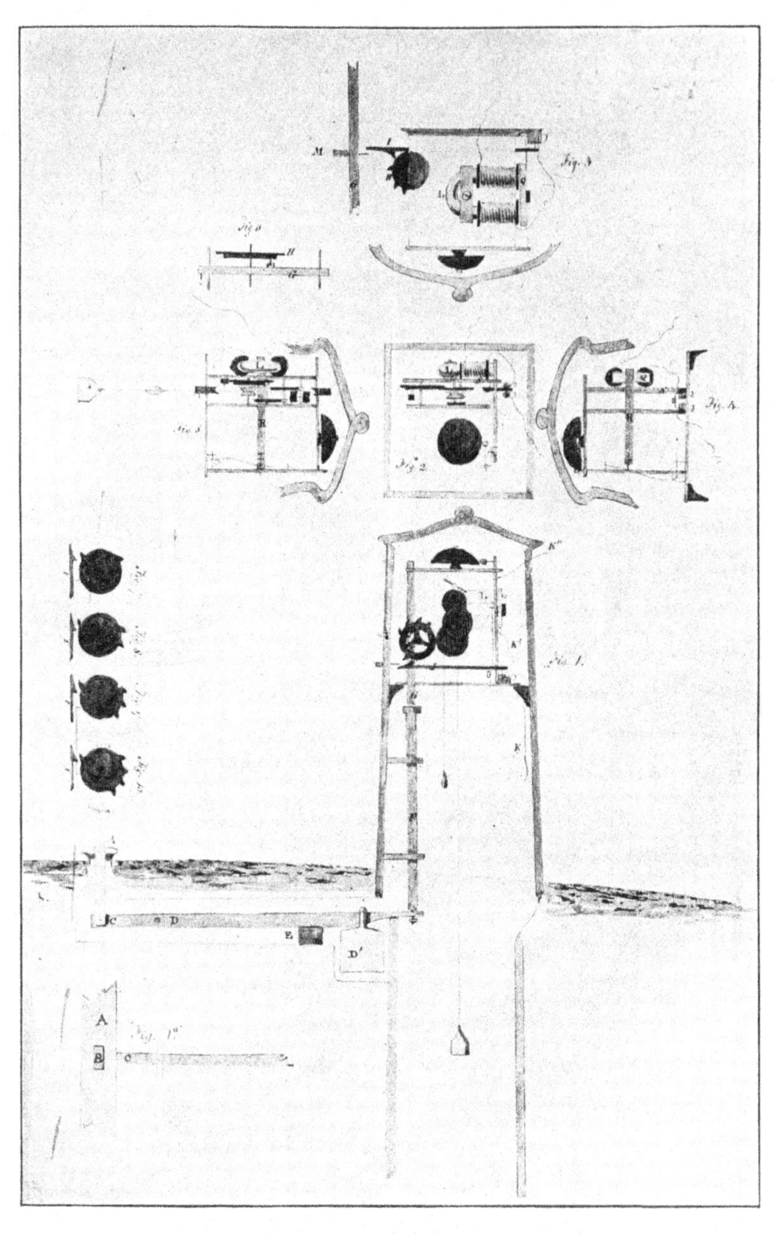

RAILWAY TELEGRAPH DRAWING BY MORSE

Patented by him in France in 1838, and embodying principle of Police and Fire Alarm Telegraph

battle-field, and gave him my reasons for believing that the army first using the facilities of the electric telegraph for military purposes would be sure of victory. He replied to me, after my answering many of his questions: —

"'Be reticent,' said he, 'on this subject for the present. I will send an officer of high rank to see and converse with you on the matter to-morrow.'

"The next day I was visited by an old Marshal of France, whose name has escaped my memory. Conversing by an interpreter, the Reverend E. N. Kirk, of Boston, I found it difficult to make the Marshal understand its practicability or its importance. The dominant idea in the Marshal's mind, which he opposed to the project, was that it involved an increase of the material of the army, for I proposed the addition of two or more light wagons, each containing in a small box the telegraph instruments and a reel of fine insulated wire to be kept in readiness at the headquarters on the field. I proposed that, when required, the wagons with the corps of operators, two or three persons, at a rapid rate should reel off the wire to the right, the centre and the left of the army, as near to these parts of the army as practicable or convenient, and thus instantaneous notice of the condition of the whole army, and of the enemy's movements, would be given at headquarters.

"To all this explanation of my plan was opposed the constant objection that it increased the material of the army. The Hon. Marshal seemed to consider that the great object to be gained by an improvement was a decrease of this material; an example of this economy which he illustrated by the case of the substitution of the leather drinking cup for the tin cup hung to the soldier's knapsack, an improvement which enabled the soldier to put his cup in his vest pocket. For this improvement, if I remember right, he said the inventor, who was a common soldier, received at the hands of the Emperor Napoleon I the cross of the Legion of Honor.

"So set was the good Marshal in his repugnance to any increase to the material of the army that, after a few moments' thought, I rebutted his position by putting to him the following case: —

"'M. Marshal,' I said, 'you are investing a fortress on the capture of which depends the success of your campaign; you have 10,000 men; on making your calculations of the chances of taking it by assault, you find that with the addition of 5000 more troops you could accomplish its capture. You have it in your power, by a simple order, to obtain from the Government these 5000 men. In this case what would you do?'

"He replied without hesitation: 'I should order the 5000, of course.'

"'But,' I rejoined, 'the material of the army would be greatly increased by such an order.'

"He comprehended the case, and, laughing heartily, abandoned the objection, but took refuge in the general skepticism of that day on the practicability of an electric telegraph. He did not believe it could ever be put in practise. This was an argument I could not then repel. Time alone could vindicate my opinion, and time has shown both its practicability and its utility."

CHAPTER XXVII

APRIL 15, 1839 — SEPTEMBER 30, 1840

Arrival in New York. — Disappointment at finding nothing done by Congress or his associates. — Letter to Professor Henry. — Henry's reply. — Correspondence with Daguerre. — Experiments with Daguerreotypes. — Professor Draper. — First group photograph of a college class. — Failure of Russian contract. — Mr. Chamberlain. — Discouragement through lack of funds. — No help from his associates. — Improvements in telegraph made by Morse. — Humorous letter.

MORSE sailed from Europe on the Great Western on the 23d of March, 1839, and reached New York, after a stormy passage, on the 15th of April. Discouraged by his lack of success in establishing a line of telegraph in Europe on a paying basis, and yet encouraged by the enthusiasm shown by the scientists of the Old World, he hoped much from what he considered the superior enterprise of his own countrymen. However, on this point he was doomed to bitter disappointment, and the next few years were destined to be the darkest through which he was to pass.

On the day after his arrival in New York he wrote to Mr. F. O. J. Smith: —

"I take the first moment of rest from the fatigues of my boisterous voyage to apprise you of my arrival yesterday in the Great Western. . . . I am quite disappointed in finding nothing done by Congress, and nothing accomplished in the way of company. I had hoped to find on my return some funds ready for prosecuting with vigor the enterprise, which I fear will suffer for the want.

"Think a moment of my situation. I left New York

for Europe to be gone three months, but have been gone eleven months. My only means of support are in my profession, which I have been compelled to abandon entirely for the present, giving my undivided time and efforts to this enterprise. I return with not a farthing in my pocket, and have to borrow even for my meals, and even worse than this, I have incurred a debt of rent by my absence which I should have avoided if I had been at home, or rather if I had been aware that I should have been obliged to stay so long abroad. I do not mention this in the way of complaint, but merely to show that I also have been compelled to make great sacrifices for the common good, and am willing to make more yet if necessary. If the enterprise is to be pursued, we must all in our various ways put the shoulder to the wheel.

"I wish much to see you and talk over all matters, for it seems to me that the present state of the enterprise in regard to Russia affects vitally the whole concern."

Thus gently did he chide one of his partners, who should have been exerting himself to forward their joint interests in America while he himself was doing what he could in Europe. The other partners, Alfred Vail and Dr. Leonard Gale, were equally lax and seem to have lost interest in the enterprise, as we learn from the following letter to Mr. Smith, of May 24, 1839: —

"You will think it strange, perhaps, that I have not answered yours of the 28th ult. sooner, but various causes have prevented an earlier attention to it. My affairs, in consequence of my protracted absence and the stagnant state of the Telegraph here at home, have caused me great embarrassment, and my whole energies have been called upon to extricate myself from the confusion in which I have been unhappily placed. You may judge a little of this when I tell you that my absence has deprived me of my usual source of income by my profession; that the state of the University is such that I shall probably leave, and shall have to move into new quarters; that my family is dispersed, requiring my care and anxieties under every disadvantage; that my engagements were such with Russia that every moment of my time was necessary to complete my arrangements to fulfill the contract in season; and, instead of finding my associates ready to sustain me with counsel and means, I find them all dispersed, leaving me without either the opportunity to consult or a cent of means, and consequently bringing everything in relation to the Telegraph to a dead stand.

"In the midst of this I am called on by the state of public opinion to defend myself against the outrageous attempt of Dr. Jackson to pirate from me my invention. The words would be harsh that are properly applicable to this man's conduct. . . .

"You see, therefore, in what a condition I found myself when I returned. I was delayed several days beyond the computed time of my arrival by the long passage of the steamer. Instead of finding any funds by a vote of Congress, or by a company, and my associates ready to back me, I find not a cent for the purpose, and my associates scattered to the four winds.

"You can easily conceive that I gave up all as it regarded Russia, and considered the whole enterprise as seriously injured if not completely destroyed. In this state of things I was hourly dreading to hear from the Russian Minister, and devising how I should save myself and the enterprise without implicating my associates in a charge of neglect; and as it has most fortunately happened for us all, the 10th of May has passed without the receipt of the promised advices, and I took advantage of this, and by the Liverpool steamer of the 18th wrote to the Baron Meyendorff, and to M. Amyot, that it was impossible to fulfill the engagement this season, since I had not received the promised advices in time to prepare."

This was, of course, before he had heard of the Czar's refusal to sign the contract, and he goes on to make plans for carrying out the Russian enterprise the next year, and concludes by saying: —

"Do think of this matter and see if means cannot be raised to keep ahead with the American Telegraph. I sometimes am astonished when I reflect how I have been able to take the stand with my Telegraph in competition with my European rivals, backed as they are with the purses of the kings and wealthy of their countries, while our own Government leaves me to fight their battles for the honor of this invention fettered hand and foot. Thanks will be due to you, not to them, if I am able to maintain the ground occupied by the American Telegraph."

Shortly after his return from abroad, on April 24, Morse wrote the following letter to Professor Henry at Princeton: —

My dear Sir, — On my return a few days since from Europe, I found directed to me, through your politeness, a copy of your valuable "Contributions," for which I beg you to accept my warmest thanks. The various cares consequent upon so long an absence from home, and which have demanded my more immediate attention, have prevented me from more than a cursory perusal of its interesting contents, yet I perceive many things of great interest to me in my telegraphic enterprise.

I was glad to learn, by a letter received in Paris from Dr. Gale, that a spool of five miles of my wire was loaned to you, and I perceive that you have already made some interesting experiments with it.

In the absence of Dr. Gale, who has gone South, I feel a great desire to consult some scientific gentleman on points of importance bearing upon my Telegraph, which I am about to establish in Russia, being under an engagement with the Russian Government agent in Paris to return to Europe for that purpose in a few weeks. I should be exceedingly happy to see you and am tempted to break away from my absorbing engagements here to find you at Princeton. In case I should be able to visit Princeton for a few days a week or two hence, how should I find you engaged? I should come as a learner and could bring no "contributions" to your stock of experiments of any value, nor any means of furthering your experiments except, perhaps, the loan of an additional five miles of wire which it may be desirable for you to have.

I have many questions to ask, but should be happy, in your reply to this letter, of an answer to this general one: Have you met with any facts in your experiments thus far that would lead you to think that my mode of telegraphic communication will prove impracticable? So far as I have consulted the savants of Paris, they have suggested no insurmountable difficulties; I have, however, quite as much confidence in your judgment, from your valuable experience, as in that of any one I have met abroad. I think that you have pursued an original course of experiments, and discovered facts of more value to me than any that have been published abroad.

Morse was too modest in saying that he could bring nothing of value to Henry in his experiments, for, as we shall see from Henry's reply, the latter had no knowledge at that time of the "relay," for bringing into use a secondary battery when the line was to stretch over long distances. This important discovery Morse had made several years before.

PRINCETON, May 6, 1839.

DEAR SIR, — Your favor of the 24th ult. came to Princeton during my absence, which will account for the long delay of my answer. I am pleased to learn that you fully sanction the loan which I obtained from Dr. Gale of your wire, and I shall be happy if any of the results are found to have a practical bearing on the electrical telegraph.

It will give me much pleasure to see you in Princeton after this week. My engagements will not then interfere with our communications on the subject of electricity. During this week I shall be almost constantly engaged with a friend in some scientific labors which we are prosecuting together.

I am acquainted with no fact which would lead me to suppose that the project of the electro-magnetic telegraph is impractical; on the contrary, I believe that science is now ripe for the application, and that there are no difficulties in the way but such as ingenuity and enterprise may obviate. But what form of the apparatus, or what application of the power will prove best, can, I believe, be only determined by careful experiment. I can say, however, that, so far as I am acquainted with the minutiæ of your plan, I see no practical difficulty in the way of its application for comparatively short distances; but, if the length of the wire between the stations is great, I think that some other modification will be found necessary in order to develop a sufficient power at the farther end of the line.

I shall, however, be happy to converse freely with you on these points when we meet. In the meantime I remain, with much respect

Yours, etc.,

JOSEPH HENRY.

I consider this letter alone a sufficient answer to those who claim that Henry was the real inventor of the telegraph. He makes no such claim himself.

In spite of the cares of various kinds which overwhelmed him during the whole of his eventful life, Morse always found time to stretch out a helping hand to others, or to do a courteous act. So now we find him writing to Daguerre on May 20, 1839: —

MY DEAR SIR, — I have the honor to enclose you the note of the Secretary of our Academy informing you of your election, at our last annual meeting, into the board of Honorary Members of our National Academy of Design. When I proposed your name it was received with enthusiasm, and the vote was *unanimous*. I hope, my dear sir, you will receive this as a testimonial, not merely of my personal esteem and deep sympathy in your late losses, but also as a proof that your genius is, in some degree, estimated on this side of the water.

Notwithstanding the efforts made in England to give to another the credit which is your due, I think I may with confidence assure you that throughout the United States your name alone will be associated with the brilliant discovery which justly bears your name. The letter I wrote from Paris, the day after your sad loss, has been published throughout this whole country in hundreds of journals, and has excited great interest. Should any attempts be made here to give to any other than yourself the honor of this discovery, my pen is ever ready for your defense.

I hope, before this reaches you, that the French Government, long and deservedly celebrated for its generosity to men of genius, will have amply supplied all your losses by a liberal sum. If, when the proper remuneration shall be secured to you in France, you should think it may be for your advantage to make an arrangement with the government to hold back the secret for six months or a year, and would consent to an exhibition of your *results* in this country for a short time, the exhibition might be managed, I think, to your pecuniary advantage. If you should think favorably of the plan, I offer you my services *gratuitously*.

To this letter Daguerre replied on July 26: —

My dear Sir,— I have received with great pleasure your kind letter by which you announce to me my election as an honorary member of the National Academy of

Design. I beg you will be so good as to express my thanks to the Academy, and to say that I am very proud of the honor which has been conferred upon me. I shall seize all opportunities of proving my gratitude for it. I am particularly indebted to you in this circumstance, and I feel very thankful for this and all other marks of interest you bestowed upon me.

The transaction with the French Government being nearly at an end, my discovery shall soon be made public. This cause, added to the immense distance between us, hinders me from taking the advantage of your good offer to get up at New York an exhibition of my results.

Believe me, my dear sir, your very devoted servant,

DAGUERRE.

A prophecy, shrewd in some particulars but rather faulty in others, of the influence of this new art upon painting, is contained in the following extracts from a letter of Morse's to his friend and master Washington Allston: —

"I had hoped to have seen you long ere this, but my many avocations have kept me constantly employed from morning till night. When I say morning I mean *half past four* in the morning! I am afraid you will think me a Goth, but really the hours from that time till twelve at noon are the richest I ever enjoy.

"You have heard of the Daguerreotype. I have the instruments on the point of completion, and if it be possible I will yet bring them with me to Boston, and show you the beautiful results of this brilliant discovery. Art is to be wonderfully enriched by this discovery.

How narrow and foolish the idea which some express that it will be the ruin of art, or rather artists, for every one will be his own painter. One effect, I think, will undoubtedly be to banish the sketchy, slovenly daubs that pass for spirited and learned; those works which possess mere general effect without detail, because, forsooth, detail destroys general effect. Nature, in the results of Daguerre's process, has taken the pencil into her own hands, and she shows that the minutest detail disturbs not the general repose. Artists will learn how to paint, and amateurs, or rather connoisseurs, how to criticise, how to look at Nature, and, therefore, how to estimate the value of true art. Our studies will now be enriched with sketches from nature which we can store up during the summer, as the bee gathers her sweets for winter, and we shall thus have rich materials for composition and an exhaustless store for the imagination to feed upon."

An interesting account of his experiences with this wonderful new discovery is contained in a letter written many years later, on the 10th of February, 1855: —

"As soon as the necessary apparatus was made I commenced experimenting with it. The greatest obstacle I had to encounter was in the quality of the plates. I obtained the common, plated copper in coils at the hardware shops, which, of course, was very thinly coated with silver, and that impure. Still I was able to verify the truth of Daguerre's revelations. The first experiment crowned with any success was a view of the Unitarian Church from the window on the staircase from the third story of the New York City University. This, of course, was before the building of the New York

Hotel. It was in September, 1839. The time, if I recollect, in which the plate was exposed to the action of light in the camera was about fifteen minutes. The instruments, chemicals, etc., were strictly in accordance with the directions in Daguerre's first book.

"An English gentleman, whose name at present escapes me, obtained a copy of Daguerre's book about the same time with myself. He commenced experimenting also. But an American of the name of Walcott was very successful with a modification of Daguerre's apparatus, substituting a metallic reflector for the lens. Previous, however, to Walcott's experiments, or rather results, my friend and colleague, Professor John W. Draper, of the New York City University, was very successful in his investigations, and with him I was engaged for a time in attempting portraits.

"In my intercourse with Daguerre I specially conversed with him in regard to the practicability of taking portraits of living persons. He expressed himself somewhat skeptical as to its practicability, only in consequence of the time necessary for the person to remain immovable. The time for taking an outdoor view was from fifteen to twenty minutes, and this he considered too long a time for any one to remain sufficiently still for a successful result. No sooner, however, had I mastered the process of Daguerre than I commenced to experiment with a view to accomplish this desirable result. I have now the results of these experiments taken in September, or beginning of October, 1839. They are full-length portraits of my daughter, single, and also in group with some of her young friends. They were taken out of doors, on the roof of a building, in the full sunlight and with the eyes closed. The time was from ten to twenty minutes.

"About the same time Professor Draper was successful in taking portraits, though whether he or myself took the first portrait successfully, I cannot say."

It was afterwards established that to Professor Draper must be accorded this honor, but I understand that it was a question of hours only between the two enthusiasts.

"Soon after we commenced together to take portraits, causing a glass building to be constructed for that purpose on the roof of the University. As our experiments had caused us considerable expense, we made a charge to those who sat for us to defray this expense. Professor Draper's other duties calling him away from the experiments, except as to their bearing on some philosophical investigations which he pursued with great ingenuity and success, I was left to pursue the artistic results of the process, as more in accordance with my profession. My expenses had been great, and for some time, five or six months, I pursued the taking of portraits by the Daguerreotype as a means of reimbursing these expenses. After this object had been attained, I abandoned the practice to give my exclusive attention to the Telegraph, which required all my time."

Before leaving the subject of the Daguerreotype, in which, as I have shown, Morse was a pioneer in this country, it will be interesting to note that he took the first group photograph of a college class. This was of the surviving members of his own class of 1810, who returned to New Haven for their thirtieth reunion in 1840.

It was not until August of the year 1839 that definite news of the failure of the Russian agreement was received, and Morse, in a letter to Smith, of August 12, comments on this and on another serious blow to his hopes: —

"I received yours of the 2d inst., and the paper accompanying it containing the notice of Mr. Chamberlain. I had previously been apprised that my forebodings were true in regard to his fate. . . . Our enterprise abroad is destined to give us anxiety, if not to end in disappointment.

"I have just received a letter from M. Amyot, who was to have been my companion to Russia, and learn from him the unwelcome news that the Emperor has decided against the Telegraph. . . . The Emperor's objections are, it seems, that 'malevolence can easily interrupt the communication.' M. Amyot scouts the idea, and writes that he refuted the objection to the satisfaction of the Baron, who, indeed, did not need the refutation for himself, for the whole matter was fully discussed between us when in Paris. The Baron, I should judge from the tone of M. Amyot's letter, was much disappointed, yet, as a faithful and obedient subject of one whose nay is nay, he will be cautious in so expressing himself as to be self-committed.

"Thus, my dear sir, prospects abroad look dark. I turn with some faint hope to my own country again. Will Congress do anything, or is my time and your generous zeal and pecuniary sacrifice to end only in disappointment? If so, I can bear it for myself, but I feel it most keenly for those who have been engaged with me; for you, for the Messrs. Vail and Dr. Gale. But I will yet hope. I don't know that our enterprise looks darker than Fulton's once appeared. There is no intrinsic difficulty; the depressing causes are extrinsic. I hope to see you soon and talk over all our affairs."

Mr. Smith, in sending a copy of the above letter to Mr. Prime, thus explains the reference to Mr. Chamberlain: —

"The allusion made in the letter just given to the fate of Mr. Chamberlain, was another depressing disappointment which occurred to the Professor comtemporaneously with those of the Russian contract. Before I left Paris we had closed a contract with Mr. Chamberlain to carry the telegraph to Austria, Prussia, the principal cities of Greece and of Egypt, and put it upon exhibition with a view to its utilization there. He was an American gentleman (from Vermont, I think) of large wealth, of eminent business capacities, of pleasing personal address and sustaining a character for strict integrity. He parted with Professor Morse in Paris to enter upon his expedition, with high expectations of both pleasure and profit, shortly after my own departure from Paris in October, 1838. He had subsequently apprised Professor Morse of very interesting exhibitions of the telegraph which he had made, and under date of Athens, January 5, 1839, wrote as follows: 'We exhibited your telegraph to the learned of Florence, much to their gratification. Yesterday evening the King and Queen of Greece were highly delighted with its performance. We have shown it also to the principal inhabitants of Athens, by all of whom it was much admired. Fame is all you will get for it in these poor countries. We think of starting in a few days for Alexandria, and hope to get something worth having from Mehemet Ali. It is, however, doubtful. Nations appear as poor as individuals, and as unwilling to risk their money upon such matters. I hope the French will avail themselves of the benefits you offer them. It is truly strange that it is not grasped at with more avidity. If I can do anything in Egypt, I will try Turkey and St. Petersburg.'"

Morse himself writes: "In another letter from Mr. Chamberlain to Mr. Lovering, dated Syra, January 9, he says: 'The pretty little Queen of Greece was delighted with Morse's telegraph. The string which carried the cannon-ball used for a weight broke, and came near falling on Her Majesty's toes, but happily missed, and we, perhaps, escaped a prison. My best respects to Mr. Morse, and say I shall ask Mehemet Ali for a purse, a beauty from his seraglio, and something else.'" And Morse concludes: "I will add that, if he will bring me the purse just now, I can dispense with the beauty and the something else."

Tragedy too often treads on the heels of comedy, and it is sad to have to relate that Mr. Chamberlain and six other gentlemen were drowned while on an excursion of pleasure on the Danube in July of 1839.

That all these disappointments, added to the necessity for making money in some way for his bare subsistence, should have weighed on the inventor's spirits, is hardly to be wondered at; the wonder is rather that he did not sink under his manifold trials. Far from this, however, he only touches on his needs in the following letter to Alfred Vail, written on November 14, 1839:—

"As to the Telegraph, I have been compelled from necessity to apply myself to those duties which yield immediate pecuniary relief. I feel the pressure as well as others, and, having several pupils at the University, I must attend to them. Nevertheless, I shall hold myself ready in case of need to go to Washington during the next session with it. The one I was constructing is completed except the rotary batteries and the pen-and-ink apparatus, which I shall soon find time to add if required.

"Mr. Smith expects me in Portland, but I have not the means to visit him. The telegraph of Wheatstone is going ahead in England, even with all its complications; so, I presume, is the one of Steinheil in Bavaria. Whether ours is to be adopted depends on the Government or on a company, and the times are not favorable for the formation of a company. Perhaps it is the part of wisdom to let the matter rest and watch for an opportunity when times look better, and which I hope will be soon."

He gives freer vent to his disappointment in a letter to Mr. Smith, of November 20, 1839:—

"I feel the want of that sum which Congress ought to have appropriated two years ago to enable me to compete with my European rivals. Wheatstone and Steinheil have money for their projects; the former by a company, and the latter by the King of Bavaria. Is there any national feeling with us on the subject? I will not say there is not until after the next session of Congress. But, if there is any cause for national exultation in being not merely *first* in the invention as to time, but *best* too, as decided by a foreign tribunal, ought the inventor to be suffered to work with his hands tied? Is it honorable to the nation to boast of its inventors, to contend for the credit of their inventions as national property, and not lift a finger to assist them to perfect that of which they boast?

"But I will not complain for myself. I can bear it, because I made up my mind from the very first for this issue, the common fate of all inventors. But I do not feel so agreeable in seeing those who have interested themselves in it, especially yourself, suffer also. Perhaps I look too much on the unfavorable side. I often thus look, not to discourage others or myself, but to check those too sanguine expectations which, with me, would rise to an inordinate height unless thus reined in and disciplined.

"Shall you not be in New York soon? I wish much to see you and to concoct plans for future operations. I am at present much straitened in means, or I should yet endeavor to see you in Portland; but I must yield to necessity and hope another season to be in different and more prosperous circumstances."

Thus the inventor, who had hoped so much from the energy and business acumen of his own countrymen, found that the conditions at home differed not much from those which he had found so exasperating abroad. Praise in plenty for the beauty and simplicity of his invention, but no money, either public or private, to enable him to put it to a practical test. His associates had left him to battle alone for his interests and theirs. F. O. J. Smith was in Portland, Maine, attending to his own affairs; Professor Gale was in the South filling a professorship; and Alfred Vail was in Philadelphia. No one of them, as far as I can ascertain, was doing anything to help in this critical period of the enterprise which was to benefit them all.

When credit is to be awarded to those who have accomplished something great, many factors must be taken into consideration. Not only must the aspirant for undying fame in the field of invention, for instance, have discovered something new, which, when properly applied, will benefit mankind, but he must prove its practical value to a world constitutionally skeptical, and he must persevere through trials and discouragements of every kind, with a sublime faith in the ultimate success of his efforts, until the fight be won. Otherwise, if he retires beaten from the field of battle, another will snatch up his sword and hew his way to victory.

It must never be forgotten that Morse won his place in the Hall of Fame, not only because of his invention of the simplest and best method of conveying intelligence by electricity, but because he, alone and unaided, carried forward the enterprise when, but for him, it would have been allowed to fail. With no thought of disparaging the others, who can hardly be blamed for their loss of faith, and who were of great assistance to him later on when the battle was nearly won, I feel that it is only just to lay emphasis on this factor in the claim of Morse to greatness.

It will not be necessary to record in detail the events of the year 1840. The inventor, always confident that success would eventually crown his efforts, lived a life of privation and constant labor in the two fields of art and science. He was still President of the National Academy of Design, and in September he was elected an honorary member of the Mercantile Library Association. He strove to keep the wolf from the door by giving lessons in painting and by practising the new art of daguerreotypy, and, in the mean time, he employed every spare moment in improving and still further simplifying his invention.

He heard occasionally from his associates. The following sentences are from a letter of Alfred Vail's, dated Philadelphia, January 13, 1840: —

FRIEND S. F. B. MORSE,

DEAR SIR, It is many a day since I last had the pleasure of seeing and conversing with you, and, if I am not mistaken, it is as long since any communications have been exchanged. However I trust it will not long be so. When I last had the pleasure of seeing you it was when on my way to Philadelphia, at which time you had the kindness to show me specimens of the greatest discovery ever made, with the exception of the Electro-Magnetic Telegraph. By the by, I have been thinking that it is time money in some way was made out of the Telegraph, and I am almost ready to order an instrument made, and to make the proposition to you to exhibit it here. What do you think of the plan? If Mr. Prosch will make me a first-rate, most perfect machine, and as speedily as possible, and will wait six or nine months for his pay, you may order one for me.

Morse's reply to this letter has not been preserved, but he probably agreed to Vail's proposition, — anything honorable to keep the telegraph in the public eye, — for, as we shall see, in a later letter he refers to the machines which Prosch was to make. Before quoting from that letter, however, I shall give the following sentences from one to Baron Meyendorff, of March 18, 1840:

"I have, since I returned to the United States, made several important improvements, which I regret my limited time will not permit me to describe or send you. . . . I have so changed the *form* of the apparatus, and condensed it into so small a compass, that you would scarcely know it for the same instrument which you saw in Paris."

This and many other allusions, in the correspondence of those years, to Morse's work in simplifying and perfecting his invention, some of which I have already noted, answer conclusively the claims of those who have said that all improvements were the work of other brains and hands.

On September 7, 1840, he writes again to Vail: —

"Your letter of 28th ult. was received several days ago, but I have not had a moment's time to give you a word in return. I am tied hand and foot during the day endeavoring to realize something from the Daguerreotype portraits. . . . As to the Telegraph, I know not what to say. The delay in finishing the apparatus on the part of Prosch is exceedingly tantalizing and vexatious. He was to have finished them more than six months ago, and I have borne with his procrastination until I utterly despair of their being completed. . . . I suppose something might be done in Washington next session if I, or some of you, could go on, but I have expended so much time in vain, there and in Europe, that I feel almost discouraged from pressing it any further; only, however, from want of funds. I have none myself, and I dislike to ask it of the rest of you. You are all so scattered that there is no consultation, and I am under the necessity of attending to duties which will give me the means of living.

"The reason of its not being in operation is not *the fault of the invention*, nor is it *my neglect*. My faith is not only unshaken in its *eventual adoption throughout the world*, but it is confirmed by every new discovery in the science of electricity."

While the future looked dark and the present was darker still, Morse maintained a cheerful exterior, and was still able to write to his friends in a light and airy vein. The following letter, dated September 30, 1840, was to a Mr. Lovering in Paris: —

"Some time since (I believe nearly a year ago) I wrote you to procure for me two lenses and some plates for the Daguerreotype process, but have never heard from you nor had any intimation that my letter was ever received. After waiting some months, I procured both lenses and plates here. Now, if I knew how to scold at you, would n't I scold.

"Well, I recollect a story of a captain who was overloaded by a great many ladies of his acquaintance with orders to procure them various articles in India, just as he was about to sail thither, all which he promised to fulfill. But, on his return, when they flocked round him for their various articles, to their surprise he had only answered the order of one of them. Upon their expressing their disappointment he addressed them thus: 'Ladies,' said he, 'I have to inform you of a most unlucky accident that occurred to your orders. I was not unmindful of them, I assure you; so one fine day I took your orders all out of my pocketbook and arranged them on the top of the companionway, but, just as they were all arranged, a sudden gust of wind took them all overboard.' 'Aye, a very good excuse,' they exclaimed. 'How happens it that Mrs. ——'s did not go overboard, too?' 'Oh!' said the captain, 'Mrs. —— had fortunately enclosed in her order some dozen doubloons which kept the wind from blowing hers away with the rest.'

"Now, friend Lovering, I have no idea of having my new order blown overboard, so I herewith send by the hands of my young friend and pupil, Mr. R. Hubbard, whom I also commend to your kind notice, ten golden half-eagles to keep my order down."

CHAPTER XXVIII

JUNE 20, 1840 — AUGUST 12, 1842

First patent issued. — Proposal of Cooke and Wheatstone to join forces rejected. — Letter to Rev. E. S. Salisbury. — Money advanced by brother artists repaid. — Poverty. — Reminiscences of General Strother, "Porte Crayon." — Other reminiscences. — Inaction in Congress. — Flattering letter of F. O. J. Smith. — Letter to Smith urging action. — Gonon and Wheatstone. — Temptation to abandon enterprise. — Partners all financially crippled. — Morse alone doing any work. — Encouraging letter from Professor Henry. — Renewed enthusiasm. — Letter to Hon. W. W. Boardman urging appropriation of $3500 by Congress. — Not even considered. — Despair of inventor.

It is only necessary to remember that the year 1840, and the years immediately preceding and following it, were seasons of great financial depression, and that in 1840 the political unrest, which always precedes a presidential election, was greatly intensified, to realize why but little encouragement was given to an enterprise so fantastic as that of an electric telegraph. Capitalists were disinclined to embark on new and untried ventures, and the members of Congress were too much absorbed in the political game to give heed to the pleadings of a mad inventor. The election of Harrison, followed by his untimely death only a month after his inauguration and the elevation of Tyler to the Presidency, prolonged the period of political uncertainty, so that Morse and his telegraph received but scant attention on Capitol Hill.

However, the year 1840 marked some progress, for on the 20th of June the first patent was issued to Morse. It may be remembered that, while his caveat and petition were filed in 1837, he had requested that action on them be deferred until after his return from Europe. He had also during the year been gradually perfecting his invention as time and means permitted.

It was during the year 1840, too, that Messrs. Wheatstone and Cooke proposed to join forces with the Morse patentees in America, but this proposition was rejected, although Morse seems to have been almost tempted, for in a letter to Smith he says: —

"I send you copies of two letters just received from England. What shall I say in answer? Can we make any arrangements with them? Need we do it? Does not our patent secure us against foreign interference, or are we to be defeated, not only in England but in our own country, by the subsequent inventions of Wheatstone?

"I feel my hands tied; I know not what to say. Do advise immediately so that I can send by the British Queen, which sails on the first prox."

Fortunately Smith advised against a combination, and the matter was dropped.

It will not be necessary to dwell at length on the events of the year 1841. The situation and aims of the inventor are best summed up in a beautiful and characteristic letter, written on February 14 of that year, to his cousin, the Reverend Edward S. Salisbury: —

"Your letter containing a draft for three hundred dollars I have received, for which accept my sincere thanks. I have hesitated about receiving it because I had begun to despair of ever being able to touch the pencil again. The blow I received from Congress, when the decision was made concerning the pictures for the Rotunda, has seriously and vitally affected my enthusiasm in my art. When that event was announced to me I was tempted to yield up all in despair, but I roused myself to resist the temptation, and, determining still to fix my mind upon the work, cast about for the means of accomplishing it in such ways as my Heavenly Father should make plain. My telegraphic enterprise was one of those means. Induced to prosecute it by the Secretary of the Treasury, and encouraged by success in every part of its progress, urged forward to complete it by the advice of the most judicious friends, I have carried the invention on my part to perfection. That is to say, so far as the invention itself is concerned. *I have done my part.* It is approved in the highest quarters — in England, France, and at home — by scientific societies and by governments, and waits only the action of the latter, or of capitalists, to carry it into operation.

"Thus after several years' expenditure of time and money in the expectation (of my friends, *never of my own* except as I yielded my own judgment to theirs) of so much at least as to leave me free to pursue my art again, I am left, humanly speaking, farther from my object than ever. I am reminded, too, that my prime is past; the snows are on my temples, the half-century of years will this year be marked against me; my eyes begin to fail, and what can I now expect to do with declining powers and habits in my art broken up by repeated disappointments?

"That prize which, through the best part of my life, animated me to sacrifice all that most men consider precious — prospects of wealth, domestic enjoyments, and, not least, the enjoyment of country — was snatched from me at the moment when it appeared to be mine beyond a doubt.

"I do not state these things to you, my dear cousin,

in the spirit of complaint of the dealings of God's Providence, for I am perfectly satisfied that, mysterious as it may seem to me, it has all been ordered in its minutest particulars in infinite wisdom, so satisfied that I can truly say I rejoice in the midst of all these trials, and in view of my Heavenly Father's hand guiding all, I have a joy of spirit which I can only express by the word 'singing.' It is not in man to direct his steps. I know I am so short-sighted that I dare not trust myself in the very next step; how then could I presume to plan for my whole life, and expect that my own wisdom had guided me into that way best for me and the universe of God's creatures?

"I have not painted a picture since that decision in Congress, and I presume that the mechanical skill I once possessed in the art has suffered by the unavoidable neglect. I may possibly recover this skill, and if anything will tend to this end, if anything can tune again an instrument so long unstrung, it is the kindness and liberality of my Cousin Edward. I would wish, therefore, the matter put on this ground that my mind may be at ease. I am at present engaged in taking portraits by the Daguerreotype. I have been at considerable expense in perfecting apparatus and the necessary fixtures, and am just reaping a little profit from it. My ultimate aim is the application of the Daguerreotype to accumulate for my studio models for my canvas. Its first application will be to the study of your picture. Yet if any accident, any unforeseen circumstances should prevent, I have made arrangements with my brother Sidney to hold the sum you have advanced subject to your order. On these conditions I accept it, and will yet indulge the hope of giving you a picture acceptable to you."

The picture was never painted, for the discouraged artist found neither time nor inclination ever to pick up his brush again; but we may be sure that the money, so generously advanced by his cousin, was repaid.

It was in the year 1841 also that, in spite of the difficulty he found in earning enough to keep him from actual starvation, he began to pay back the sums which had been advanced to him by his friends for the painting of a historical picture, which should, in a measure, atone to him for the undeserved slight of Congress. In a circular addressed to each of the subscribers he gives the history of the matter and explains why he had hoped that the telegraph would supply him with the means to paint the picture, and then he adds: —

"I have, as yet, not realized one cent, and thus I find myself farther from my object than ever. Upon deliberately considering the matter the last winter and spring, I came to the determination, in the first place, to free myself from the pecuniary obligation under which I had so long lain to my friends of the Association, and I commenced a system of economy and retrenchment by which I hoped gradually to amass the necessary sum for that purpose, which sum, it will be seen, amounts in the aggregate to $510. Three hundred dollars of this sum I had already laid aside, when an article in the New York 'Mirror,' of the 16th October, determined me at once to commence the refunding of the sums received."

What the substance of the article in the "Mirror" was, I do not know, but it was probably one of those scurrilous and defamatory attacks, from many of which he suffered in common with other persons of prominence, and which was called forth, perhaps, by his activity in the politics of the day.

That I have not exaggerated in saying that he was almost on the verge of starvation during these dark years is evidenced by the following word picture from the pen of General Strother, of Virginia, known in the world of literature under the pen name of "Porte Crayon": —

"I engaged to become Morse's pupil, and subsequently went to New York and found him in a room in University Place. He had three other pupils, and I soon found that our professor had very little patronage. I paid my fifty dollars that settled for one quarter's instruction. Morse was a faithful teacher, and took as much interest in our progress — more indeed than — we did ourselves. But he was very poor. I remember that when my second quarter's pay was due my remittance from home did not come as expected, and one day the professor came in and said, courteously: —

"'Well, Strother my boy, how are we off for money?'

"'Why, Professor,' I answered, 'I am sorry to say I have been disappointed; but I expect a remittance next week.'

"'Next week!' he repeated sadly. 'I shall be dead by that time.'

"'Dead, Sir?'

"'Yes, dead by starvation.'

"I was distressed and astonished. I said hurriedly: —

"'Would ten dollars be of any service?'

"'Ten dollars would save my life; that is all it would do.'

"I paid the money, all that I had, and we dined together. It was a modest meal but good, and, after he had finished, he said: —

"'This is my first meal for twenty-four hours. Strother, don't be an artist. It means beggary. Your life depends upon people who know nothing of your art and care nothing for you. A house-dog lives better, and the very sensitiveness that stimulates an artist to work keeps him alive to suffering.'"

Another artist describes the conditions in 1841 in the following words: —

"In the spring of 1841 I was searching for a studio in which to set up my easel. My 'house-hunting' ended at the New York University, where I found what I wanted in one of the turrets of that stately edifice. When I had fixed my choice, the janitor, who accompanied me in my examination of the rooms, threw open a door on the opposite side of the hall and invited me to enter. I found myself in what was evidently an artist's studio, but every object in it bore indubitable signs of unthrift and neglect. The statuettes, busts, and models of various kinds were covered with dust and cobwebs; dusty canvases were faced to the wall, and stumps of brushes and scraps of paper littered the floor. The only signs of industry consisted of a few masterly crayon drawings, and little luscious studies of color pinned to the wall.

"'You will have an artist for a neighbor,' said the janitor, 'though he is not here much of late; he seems to be getting rather shiftless; he is wasting his time over some silly invention, a machine by which he expects to send messages from one place to another. He is a very good painter, and might do well if he would only stick to his business; but, Lord!' he added with a sneer of contempt, 'the idea of telling by a little streak of lightning what a body is saying at the other end of it.'

"Judge of my astonishment when he informed me that the 'shiftless individual' whose foolish waste of time so much excited his commiseration, was none other than the President of the National Academy of Design — the most exalted position, in my youthful artistic fancy, it was possible for mortal to attain — S. F. B. Morse, since better known as the inventor of the Electric Telegraph. But a little while after this his fame was flashing through the world, and the unbelievers who voted him insane were forced to confess that there was, at least, 'method in his madness.'"

The spring and summer of 1841 wore away and nothing was accomplished. On August 16 Morse writes to Smith: —

"Our Telegraph matters are in a situation to do none of us any good, unless some understanding can be entered into among the proprietors. I have recently received a letter from Mr. Isaac N. Coffin, from Washington, with a commendatory letter from Hon. R. McClellan, of the House. Mr. Coffin proposes to take upon himself the labor of urging through the two houses the bill relating to my Telegraph, which you know has long been before Congress. He will press it and let his compensation depend on his success."

This Mr. Coffin wrote many long letters telling, in vivid language, of the great difficulties which beset the passage of a bill through both houses of Congress, and of how skilled he was in all the diplomatic moves necessary to success, and finally, after a long delay, occasioned by the difficulty of getting powers of attorney from all the proprietors, he was authorized to go ahead. The sanguine inventor hoped much from this unsolicited offer of assistance, but he was again doomed to disappointment, for Mr. Coffin's glowing promises amounted to nothing at all, and the session of 1841–42 ended with no action taken on the bill.

In view of the fact, alluded to in a former chapter, that Francis O. J. Smith later became a bitter enemy of Morse's, and was responsible for many of the virulent attacks upon him, going so far as to say that most, if not all, of the essentials of the telegraph had been invented by others, it may be well to quote the following sentences from a letter of August 21, 1841, in reply to Morse's of August 16: —

"I shall be in Washington more next winter, and will lend all aid in my power, of course, to any agent we may have there. My expenditures in the affair, as you know, have been large and liberal, and have somewhat embarrassed me. Hence I cannot incur more outlay. I am, however, extremely solicitous for the double purpose of having you witness with your own eyes and in your own lifetime the consummation in actual, practical, national utility [of] this beautiful and wonderful offspring of your mechanical and philosophical genius, and know that you have not overestimated the service you have been ambitious of rendering to your country and the world."

On December 3, 1841, Morse again urges Smith to action: —

"Indeed, my dear sir, something ought to be done to carry forward this enterprise that we may all receive what I think we all deserve. The whole labor and expense of moving at all devolve on me, and I have nothing in the world. Completely crippled in means I have scarcely (indeed, I have not at all) the means even to pay the postage of letters on the subject. I feel it most tantalizing to find that there is a movement in Washington on the subject; to know that telegraphs will be before Congress this session, and from the means possessed by Gonon and Wheatstone!! (yes, Wheatstone who successfully headed us off in England), one or the other of their two plans will probably be adopted. Wheatstone, I suppose you know, has a patent here, and has expended $1000 to get everything prepared for a campaign to carry his project into operation, and more than that, his patent is dated *before mine!*

"My dear sir, to speak as I feel, I am sick at heart to perceive how easily others, *foreigners*, can manage our Congress, and can contrive to cheat our country out of the honor of a discovery of which the country boasts, and our countrymen out of the profits which are our due; to perceive how easily they can find men and means to help them in their plans, and how difficult, nay, *impossible*, for us to find either. Is it really so, or am I deceived? What can be done? Do write immediately and propose something. Will you not be in Washington this winter? Will you not call on me as you pass through New York, if you do go?

"Gonon has his telegraph on the Capitol, and a committee of the Senate reported in favor of trying his for a short distance, and will pass a bill this session if we are not doing something. Some means, somehow, must be raised. I have been compelled to stop my machine just at the moment of completion. I cannot move a step without running in debt, and that I cannot do.

"As to the company that was thought of to carry the Telegraph into operation here, it is another of those *ignes fatui* that have just led me on to waste a little more time, money, and patience, and then vanished. The gentleman who proposed the matter was, doubtless, friendly disposed, but he lacks judgment and perseverance in a matter of this sort.

"If Congress would but pass the bill of $30,000 before them, there would be no difficulty. There is no difficulty in the scientific or mechanical part of the matter; *that is a problem solved.* The only difficulty that remains is obtaining funds, which Congress can furnish, to carry it into execution. I have a great deal to say, but must stop for want of time to write more."

But he does not stop. He is so full of his subject that he continues at some length: —

"Everything done by me in regard to the Telegraph is at arm's length. I can do nothing without consultation, and when I wish to consult on the most trivial thing I have three letters to write, and a week or ten days to wait before I can receive an answer.

"I feel at times almost ready to cast the whole matter to the winds, and turn my attention forever from the subject. Indeed, I feel almost inclined, at times, to destroy the evidences of priority of invention in my possession and let Wheatstone and England take the credit of it. For it is tantalizing in the highest degree to find the papers and the lecturers boasting of the invention as one of the greatest of the age, and as an honor to America, and yet to have the nation by its representatives leave the inventor without the means either to put his invention fairly before his countrymen, or to defend himself against foreign attack.

"If I had the means in any way of support in Washington this winter, I would go on in the middle of January and push the matter, but I cannot run the risk. I would write a detailed history of the invention, which would be an interesting document to have printed in the Congressional documents, and establish beyond contradiction both priority and superiority of my invention. Has not the Postmaster-General, or Secretary of War or Treasury, the power to pay a few hundred dollars from a contingent fund for such purposes?

"Whatever becomes of the invention through the neglect of those who could but would not lend a helping hand, *you*, my dear sir, will have the reflection that you did all in your power to aid me, and I am deterred from giving up the matter as desperate most of all for the consideration that those who kindly lent their aid when the invention was in its infancy would suffer, and that, therefore, I should not be dealing right by them. If this is a little *blue*, forgive it."

It appears from this letter that Morse bore no ill-will towards his partners for not coming to his assistance at this critical stage of the enterprise, so that it behooves us not to be too harsh in our judgment. Perhaps I have not sufficiently emphasized the fact that, owing to the great financial depression which prevailed at that time, Mr. Smith and the Vails were seriously crippled in their means, and were not able to advance any more money, and Professor Gale had never been called upon to contribute money. This does not alter my main contention, however, for it still remains true that, if it had not been for Morse's dogged persistence during these dark years, the enterprise would, in all probability, have failed. With the others it was merely an incident, with him it had become his whole life.

The same refrain runs through all the letters of 1841 and 1842; discouragement at the slow progress which is being made, and yet a sincere conviction that eventually the cause will triumph. On December 13, 1841, he says in a letter to Vail: —

"We are all somewhat crippled, and I most of all, being obliged to superintend the getting up of a set of machinery complete, and to make the greater part myself, and without a cent of money. . . . All the burden now rests on my shoulders after years of time devoted to the enterprise, and I am willing, as far as I am able, to bear my share if the other proprietors will lend a helping hand, and give me facilities to act and a reasonable recompense for my services in case of success."

Vail, replying to this letter on December 15, says: "I have recently given considerable thought to the subject of the Telegraph, and was intending to get permission of you, if there is anything to the contrary in our articles of agreement, to build for myself and my private use a Telegraph upon your plan."

In answering this letter, on December 18, Morse again urges Vail to give him a power of attorney, and adds: —

"You can see in a moment that, if I have to write to all the scattered proprietors of the Telegraph every time any movement is made, what a burden falls upon me both of expense of time and money which I cannot afford. In acting for my own interest in this matter I, of course, act for the interest of all. If we can get that thirty thousand dollars bill through Congress, the experiment (if it can any longer be called such) can then be tried on such a scale as to insure its success.

"You ask permission to make a Telegraph for your own use. I have no objection, but, before you commence one, you had better see me and the improvements which I have made, and I can suggest a few more, rather of an ornamental character, and some economical arrangements which may be of use to you.

"I thank you for your kind invitation, and, when I come to Philadelphia, shall *A. Vail* myself of your politeness. I suppose by this time you have a brood of chickens around you. Well, go on and prosper. As for me, I am not well; am much depressed at times, and have many cares, anxieties, and disappointments, in which I am aware I am not alone. But all will work for the best if we only look through the cloud and see a kind Parent directing all. This reflection alone cheers me and gives me renewed strength."

Conditions remained practically unchanged during the early part of the year 1842. If it had not been for occasional bits of encouragement from different quarters the inventor would probably have yielded to the temptation to abandon all and depend on his brush again for a living. Perhaps the ray of greatest encouragement which lightened the gloom of this depressing period was the following letter from Professor Henry, dated February 24, 1842: —

My dear Sir — I am pleased to learn that you have again petitioned Congress in reference to your telegraph, and I most sincerely hope you will succeed in convincing our representatives of the importance of the invention. In this you may, perhaps, find some difficulty, since, in the minds of many, the electro-magnetic telegraph is associated with the various chimerical projects constantly presented to the public, and particularly with the schemes so popular a year or two ago for the application of electricity as a moving power in the arts. I have asserted, from the first, that all attempts of this kind are premature and made without a proper knowledge of scientific principles. The case is, however, entirely different in regard to the electro-magnetic telegraph. Science is now fully ripe for this application, and I have not the least doubt, if proper means be afforded, of the perfect success of the invention.

The idea of transmitting intelligence to a distance by means of electrical action, has been suggested by various persons, from the time of Franklin to the present; but, until the last few years, or since the principal discoveries in electro-magnetism, all attempts to reduce it to practice were, necessarily, unsuccessful. The mere suggestion however, of a scheme of this kind is a matter for which little credit can be claimed, since it is one which would naturally arise in the mind of almost any person familiar with the phenomena of electricity; but the bringing it forward at the proper moment, when the developments of science are able to furnish the means of certain success, and the devising a plan for carrying it into practical operation, are the grounds of a just claim to scientific reputation, as well as to public patronage.

About the same time with yourself Professor Wheatstone, of London, and Dr. Steinheil, of Germany, proposed plans of the electro-magnetic telegraph, but these differ as much from yours as the nature of the common principle would well permit; and, unless some essential improvements have lately been made in these European plans, *I should prefer the one invented by yourself.*

With my best wishes for your success I remain, with much esteem

Yours truly

JOSEPH HENRY.

I consider this one of the most important bits of contemporary evidence that has come down to us. Professor Henry, perfectly conversant with all the minutiæ of science and invention, practically gives to Morse all the credit which the inventor himself at any time claimed. He dismisses the claims of those who merely suggested a telegraph, or even made unsuccessful attempts to reduce one to practice, unsuccessful because the time was not yet ripe; and he awards Morse scientific as well as popular reputation. Furthermore Professor Henry, with the clear vision of a trained mind, points out that advances in discovery and invention are necessarily slow and dependent upon the labors of many in the same field. His cordial endorsement of the invention, in this letter and later, so pleased and encouraged Morse that he refers to it several times in his correspondence. To Mr. Smith, on July 16, 1842, he writes: —

"Professor Henry visited me a day or two ago; he knew the principles of the Telegraph, but had never before seen it. He told a gentleman, who mentioned it again to me, that without exception it was the most beautiful and ingenious instrument he had ever seen.

He says mine is the only truly practicable plan. He has been experimenting and making discoveries on celestial electricity, and he says that Wheatstone's and Steinheil's telegraphs must be so influenced in a highly electrical state of the atmosphere as at times to be useless, they using the deflection of the needle, while mine, from the use of the magnet, is not subject to this disturbing influence. I believe, if the truth were known, some such cause is operating to prevent our hearing more of these telegraphs."

In this same letter he tells of the application of a certain Mr. John P. Manrow for permission to form a company, but, as nothing came of it, it will not be necessary to particularize. Mr. Manrow, however, was a successful contractor on the New York and Erie Railroad, and it was a most encouraging sign to have practical business men begin to take notice of the invention.

So cheered was the ever-hopeful inventor by the praise of Professor Henry, that he redoubled his efforts to get the matter properly before Congress; and in this he worked alone, for, in the letter to Smith just quoted from, he says: "I have not heard a word from Mr. Coffin at Washington since I saw you. I presume he has abandoned the idea of doing anything on the terms we proposed, and so has given it up. Well, so be it; I am content."

Taking advantage of the fact that he was personally acquainted with many members of Congress, he wrote to several of them on the subject. In some of the letters he treats exhaustively of the history and scientific principles of his telegraph, but I have selected the following, addressed to the Honorable W. W. Boardman, as containing the most essential facts in the most concise form: —

August 10, 1842.

My dear Sir, — I enclose you a copy of the "Tribune" in which you will see a notice of my Telegraph. I have showed its operation to a few friends occasionally within a few weeks, among others to Professor Henry, of Princeton (a copy of whose letter to me on this subject I sent you some time since). He had never seen it in operation, but had only learned from description the principle on which it is founded. He is not of an enthusiastic temperament, but exceedingly cautious in giving an opinion on scientific inventions, yet in this case he expressed himself in the warmest terms, and told my friend Dr. Chilton (who informed me of it) that he had just been witnessing "the operation of the most beautiful and ingenious instrument he had ever seen."

Indeed, since I last wrote you, I have been wholly occupied in perfecting its details and making myself familiar with the whole system. There is not a shadow of a doubt as to its performing all that I have promised in regard to it, and, indeed, all that has been conceived of it. Few can understand the obstacles arising from want of pecuniary means that I have had to encounter the past winter. To avoid debt (which I will never incur) I have been compelled to make with my own hands a great part of my machinery, but at an expense of time of very serious consideration to me. I have executed in six months what a good machinist, if I had the means to employ him, would have performed in as many weeks, and performed much better.

I had hoped to be able to show my perfected instrument in Washington long before this, and was (until this morning) contemplating its transportation thither next week. The news, just arrived, of the proposed adjournment of Congress has stopped my preparations, and interposes, I fear, another year of anxious suspense.

Now, my dear sir, as your time is precious, I will state in few words what I desire. The Government will eventually, without doubt, become possessed of this invention, for it will be necessary from many considerations; not merely as a direct advantage to the Government and public at large if regulated by the Government, but as a preventive of the evil effects which must result if it be a monopoly of a company. To this latter mode of remunerating myself I shall be compelled to resort if the Government should not eventually act upon it.

You were so good as to call the attention of the House to the subject by a resolution of inquiry early in the session. I wrote you some time after requesting a stay of action on the part of the committee, in the hope that, long before this, I could show them the Telegraph in Washington; but, just as I am ready, I find that Congress will adjourn before I can reach Washington and put the instrument in order for their inspection.

Will it be possible, before Congress rises, to appropriate a small sum, say $3500, under the direction of the Secretary of the Treasury, to put my Telegraph in operation for the inspection of Congress the next session? If Congress will grant this sum, I will engage to have a complete Telegraph on my Electro-Magnetic plan between the President's house, or one of the Departments, and the Capitol and the Navy Yard, so that instantaneous communication can be held between these three points at pleasure, at any time of day or night, at any season, in clear or rainy weather, and ready for their examination during the next session of Congress, so that the whole subject may be fairly understood.

I believe that, did the great majority of Congress but consider seriously the results of this invention of the Electric Telegraph on all the interests of society; did they suffer themselves to dwell but for a moment on the vast consequences of the instantaneous communication of intelligence from one part to the other of the land in a commercial point of view, and as facilitating the defenses of the country, which my invention renders certain; they would not hesitate to pass all the acts necessary to secure its control to the Government. I ask not this until they have thoroughly examined its merits, but will they not assist me in placing the matter fairly before them? Surely so small a sum to the Government for so great an object cannot reasonably be denied.

I hardly know in what form this request of mine should be made. Should it be by petition to Congress, or will this letter handed in to the committee be sufficient? If a petition is required, for form's sake, to be referred to the committee to report, shall I ask the favor of you to make such petition in proper form?

You know, my dear sir, just what I wish, and I know, from the kind and friendly feeling you have shown toward my invention, I may count on your aid. If, on your return, you stop a day or two in New York, I shall be glad to show you the operation of the Telegraph as it is.

This modest request of the inventor was doomed, like so many of his hopes, to be shattered, as we learn from the courteous reply of Mr. Boardman, dated August 12: —

DEAR SIR, — Yours of the 10th is received. I had already seen the notice of your Telegraph in the "Tribune," and was prepared for such a report. This is not the time to commence any new project before Congress. We are, I trust, within ten days of adjournment. There is no prospect of a tariff at this session, and, as that matter appears settled, the sooner Congress adjourns the better. The subject of your Telegraph was some months ago, as you know, referred to the Committee on Commerce, and by that committee it was referred to Mr. Ferris, one of the members of that committee, from the city of New York, and who, by-the-way, is now at home in the city and will be glad to see you on the subject. I cannot give you his address, but you can easily find him.

The Treasury and the Government are both bankrupt, and that foolish Tyler has vetoed the tariff bill; the House is in bad humor and nothing of the kind you propose could be done. The only chance would be for the Committee on Commerce to report such a plan, but there would be little or no chance of getting such an appropriation through this session. I have much faith in your plan, and hope you will continue to push it toward Congress.

This was almost the last straw, and it is not strange that the long-suffering inventor should have been on the point of giving up in despair, nor that he should have given vent to his despondency in the following letter to Smith: —

"While, so far as the invention itself is concerned, everything is favorable, I find myself without sympathy or help from any who are associated with me, whose interest, one would think, would impel them at least to inquire if they could render some assistance. For two years past I have devoted all my time and scanty means, living on a mere pittance, denying myself all pleasures and even necessary food, that I might have a sum to put my Telegraph into such a position before Congress as to insure success to the common enterprise.

"I am crushed for want of means, and means of so trivial a character, too, that they who know how to ask (which I do not) could obtain in a few hours. One more year has gone for want of these means. I have now ascertained that, however unpromising were the times last session, if I could but have gone to Washington, I could have got some aid to enable me to insure success at the next session."

The other projects for telegraphs must have been abandoned, for he goes on to say:—

"As it is, although everything is favorable, although I have no competition and no opposition — on the contrary, although every member of Congress, as far as I can learn, is favorable — yet I fear all will fail because I am too poor to risk the trifling expense which my journey and residence in Washington will occasion me. *I will not run in debt* if I lose the whole matter. So, unless I have the means from some source, I shall be compelled, however reluctantly, to leave it, and, if I get once engaged in my proper profession again, the Telegraph and its proprietors will urge me from it in vain.

"No one can tell the days and months of anxiety and

labor I have had in perfecting my telegraphic apparatus. For want of means I have been compelled to make with my own hands (and to labor for weeks) a piece of mechanism which could be made much better, and in a tenth part of the time, by a good mechanician, thus wasting *time* — time which I cannot recall and which seems double-winged to me.

"'Hope deferred maketh the heart sick.' It is true and I have known the full meaning of it. Nothing but the consciousness that I have an invention which is to mark an era in human civilization, and which is to contribute to the happiness of millions, would have sustained me through so many and such lengthened trials of patience in perfecting it."

CHAPTER XXIX

JULY 16, 1842 — MARCH 26, 1843

Continued discouragements. — Working on improvements. — First submarine cable from Battery to Governor's Island. — The Vails refuse to give financial assistance. — Goes to Washington. — Experiments conducted at the Capitol. — First to discover duplex and wireless telegraphy. — Dr. Fisher. — Friends in Congress. — Finds his statuette of Dying Hercules in basement of Capitol. — Alternately hopes and despairs of bill passing Congress. — Bill favorably reported from committee. — Clouds breaking. — Ridicule in Congress. — Bill passes House by narrow majority. — Long delay in Senate. — Last day of session. — Despair. — Bill passes. — Victory at last.

SLOWLY the mills of the gods had been grinding, so slowly that one marvels at their leaden pace, and wonders why the dream of the man so eager to benefit his fellowmen could not have been realized sooner. We are forced to echo the words of the inventor himself in a previously quoted letter: "I am perfectly satisfied that, mysterious as it may seem to me, it has all been ordered in its minutest particulars in infinite wisdom." He enlarges on this point in the letter to Smith of July 16, 1842. Referring to the difficulties he has encountered through lack of means, he says: —

"I have oftentimes risen in the morning not knowing where the means were to come from for the common expenses of the day. Reflect one moment on my situation in regard to the invention. Compelled from the first, from my want of the means to carry out the invention to a practical result, to ask assistance from those who had means, I associated with me the Messrs. Vail and Dr. Gale, by making over to them, on certain conditions, a portion of the patent right. These means enabled me to carry it successfully forward to a certain point. At this point you were also admitted into a share of the patent on certain conditions, which carried the enterprise forward successfully still further. Since then disappointments have occurred and disasters to the property of every one concerned in the enterprise, but of a character not touching the intrinsic merits of the invention in the least, yet bearing on its progress so fatally as for several years to paralyze all attempts to proceed.

"The depressed situation of all my associates in the invention has thrown the whole burden of again attempting a movement entirely on me. With the trifling sum of five hundred dollars I could have had my instruments perfected and before Congress six months ago, but I was unable to run the risk, and I therefore chose to go forward more slowly, but at a great waste of time.

"In all these remarks understand me as not throwing the least blame on any individual. I believe that the situation in which you all are thrown is altogether providential — that human foresight could not avert it, and I firmly believe, too, that the delays, tantalizing and trying as they have been, will, in the end, turn out to be beneficial."

I have hazarded the opinion that it was a kindly fate which frustrated the consummation of the Russian contract, and here again I venture to say that the Fates were kind, that Morse was right in saying that the "delays" would "turn out to be beneficial." And why? Because it needed all these years of careful thought and experiment on the part of the inventor to bring his instruments to the perfection necessary to complete success, and because the period of financial depression, through which the country was then passing, was unfavorable to an enterprise of this character. The history of all inventions proves that, no matter how clear a vision of the future some enthusiasts may have had, the dream was never actually realized until all the conditions were favorable and the psychological moment had arrived. Professor Henry showed, in his letter of February 24, that he realized that some day electricity would be used as a motive power, but that much remained yet to be discovered and invented before this could be actually and practically accomplished. So, too, the conquest of the air remained a dream for centuries until, to use Professor Henry's words, "science" was "ripe for its application." Therefore I think we can conclude that, however confident Morse may have been that his invention could have stood the test of actual commercial use during those years of discouragement, it needed the perfection which he himself gave it during those same years to enable it to prove its superiority over other methods.

Among the other improvements made by Morse at this time, the following is mentioned in the letter to Smith of July 16, 1842, just quoted from: "I have invented a battery which will delight you; it is the most powerful of its size ever invented, and this part of my telegraphic apparatus the results of experiments have enabled me to simplify and truly to perfect."

Another most important development of the invention was made in the year 1842. The problem of crossing wide bodies of water had, naturally, presented itself to the mind of the inventor at an early date, and during the most of this year he had devoted himself seriously to its solution. He laboriously insulated about two miles of copper wire with pitch, tar, and rubber, and, on the evening of October 18, 1842, he carried it, wound on a reel, to the Battery in New York and hired a row-boat with a man to row him while he paid out his "cable." Tradition says that it was a beautiful moonlight night and that the strollers on the Battery were mystified, and wondered what kind of fish were being trolled for. The next day the following editorial notice appeared in the New York "Herald": —

MORSE'S ELECTRO-MAGNETIC TELEGRAPH

This important invention is to be exhibited in operation at Castle Garden between the hours of twelve and one o'clock to-day. One telegraph will be erected on Governor's Island, and one at the Castle, and messages will be interchanged and orders transmitted during the day.

Many have been incredulous as to the powers of this wonderful triumph of science and art. All such may now have an opportunity of fairly testing it. *It is destined to work a complete revolution in the mode of transmitting intelligence throughout the civilized world.*

Before the appointed hour on the morning of the 19th, Morse hastened to the Battery, and found a curious crowd already assembled to witness this new marvel. With confidence he seated himself at the instrument and had succeeded in exchanging a few signals between himself and Professor Gale at the other end on Governor's Island, when suddenly the receiving instrument was dumb. Looking out across the waters of the bay, he soon saw the cause of the interruption. Six or seven vessels were anchored along the line of his cable, and one of them, in raising her anchor, had fouled the cable and pulled it up. Not knowing what it was, the sailors hauled in about two hundred feet of it; then, finding no end, they cut the cable and sailed away, ignorant of the blow they had inflicted on the mortified inventor. The crowd, thinking they had been hoaxed, turned away with jeers, and Morse was left alone to bear his disappointment as philosophically as he could.

Later, in December, the experiment was repeated across the canal at Washington, and this time with perfect success.

Still cramped for means, chafing under the delay which this necessitated, he turned to his good friends the Vails, hoping that they might be able to help him. While he shrank from borrowing money he considered that, as they were financially interested in the success of the invention, he could with propriety ask for an advance to enable him to go to Washington.

To his request he received the following answer from the Honorable George Vail: —

SPEEDWELL IRON WORKS,
December 31, 1842.

S. F. B. MORSE, ESQ.,

DEAR SIR, — Your favor is at hand. I had expected that my father would visit you, but he could not go out in the snow-storm of Wednesday, and, if he had, I do not think anything could induce him to raise the needful for the prosecution of our object. He says: "Tell Mr. Morse that there is no one I would sooner assist than him if I could, but, in the present posture of my affairs,

I am not warranted in undertaking anything more than to make my payments as they become due, of which there are not a few."

He thinks that Mr. S—— might soon learn how to manage it, and, as he is there, it would save a great expense. I do not myself know that he could learn; but, as my means are nothing at the present time, I can only wish you success, if you go on.

Of course Mr. Vail meant "if you go on to Washington," but to the sensitive mind of the inventor the words must have seemed to imply a doubt of the advisability of going on with the enterprise. However, he was not daunted, but in some way he procured the means to defray his expenses, perhaps from his good brother Sidney, for the next letter to Mr. Vail is from Washington, on December 18, 1842: —

"I have not written you since my arrival as I had nothing special to say, nor have I now anything very decided to communicate in relation to my enterprise, except that it is in a very favorable train. The Telegraph, as you will see by Thursday or Friday's 'Intelligencer,' is established between two of the committee rooms in the Capitol, and excites universal admiration. I am told from all quarters that there is but one sentiment in Congress respecting it, and that the appropriation will unquestionably pass.

"The discovery I made with Dr. Fisher, just before leaving New York, of the fact that two or more currents will pass, without interference, at the same time, on the same wire, excites the wonder of all the scientific in and out of Congress here, and when I show them the certainty of it, in the practical application of it to simplify my Telegraph, their admiration is loudly expressed, and it has created a feeling highly advantageous to me.

"I believe I drew for you a method by which I thought I could pass rivers, *without any wires*, through the water. I tried the experiment across the canal here on Friday afternoon *with perfect success*. This also has added a fresh interest in my favor, and I begin to hope that I am on the eve of realizing something in the shape of compensation for my time and means expended in bringing my invention to its present state. I dare not be sanguine, however, for I have had too much experience of delusive hopes to indulge in any premature exultation. Now there is no opposition, but it may spring up unexpectedly and defeat all. . . .

"I find Dr. Fisher a great help. He is acquainted with a great many of the members, and he is round among them and creating an interest for the Telegraph. Mr. Smith has not yet made his appearance, and, if he does not come soon, everything will be accomplished without him. My associate proprietors, indeed, are at present broken reeds, yet I am aware they are disabled in various ways from helping me, and I ought to remember that their help in the commencement of the enterprise was essential in putting the Telegraph into the position it now is [in]; therefore, although they give me now no aid, it is not from unwillingness but from inability, and I shall not grudge them their proportion of its profits, nor do I believe they will be unwilling to reimburse me my expenses, should the Telegraph eventually be purchased by the Government.

"Mr. Ferris, our representative, is very much interested in understanding the scientific principles on which my Telegraph is based, and has exerted himself very strongly in my behalf; so has Mr. Boardman, and, in a special manner, Dr. Aycrigg, of New Jersey, the latter of whom is determined the bill shall pass by acclamation. Mr. Huntington, of the Senate, Mr. Woodbury and Mr. Wright are also very strongly friendly to the Telegraph."

This letter, to the best of my knowledge, has never before been published, and yet it contains statements of the utmost interest. The discovery of duplex telegraphy, or the possibility of sending two or more messages over the same wire at the same time has been credited by various authorities to different persons; by some to Moses G. Farmer in 1852, by others to Gintl, of Vienna, in 1853, or to Frischen or Siemens and Halske in 1854. Yet we see from this letter that Morse and his assistant Dr. Fisher not only made the discovery ten years earlier, in 1842, but demonstrated its practicability to the scientists and others in Washington at that date. Why this fact should have been lost sight of I cannot tell, but I am glad to be able to bring forward the proof of the paternity of this brilliant discovery even at this late day.

Still another scientific principle was established by Morse at this early period, as we learn from this letter, and that is the possibility of wireless telegraphy; but, as he has been generally credited with the first suggestion of what has now become one of the greatest boons to humanity, it will not be necessary to enlarge on it.

A brighter day seemed at last to be dawning, and a most curious happening, just at this time, came to the inventor as an auspicious omen. In stringing his wires between the two committee rooms he had to descend into a vault beneath them which had been long unused. A workman, who was helping him, went ahead and carried a lamp, and, as he glanced around the chamber, Morse noticed something white on a shelf at one side. Curious to see what this could be, he went up to it, when what was his amazement to find that it was a plaster cast of that little statuette of the Dying Hercules which had won for him the Adelphi Gold Medal so many years before in London. There was the token of his first artistic success appearing to him out of the gloom as the harbinger of another success which he hoped would also soon emerge from behind the lowering clouds.

The apparently mysterious presence of the little demigod in such an out-of-the-way place was easily explained. Six casts of the clay model had been made before the original was broken up. One of these Morse had kept for himself, four had been given to various institutions, and one to his friend Charles Bulfinch, who succeeded Latrobe as the architect of the Capitol. A sinister fate seemed to pursue these little effigies, for his own, and the four he had presented to different institutions, were all destroyed in one way and another. After tracing each one of these five to its untimely end, he came to the conclusion that this evidence of his youthful genius had perished from the earth; but here, at last, the only remaining copy was providentially revealed to the eyes of its creator, having undoubtedly been placed in the vault for safe-keeping and overlooked. It was cheerfully returned to him. By him it was given to his friend, the Reverend E. Goodrich Smith, and by the latter presented to Yale University, where it now rests in the Fine Arts Building.

So ended the year 1842, a decade since the first conception of the telegraph on board the Sully, and it found the inventor making his last stand for recognition from that Government to which he had been so loyal, and upon which he wished to bestow a priceless gift. With the dawn of the new year, a year destined to mark an epoch in the history of civilization, his flagging spirits were revived, and he entered with zest on what proved to be his final and successful struggle.

It passes belief that with so many ocular demonstrations of the practicability of the Morse telegraph, and with the reports of the success of other telegraphs abroad, the popular mind, as reflected in its representatives in Congress, should have remained so incredulous. Morse had been led to hope that his bill was going to pass by acclamation, but in this he was rudely disappointed. Still he had many warm friends who believed in him and his invention. First and foremost should be mentioned his classmate, Henry L. Ellsworth, the Commissioner of Patents, at whose hospitable home the inventor stayed during some of these anxious days, and who, with his family, cheered him with encouraging words and help. Among the members of Congress who were energetic in support of the bill especially worthy of mention are — Kennedy, of Maryland; Mason, of Ohio; Wallace, of Indiana; Ferris and Boardman, of New York; Holmes, of South Carolina; and Aycrigg, of New Jersey.

The alternating moods of hope and despair, through

which the inventor passed during the next few weeks, are best pictured forth by himself in brief extracts from letters to his brother Sidney: —

"*January 6, 1843.* I sent you a copy of the Report on the Telegraph a day or two since. I was in hopes of having it called up to-day, but the House refused to go into Committee of the Whole on the State of the Union, so it is deferred. The first time they go into Committee of the Whole on the State of the Union it will probably be called up and be decided upon.

"Everything looks favorable, but I do not suffer myself to be sanguine, for I do not know what may be doing secretly against it. I shall believe it passed when the signature of the President is affixed to it, and not before."

"*January 16.* I snatch the moments of waiting for company in the Committee Room of Commerce to write a few lines. Patience is a virtue much needed and much tried here. So far as opinion goes everything is favorable to my bill. I hear of no opposition, but should not be surprised if it met with some. The great difficulty is to get it up before the House; there are so many who must '*define their position*,' as the term is, so many who must say something to 'Bunkum,' that a great deal of the people's time is wasted in mere idle, unprofitable speechifying. I hope something may be done this week that shall be decisive, so that I may know what to do. . . . This waiting at so much risk makes me question myself: am I in the path of duty? When I think that the little money I brought with me is nearly gone, that, if nothing should be done by Congress, I shall be in a destitute state; that perhaps I shall have again to be a burden to friends until I know to what to turn my hands, I feel low-spirited. I am only relieved by naked trust in God, and it is right that this should be so."

"*January 20.* My patience is still tried in waiting for the action of Congress on my bill. With so much at stake you may easily conceive how tantalizing is this state of suspense. I wish to feel right on this subject; not to be impatient, nor distrustful, nor fretful, and yet to be prepared for the worst. I find my funds exhausting, my clothing wearing out, my time, especially, rapidly waning, and my affairs at home requiring some little looking after; and then, if I should after all be disappointed, the alternative looks dark, and to human eyes disastrous in the extreme.

"I hardly dare contemplate this side of the matter, and yet I ought so far to consider it as to provide, if possible, against being struck down by such a blow. At times, after waiting all day and day after day, in the hope that my bill may be called up, and in vain, I feel heart-sick, and finding nothing accomplished, that no progress is made, that *precious time* flies, I am depressed and begin to question whether I am in the way of duty. But when I feel that I have done all in my power, and that this delay may be designed by the wise disposer of all events for a trial of patience, I find relief and a disposition quietly to wait such issue as he shall direct, knowing that, if I sincerely have put my trust in him, he will not lead me astray, and my way will, in any event, be made plain."

"*January 25.* I am still *waiting, waiting.* I know not what the issue will be and wish to be prepared, and have you all prepared, for the worst in regard to the bill.

Although I learn of no opposition yet I have seen enough of the modes of business in the House to know that everything there is more than in ordinary matters uncertain. It will be the end of the session, probably, before I return. I will not have to reproach myself, or be reproached by others, for any neglect, but under all circumstances I am exceedingly tried. I am too foreboding probably, and ought not so to look ahead as to be distrustful. I fear that I have no right feelings in this state of suspense. It is easier to say 'Thy will be done' than at all times to feel it, yet I can pray that God's will may be done whatever becomes of me and mine."

"*January 30.* I am still kept in suspense which is becoming more and more tantalizing and painful. But I endeavor to exercise patience."

"*February 21.* I think the clouds begin to break away and a little sunlight begins to cheer me. The House in Committee of the Whole on the State of the Union have just passed my bill through committee to report to the House. There was an attempt made to cast ridicule upon it by a very few headed by Mr. Cave Johnson, who proposed an amendment that half the sum should be appropriated to mesmeric experiments. Only 26 supported him and it was laid aside to be reported to the House without amendment and without division.

"I was immediately surrounded by my friends in the House, congratulating me and telling me that the crisis is passed, and that the bill will pass the House by a large majority. Mr. Kennedy, chairman of the Committee on Commerce, has put the bill on the Speaker's calendar for Thursday morning, when the final vote in the House will be taken. It then has to go to the Senate, where I have reason to believe it will meet with a favorable reception. Then to the President, and, if signed by him, I shall return with renovated spirits, for I assure you I have for some time been at the lowest ebb, and can now scarcely realize that a turn has occurred in my favor. I don't know when I have been so much tried as in the tedious delays of the last two months, but I see a reason for it in the Providence of God. He has been pleased to try my patience, and not until my impatience had yielded unreservedly to submission has He relieved me by granting light upon my path. Praised be His name, for to Him alone belongs all the glory.

"I write with a dreadful headache caused by over-excitement in the House, but hope to be better after a night's rest. I have written in haste just to inform you of the first symptoms of success."

On the same date as that of the preceding letter, February 21, the following appeared in the "Congressional Globe," and its very curtness and flippancy is indicative of the indifference of the public in general to this great invention, and the proceedings which are summarized cast discredit on the intelligence of our national law-makers: —

ELECTRO AND ANIMAL MAGNETISM

On motion of Mr. Kennedy of Maryland, the committee took up the bill to authorize a series of experiments to be made in order to test the merits of Morse's electro-magnetic telegraph. The bill appropriates $30,000, to be expended under the direction of the Postmaster-General.

On motion of Mr. Kennedy, the words "Postmaster-

General" were stricken out and "Secretary of the Treasury" inserted.

Mr. Cave Johnson wished to have a word to say upon the bill. As the present Congress had done much to encourage science, he did not wish to see the science of mesmerism neglected and overlooked. He therefore proposed that one half of the appropriation be given to Mr. Fisk, to enable him to carry on experiments, as well as Professor Morse.

Mr. Houston thought that Millerism should also be included in the benefits of the appropriation.

Mr. Stanly said he should have no objection to the appropriation for mesmeric experiments, provided the gentleman from Tennessee [Mr. Cave Johnson] was the subject. [A laugh.]

Mr. Cave Johnson said he should have no objection provided the gentleman from North Carolina [Mr. Stanly] was the operator. [Great laughter.]

Several gentlemen called for the reading of the amendment, and it was read by the Clerk, as follows: —

"*Provided*, That one half of the said sum shall be appropriated for trying mesmeric experiments under the direction of the Secretary of the Treasury."

Mr. S. Mason rose to a question of order. He maintained that the amendment was not *bona fide*, and that such amendments were calculated to injure the character of the House. He appealed to the chair to rule the amendment out of order.

The Chairman said it was not for him to judge of the motives of members in offering amendments, and he could not, therefore, undertake to pronounce the amendment not *bona fide*. Objections might be raised to it on the ground that it was not sufficiently analogous in character to the bill under consideration, but, in the opinion of the Chair, it would require a scientific analysis to determine how far the magnetism of mesmerism was analogous to that to be employed in telegraphs. [Laughter.] He therefore ruled the amendment in order.

On taking the vote, the amendment was rejected — ayes 22, noes not counted.

The bill was then laid aside to be reported.

On February 23, the once more hopeful inventor sent off the following hurriedly written letter to his brother: —

"You will perceive by the proceedings of the House to-day that *my bill has passed the House by a vote of 89 to 80.* A close vote after the expectations raised by some of my friends in the early part of the session, but enough is as good as a feast, and it is safe so far as the House is concerned. I will advise you of the progress of it through the Senate. All my anxieties are now centred there. I write in great haste."

A revised record of the voting showed that the margin of victory was even slighter, for in a letter to Smith, Morse says: —

"The long agony (truly agony to me) is over, for you will perceive by the papers of to-morrow that, so far as the House is concerned, the matter is decided. *My bill has passed by a vote of eighty-nine to eighty-three.* A close vote, you will say, but explained upon several grounds not affecting the disposition of many individual members, who voted against it, to the invention. In this matter six votes are as good as a thousand, so far as the appropriation is concerned.

"The yeas and nays will tell you who were friendly and who adverse to the bill. I shall now bend all my attention to the Senate. There is a good disposition there and I am now strongly encouraged to think that my invention will be placed before the country in such a position as to be properly appreciated, and to yield to all its proprietors a proper compensation.

"I have no desire to vaunt my exertions, but I can truly say that I have never passed so trying a period as the last two months. Professor Fisher (who has been of the greatest service to me) and I have been busy from morning till night every day since we have been here. I have brought him on with me at my expense, and he will be one of the first assistants in the first experimental line, if the bill passes. . . . My feelings at the prospect of success are of a joyous character, as you may well believe, and one of the principal elements of my joy is that I shall be enabled to contribute to the happiness of all who formerly assisted me, some of whom are, at present, specially depressed."

Writing to Alfred Vail on the same day, he says after telling of the passage of the bill: —

"You can have but a faint idea of the sacrifices and trials I have had in getting the Telegraph thus far before the country and the world. I cannot detail them here; I can only say that, for two years, I have labored all my time and at my own expense, without assistance from the other proprietors (except in obtaining the iron of the magnets for the last instruments obtained of you) to forward our enterprise. My means to defray my expenses, to meet which every cent I owned in the world was collected, are nearly all gone, and if, by any means, the bill should fail in the Senate, I shall return to New York with the *fraction of a dollar* in my pocket."

And now the final struggle which meant success or failure was on. Only eight days of the session remained and the calendar was, as usual, crowded. The inventor, his nerves stretched to the breaking point, hoped and yet feared. He had every reason to believe that the Senate would show more broad-minded enlightenment than the House, and yet he had been told that his bill would pass the House by acclamation, while the event proved that it had barely squeezed through by a beggarly majority of six. He heard disquieting rumors of a determination on the part of some of the House members to procure the defeat of the bill in the Senate. Would they succeed, would the victory, almost won, be snatched from him at the last moment, or would his faith in an overruling Providence, and in his own mission as an instrument of that Providence, be justified at last?

Every day of that fateful week saw him in his place in the gallery of the Senate chamber, and all day long he sat there, listening, as we can well imagine, with growing impatience to the senatorial oratory on the merits or demerits of bills which to him were of such minor importance, however heavily freighted with the destinies of the nation they may have been. And every night he returned to his room with the sad reflection that one more of the precious days had passed and his bill had not been reached. And then came the last day, March 3, that day when the session of the Senate is prolonged till midnight, when the President, leaving the White House, sits in the room provided for him at the Capitol, ready to sign the bills which are passed in these last few hurried hours, if they meet with his approval, or to consign them to oblivion if they do not.

The now despairing inventor clung to his post in the gallery almost to the end, but, being assured by his senatorial friends that there was no possibility of the bill being reached, and unable to bear the final blow of hearing the gavel fall which should signalize his defeat, shrinking from the well-meant condolences of his friends, he returned almost broken-hearted to his room.

The future must have looked black indeed. He had staked his all and lost, and he was resolved to abandon all further efforts to press his invention on an unfeeling and a thankless world. He must pick up his brush again; he must again woo the fickle goddess of art, who had deserted him before, and who would, in all probability, be chary of her favors now. In that dark hour it would not have been strange if his trust in God had wavered, if he had doubted the goodness of that Providence to whose mysterious workings he had always submissively bowed. But his faith seems to have risen triumphant even under this crushing stroke, for he thus describes the events of that fateful night, and of the next morning, in a letter to Bishop Stevens, of Pennsylvania, written many years later: —

"The last days of the last session of that Congress were about to close. A bill appropriating thirty thousand dollars for my purpose had passed the House, and was before the Senate for concurrence. On the last day of the session [3d of March, 1843] I had spent the whole day and part of the evening in the Senate chamber, anxiously watching the progress of the passing of the various bills, of which there were, in the morning of that day, over one hundred and forty to be acted upon before the one in which I was interested would be reached; and a resolution had a few days before been passed to proceed with the bills on the calendar in their regular order, forbidding any bill to be taken up out of its regular place.

"As evening approached there seemed to be but little chance that the Telegraph Bill would be reached before the adjournment, and consequently I had the prospect of the delay of another year, with the loss of time, and all my means already expended. In my anxiety I consulted with two of my senatorial friends — Senator Huntington, of Connecticut, and Senator Wright, of New York — asking their opinion of the probability of reaching the bill before the close of the session. Their answers were discouraging, and their advice was to prepare myself for disappointment. In this state of mind I retired to my chamber and made all my arrangements for leaving Washington the next day. Painful as was this prospect of renewed disappointment, you, my dear sir, will understand me when I say that, knowing from experience whence my help must come in any difficulty, I soon disposed of my cares, and slept as quietly as a child.

"In the morning, as I had just gone into the breakfast-room, the servant called me out, announcing that a young lady was in the parlor wishing to speak with me. I was at once greeted with the smiling face of my young friend, the daughter of my old and valued friend and classmate, the Honorable H. L. Ellsworth, the Commissioner of Patents. On my expressing surprise at so early a call, she said: —

"'I have come to congratulate you.'

"'Indeed, for what?'

"'On the passage of your bill.'

"'Oh! no, my young friend, you are mistaken; I was in the Senate chamber till after the lamps were lighted, and my senatorial friends assured me there was no chance for me.'

"'But,' she replied, 'it is you that are mistaken. Father was there at the adjournment at midnight, and saw the President put his name to your bill, and I asked father if I might come and tell you, and he gave me leave. Am I the first to tell you?'

"The news was so unexpected that for some moments I could not speak. At length I replied: —

"'Yes, Annie, you are the first to inform me, and now I am going to make you a promise; the first dispatch on the completed line from Washington to Baltimore shall be yours.'

"'Well,' said she, 'I shall hold you to your promise.'"

This was the second great moment in the history of the Morse Telegraph. The first was when the inspiration came to him on board the Sully, more than a decade before, and now, after years of heart-breaking struggles with poverty and discouragements of all kinds, the faith in God and in himself, which had upheld him through all, was justified, and he saw the dawning of a brighter day.

On what slight threads do hang our destinies! The change of a few votes in the House, the delay of a few minutes in the Senate, would have doomed Morse to failure, for it is doubtful whether he would have had the heart, the means, or the encouragement to prosecute the enterprise further.

He lost no time in informing his associates of the happy turn in their affairs, and, in the excitement of the moment, he not only dated his letter to Smith March 3, instead of March 4, but he seems not to have understood that the bill had already been signed by the President, and had become a law: —

"Well, my dear Sir, the matter is decided. *The Senate has just passed my bill without division and without opposition*, and it will probably be signed by the President in a few hours. This, I think, is news enough for you at present, and, as I have other letters that I must write before the mail closes, I must say good-bye until I see you or hear from you. Write to me in New York, where I hope to be by the latter part of next week."

And to Vail he wrote on the same day: —

"You will be glad to learn, doubtless, that my bill has passed the Senate without a division and without opposition, so that now the telegraphic enterprise begins to look bright. I shall want to see you in New York after my return, which will probably be the latter part of next week. I have other letters to write, so excuse the shortness of this, which, IF SHORT, IS SWEET, at least. My kind regards to your father, mother, brothers, sisters, and wife. The whole delegation of your State, without exception, deserve the highest gratitude of us all."

The Representatives from the State of New Jersey in the House voted unanimously for the bill, those of every other State were divided between the yeas and the nays and those not voting.

Congratulations now poured in on him from all sides,

and the one he, perhaps, prized the most was from his friend and master, Washington Allston, then living in Boston: —

"*March 24, 1843.* All your friends here join me in rejoicing at the passing of the act of Congress appropriating thirty thousand dollars toward carrying out your Electro-Magnetic Telegraph. I congratulate you with all my heart. Shakespeare says: 'There is a tide in the affairs of men that, taken at the flood, leads on to fortune.' You are now fairly launched on what I hope will prove to you another Pactolus. *I pede fausto!*

"This has been but a melancholy year to me. I have been ill with one complaint or another nearly the whole time; the last disorder the erysipelas, but this has now nearly disappeared. I hope this letter will meet you as well in health as I take it you are now in spirits."

Morse lost no time in replying: —

"I thank you, my dear sir, for your congratulations in regard to my telegraphic enterprise. I hope I shall not disappoint the expectations of my friends. I shall exert all my energies to show a complete and satisfactory result. When I last wrote you from Washington, I wrote under the apprehension that my bill would not be acted upon, and consequently I wrote in very low spirits.

"'What has become of painting?' I think I hear you ask. Ah, my dear sir, when I have diligently and perseveringly wooed the coquettish jade for twenty years, and she then jilts me, what can I do? But I do her injustice, she is not to blame, but her guardian for the time being. I shall not give her up yet in despair, but pursue her even with lightning, and so overtake her at last.

"I am now absorbed in my arrangements for fulfilling my designs with the Telegraph in accordance with the act of Congress. I know not that I shall be able to complete my experiment before Congress meets again, but I shall endeavor to show it to them at their next session."

CHAPTER XXX

MARCH 15, 1843 — JUNE 13, 1844

Work on first telegraph line begun. — Gale, Fisher, and Vail appointed assistants. — F. O. J. Smith to secure contract for trenching. — Morse not satisfied with contract. — Death of Washington Allston. — Reports to Secretary of the Treasury. — Prophesies Atlantic cable. — Failure of underground wires. — Carelessness of Fisher. — F. O. J. Smith shows cloven hoof. — Ezra Cornell solves a difficult problem. — Cornell's plan for insulation endorsed by Professor Henry. — Many discouragements. — Work finally progresses favorably. — Frelinghuysen's nomination as Vice-President reported by telegraph. — Line to Baltimore completed. — First message. — Triumph. — Reports of Democratic Convention. — First long-distance conversation. — Utility of telegraph established. — Offer to sell to Government.

OUT of the darkness of despair into which he had been plunged, Morse had at last emerged into the sunlight of success. For a little while he basked in its rays with no cloud to obscure the horizon, but his respite was short, for new difficulties soon arose, and new trials and sorrows soon darkened his path.

Immediately after the telegraph bill had become a law he set to work with energy to carry out its provisions. He decided, after consultation with the Secretary of the Treasury, Hon. J. C. Spencer, to erect the experimental line between Washington and Baltimore, along the line of railway, and all the preliminaries and details were carefully planned. With the sanction of the Secretary he appointed Professors Gale and Fisher as his assistants, and soon after added Mr. Alfred Vail to their number. He returned to New York, and from there wrote to Vail on March 15: —

"You will not fail, with your brother and, if possible, your father, to be in New York on Tuesday the 21st, to meet the proprietors of the Telegraph. I was on the point of coming out this afternoon with young Mr. Serrell, the patentee of the lead-pipe machine, which I think promises to be the best for our purposes of all that have been invented, as to it can be applied '*a mode of filling lead-pipe with wire,*' for which Professor Fisher and myself have entered a caveat at the Patent Office."

Vail gladly agreed to serve as assistant in the construction of the line, and, on March 21 signed the following agreement: —

PROFESSOR MORSE, — As an assistant in the telegraphic experiment contemplated by the Act of Congress lately passed, I can superintend and procure the making of the *Instruments complete* according to your direction, namely: the registers, the correspondents with their magnets, the batteries, the reels, and the paper, and will attend to the procuring of the acids, the ink, and the preparation of the various stations. I will assist in filling the tubes with wire, and the resinous coating, and I will devote my whole time and attention to the business so as to secure a favorable result, and should you wish to devolve upon me any other business connected with the Telegraph, I will cheerfully undertake it.

Three dollars per diem, with travelling expenses, I shall deem a satisfactory salary.

Very respectfully, your ob't ser't,

ALFRED VAIL.

Professor Fisher was detailed to superintend the manufacture of the wire, its insulation and its insertion in the lead tubes, and Professor Gale's scientific knowledge was to be placed at the disposal of the patentees wherever and whenever it should be necessary. F. O. J. Smith undertook to secure a favorable contract for the trenching, which was necessary to carry out the first idea of placing the wires underground, and Morse himself was, of course, to be general superintendent of the whole enterprise.

In advertising for lead pipe the following quaint answer was received from Morris, Tasker & Morris, of Philadelphia: —

"Thy advertisements for about one hundred and twenty miles of ½ in. lead tube, for Electro Magnetic Telegraphic purposes, has induced us to forward thee some samples of Iron Tube for thy inspection. The quantity required and the terms of payment are the inducement to offer it to thee at the exceeding low price here stated, which thou wilt please keep *to thyself undivulged to other person*, etc., etc."

As iron tubing would not have answered Morse's purpose, this decorous solicitation was declined with thanks.

During the first few months everything worked smoothly, and the prospect of an early completion of the line was bright. Morse kept all his accounts in the most businesslike manner, and his monthly accounts to the Secretary of the Treasury were models of accuracy and a conscientious regard for the public interest.

One small cloud appeared above the horizon, so small that the unsuspecting inventor hardly noticed it, and yet it was destined to develop into a storm of portentous dimensions. On May 17, he wrote to F. O. J. Smith from New York: —

"Yours of the 27th April I have this morning received enclosing the contracts for trenching. I have examined the contract and I must say I am not exactly pleased with the terms. If I understood you right, before you left for Boston, you were confident a contract could be made far within the estimates given in to the Government, and I had hoped that something could be saved from that estimate as from the others, so as to present the experiment before the country in as cheap a form as possible.

"I have taken a pride in showing to Government how cheaply the Telegraph could be laid, since the main objection, and the one most likely to defeat our ulterior plans, is its great expense. I have in my other contracts been able to be far within my estimates to Government, and I had hoped to be able to present to the Secretary the contract for trenching likewise reduced. There are plenty of applicants here who will do it for much less, and one even said he thought for one half. I shall do nothing in regard to the matter until I see you."

A great personal sorrow came to him also, a short time after this, to dim the brilliance of success. On July 9, 1843, his dearly loved friend and master, Washington Allston, died in Boston after months of suffering. Morse immediately dropped everything and hastened to Boston to pay the last tributes of respect to him whom he regarded as his best friend. He obtained as a memento one of the brushes, still wet with paint, which Allston was using on his last unfinished work, "The Feast of Belshazzar," when he was suddenly stricken. This brush he afterwards presented to the National Academy of Design, where it is, I believe, still preserved.

Sorrowfully he returned to his work in Washington, but with the comforting thought that his friend had lived to see his triumph, the justification for his deserting that art which had been the bond to first bring them together.

On July 24, in his report to the Secretary of the Treasury, he says: —

"I have also the gratification to report that the contract for the wire has been faithfully fulfilled on the part of Aaron Benedict, the contractor; that the first covering with cotton and two varnishings of the whole one hundred and sixty miles is also completed; that experiments made upon forty-three miles have resulted in the most satisfactory manner, and that the whole work is proceeding with every prospect of a successful issue."

It was at first thought necessary to insulate the whole length of the wire, and it was not until some time afterwards that it was discovered that naked wires could be successfully employed.

On August 10, in his report to the Secretary, he indulges in a prophecy which must have seemed in the highest degree visionary in those early days: —

"Some careful experiments on the decomposing power at various distances were made from which the law of propulsion has been deduced, verifying the results of Ohm and those which I made in the summer of 1842, and alluded to in my letter to the Honorable C. G. Ferris, published in the House Report, No. 17, of the last Congress.

"The practical inference from this law is that a telegraphic communication on my plan may with certainty be established across the Atlantic!

"Startling as this may seem now, the time will come when this project will be realized."

On September 11, he reports an item of saving to the Government which illustrates his characteristic honesty in all business dealings: —

"I would also direct the attention of the Honorable Secretary to the payment in full of Mr. Chase, (voucher 215), for covering the wire according to the contract with him. The sum of $1010 was to be paid him. In the course of the preparation of the wire several improvements occurred to me of an economical character, in which Mr. Chase cheerfully concurred, although at a considerable loss to him of labor contracted for; so that my wire has been prepared at a cost of $551.25, which is receipted in full, instead of $1010, producing an economy of $458.75."

The work of trenching was commenced on Saturday, October 21, at 8 A.M., and then his troubles began. Describing them at a later date he says: —

"Much time and expense were lost in consequence of my following the plan adopted in England of laying the conductors beneath the ground. At the time the Telegraph bill was passed there had been about thirteen miles of telegraph conductors, for Professor Wheatstone's telegraph system in England, put into tubes and interred in the earth, and there was no hint publicly given that that mode was not perfectly successful. I did not feel, therefore, at liberty to expend the public moneys in useless experiments on a plan which seemed to be already settled as effective in England. Hence I fixed upon this mode as one supposed to be the best. It was prosecuted till the winter of 1843–44. It was abandoned, among other reasons, in consequence of ascertaining that, in the process of inserting the wire into the leaden tubes (which was at the moment of forming the tube from the lead at melting heat), the insulating covering of the wires had become charred, at various and numerous points of the line, to such an extent that greater delay and expense would be necessary to repair the damage than to put the wire on posts.

"In my letter to the Secretary of the Treasury, of September 27, 1837, one of the modes of laying the conductors for the Telegraph was the present almost universal one of extending them on posts set about two hundred feet apart. This mode was adopted with success."

The sentence in the letter of September 27, 1837, just referred to, reads as follows: "If the circuit is laid through the air, the first cost would, doubtless, be much lessened. Stout spars, of some thirty feet in height, well planted in the ground and placed about three hundred and fifty feet apart, would in this case be required, along the tops of which the circuit might be stretched."

A rough drawing of this plan also appears in the 1832 sketch-book.

It would seem, from a voluminous correspondence, that Professor Fisher was responsible for the failure of the underground system, inasmuch as he did not properly test the wires after they had been inserted in the lead pipe. Carelessness of this sort Morse could never brook, and he was reluctantly compelled to dispense with the services of one who had been of great use to him previously. He refers to this in a letter to his brother Sidney of December 16, 1843: —

"The season is against all my operations, and I expect to resume in the spring. I have difficulties and trouble in my work, but none of a nature as yet to discourage; they arise from neglect and unfaithfulness (*inter nos*) on the part of Fisher, whom I shall probably dismiss, although on many accounts I shall do it reluctantly. I shall give him an opportunity to excuse himself, if he ever gets here. I have been expecting both him and Gale for three weeks, and written, but without bringing either of them. They may have a good excuse. We shall see."

The few months of sunshine were now past, and the clouds began again to gather: —

December 18, 1843.

DEAR SIDNEY, — I have made every effort to try and visit New York. Twice I have been ready with my baggage in hand, but am prevented by a pressure of difficulties which you cannot conceive. I was never so tried and never needed more your prayers and those of Christians for me. Troubles cluster in such various shapes that I am almost overwhelmed.

And then the storm of which the little cloud was the forerunner burst in fury: —

December 30, 1843.

DEAR SIDNEY, — I have no heart to give you the details of the troubles which almost crush me, and which have unexpectedly arisen to throw a cloud over all my prospects. It must suffice at present to say that the unfaithfulness of Dr. Fisher in his inspection of the wires, and connected with Serrell's bad pipe, is the main origin of my difficulties.

The trenching is stopped in consequence of this among other reasons, and has brought the contractor upon me for damages (that is, upon the Government). Mr. Smith is the contractor, and where I expected to find a *friend* I find a FIEND. The word is not too strong, as I may one day show you. I have been compelled to dismiss Fisher, and have received a very insolent letter from him in reply. The lead-pipe contract will be litigated, and Smith has written a letter full of the bitterest malignity against me to the Secretary of the Treasury. He seems perfectly reckless and acts like a madman, and all for what? Because the condition of my pipe and the imperfect insulation of my wires were such that it became necessary to stop trenching on this account alone, but, taken in connection with the advanced state of the season, when it was impossible to carry on my operations out of doors, I was compelled to stop any further trenching. This causes him to lose his profit on the contract. *Hinc illæ lachrymæ.* And because I refused to accede to terms which, as a public officer, I could not do without dishonor and violation of trust, he pursues me thus malignantly.

Blessed be God, I have escaped snares set for me by this arch-fiend, one of which a simple inquiry from you was the means of detecting. You remember I told you that Mr. Smith had made an advantageous contract with Tatham & Brothers for pipe, and had divided the profits with me by which I should gain five hundred dollars. You asked if it was all right and, if it should be made public, it would be considered so. I replied, 'Oh! yes; Mr. Smith says it is all perfectly fair' (for I had the utmost confidence in his fair dealing and uprightness).

But your remark led me to think of the matter, and I determined at once that, since there was a doubt, I would not touch it for myself, but credit it to the Government, and I accordingly credited it as so much saved to the Government from the contract.

And now, will you believe it! the man who would have persuaded me that all was right in that matter, turns upon me and accuses me to the Secretary as dealing in bad faith to the Government, citing this very transaction in proof. But, providentially, my friend Ellsworth, and also a clerk in the Treasury Department, are witnesses that that sum was credited to the Government before any difficulties arose on the part of Smith.

But I leave this unpleasant matter. The enterprise yet looks lowering, but I know who can bring light out of darkness, and in Him I trust as a sure refuge till these calamities be overpast. . . . Oh! how these troubles drive all thought of children and brothers and all relatives out of my mind except in the wakeful hours of the night, and then I think of you all with sadness, that I cannot add to your enjoyment but only to your anxiety. . . . Love to all. Specially remember me in your prayers that I may have wisdom from above to act wisely and justly and calmly in this sore trial.

While thus some of those on whom he had relied failed him at a critical moment, new helpers were at hand to assist him in carrying on the work. On December 27, he writes to the Secretary of the Treasury: "I have the honor to report that I have dismissed Professor James C. Fisher, one of my assistants, whose salary was $1500 per annum. . . . My present labors require the services of an efficient mechanical assistant whom I believe I have found in Mr. Ezra Cornell, and whom I present for the approval of the Honorable Secretary, with a compensation at the rate of $1000 per annum from December 27, 1843."

Cornell proved himself, indeed, an efficient assistant, and much of the success of the enterprise, from that time forward, was due to his energy, quick-wittedness, and faithfulness.

Mr. Prime, in his biography of Morse, thus describes a dramatic episode of those trying days: —

"When the pipe had been laid as far as the Relay House, Professor Morse came to Mr. Cornell and expressed a desire to have the work arrested until he could try further experiments, but he was very anxious that nothing should be said or done to give to the public the impression that the enterprise had failed. Mr. Cornell said he could easily manage it, and, stepping up to the machine, which was drawn by a team of eight mules, he cried out: 'Hurrah, boys! we must lay another length of pipe before we quit.' The teamsters cracked their whips over the mules and they started on a lively pace. Mr. Cornell grasped the handles of the plough, and, watching an opportunity, canted it so as to catch the point of a rock, and broke it to pieces while Professor Morse stood looking on.

"Consultations long and painful followed. The anxiety of Professor Morse at this period was greater than at any previous hour known in the history of the invention. Some that were around him had serious apprehensions that he would not stand up under the pressure."

Cornell having thus cleverly cut the Gordian knot,

it was decided to string wires on poles, and Cornell himself thus describes the solution of the insulation problem: —

"In the latter part of March Professor Morse gave me the order to put the wires on poles, and the question at once arose as to the mode of *fastening the wires to the poles*, and the insulation of them at the point of fastening. I submitted a plan to the Professor which I was confident would be successful as an insulating medium, and which was easily available then and inexpensive. Mr. Vail also submitted a plan for the same purpose, which involved the necessity of going to New York or New Jersey to get it executed. Professor Morse gave preference to Mr. Vail's plan, and started for New York to get the fixtures, directing me to get the wire ready for use and arrange for setting the poles.

"At the end of a week Professor Morse returned from New York and came to the shop where I was at work, and said he wanted to provide the insulators for putting the wires on the poles upon the plan I had suggested; to which I responded: 'How is that, Professor; I thought you had decided to use Mr. Vail's plan?' Professor Morse replied: 'Yes, I did so decide, and on my way to New York, where I went to order the fixtures, I stopped at Princeton and called on my old friend, Professor Henry, who inquired how I was getting along with my Telegraph.

"'I explained to him the failure of the insulation in the pipes, and stated that I had decided to place the wires on poles in the air. He then inquired how I proposed to insulate the wires when they were attached to the poles. I showed him the model I had of Mr. Vail's plan, and he said, "It will not do; you will meet the same difficulty you had in the pipes." I then explained to him your plan which he said would answer.'"

However, before the enterprise had reached this point in March, 1844, many dark and discouraging days and weeks had to be passed, which we can partially follow by the following extracts from letters to his brother Sidney and others. To his brother he writes on January 9, 1844: —

"I thank you for your kind and sympathizing letter, which, I assure you, helped to mitigate the acuteness of my mental sufferings from the then disastrous aspect of my whole enterprise. God works by instrumentalities, and he has wonderfully thus far interposed in keeping evils that I feared in abeyance. All, I trust, will yet be well, but I have great difficulties to encounter and overcome, with the details of which I need not now trouble you. I think I see light ahead, and the great result of these difficulties, I am persuaded, will be a great economy in laying the telegraphic conductors. . . . I am well in health but have sleepless nights from the great anxieties and cares which weigh me down."

"*January 13*. I am working to retrieve myself under every disadvantage and amidst accumulated and most diversified trials, but I have strength from the source of strength, and courage to go forward. Fisher I have dismissed for unfaithfulness; Dr. Gale has resigned from ill-health; Smith has become a malignant enemy, and Vail only remains true at his post. All my pipe is useless as the wires are all injured by the *hot process* of manufacture. I am preparing (as I said before, under every disadvantage) a short distance between the Patent

Office and Capitol, which I am desirous of having completed as soon as possible, and by means of it relieving the enterprise from the heavy weight which now threatens it."

To his good friend, Commissioner Ellsworth, he writes from Baltimore on February 7: —

"In complying with your kind request that I would write you, I cannot refrain from expressing my warm thanks for the words of sympathy and the promise of a welcome on my return, which you gave me as I was leaving the door. I find that, brace myself as I will against trouble, the spirit so sympathises with the body that its moods are in sad bondage to the physical health; the latter vanquishing the former. For the spirit is often willing and submits, while the flesh is weak and rebels.

"I am fully aware that of late I have evinced an unusual sensitiveness, and exposed myself to the charge of great weakness, which would give me the more distress were I not persuaded that I have been among real friends who will make every allowance. My temperament, naturally sensitive, has lately been made more so by the combination of attacks from deceitful associates without and bodily illness within, so that even the kind attentions of the dear friends at your house, and who have so warmly rallied around me, have scarcely been able to restore me to my usual buoyancy of spirit, and I feel, amidst other oppressive thoughts, that I have not been grateful enough for your friendship. But I hope yet to make amends for the past. . . . I have no time to add more than that I desire sincere love to dear Annie, to whom please present for me the accompanying piece from my favorite Bellini, and the book on Etiquette, after it shall have passed the ordeal of a mother's examination, as I have not had time to read it myself."

On March 4, he writes to his brother: —

"I have nothing new. Smith continues to annoy me, but I think I have got him in check by a demand for compensation for my services for seven months, for doing that for him in Paris which he was bound to do. The agreement stipulates that I give my services for '*three months and no longer*,' but, at his earnest solicitation, I remained seven months longer and was his agent in 'negotiating the sale of rights,' which by the articles he was obliged to do; consequently I have a right to compensation, and Mr. E. and others think my claim a valid one. If it is sustained the tables are completely turned on him, and he is debtor to me to the amount of six or seven hundred dollars. I have commenced my operations with posts which promise well at present."

"*March 23.* My Telegraph labors go on well at present. The whole matter is now critical, or, as our good father used to say, 'a crisis is at hand.' I hope for the best while I endeavor to prepare my mind for the worst. Smith, if he goes forward with his claim, is a ruined man in reputation, but he may sink the Telegraph also in his passion; but, when he returns from the East, where he fortunately is now, we hope through his friends to persuade him to withdraw it, which he may do from fear of the consequences. As to his claims privately on me, I think I have him in check, but he is a man of consummate art and unprincipled; he will, therefore, doubtless give me trouble."

"*April 10.* A brighter day is dawning upon me. I

send you the Intelligencer of to-day, in which you will see that the Telegraph is successfully under way. Through six miles the experiment has been most gratifying. In a few days I hope to advise you of more respecting it. I have preferred reserve until I could state something positive. I have my posts set to Beltsville, twelve miles, and you will see by the Intelligencer that I am prepared to go directly on to Baltimore and hope to reach there by the middle of May."

"*May 7.* Let me know when Susan and the two Charles arrive [his son and his grandson] for, if they come within the next fortnight, I think I can contrive to run on and pay a visit of two or three days, unless my marplot Smith should prevent again, as he is likely to do if he comes on here. As yet there is no settlement of that matter, and he seems determined (*inter nos*) to be as ugly as he can and defeat all application for an appropriation if I am to have the management of it. He chafes like a wild boar, but, when he finds that he can effect nothing by such a temper, self-interest may soften him into terms.

"You will see by the papers that the Telegraph is in successful operation for twenty-two miles, to the Junction of the Annapolis road with the Baltimore and Washington road. The nomination of Mr. Frelinghuysen as Vice-President was written, sent on, and the receipt acknowledged back in two minutes and one second, a distance of forty-four miles. The news was spread all over Washington one hour and four minutes before the cars containing the news by express arrived. In about a fortnight I hope to be in Baltimore, and a communication will be established between the two cities. Good-bye. I am almost asleep from exhaustion, so excuse abrupt closing."

This was the first great triumph of the telegraph. Morse and Vail and Cornell had worked day and night to get the line in readiness as far as the Junction so that the proceedings of the Whig Convention could be reported from that point. Many difficulties were encountered — crossing of wires, breaks, injury from thunder storms, and the natural errors incidental to writing and reading what was virtually a new language. But all obstacles were overcome in time, and the day before the convention met, Morse wrote to Vail: —

"Get everything ready in the morning for the day, and do not be out of hearing of your bell. When you learn the name of the candidate nominated, see if you cannot give it to me and receive an acknowledgment of its receipt before the cars leave you. If you can it will do more to excite the wonder of those in the cars than the mere announcement that the news is gone to Washington."

The next day's report was most encouraging: —

"Things went well to-day. Your last writing was good. You did not correct your error of running your letters together until some time. Better be deliberate; we have time to spare, since we do not spend upon our stock. Get ready to-morrow (Thursday) as to-day. There is great excitement about the Telegraph and my room is thronged, therefore it is important to have it in action during the hours named. I may have some of the Cabinet to-morrow. . . . Get from the passengers in the cars from Baltimore, or elsewhere, all the news you can and transmit. A good way of exciting wonder will be to tell the passengers to give you some short sentence to send me; let them note time and call at the Capitol to verify the time I received it. Before transmitting notify me with (48). Your message to-day that 'the passengers in the cars gave three cheers for Henry Clay,' excited the highest wonder in the passenger who gave it to you to send when he found it verified at the Capitol."

In a letter to his friend, Dr. Aycrigg of New Jersey, written on May 8, and telling of these successful demonstrations, this interesting sentence occurs: "I find that the ground, in conformity with the results of experiments of Dr. Franklin, can be made a part of the circuit, and I have used one wire and the ground with better effect for one circuit than two wires."

On the 11th of May he again cautions Vail about his writing: " Everything worked well yesterday, but there is one defect in your writing. Make a *longer* space between each letter and a still longer space between each word. I shall have a great crowd to-day and wish all things to go off well. Many M.C.s will be present, perhaps Mr. Clay. Give me news by the cars. When the cars come along, try and get a newspaper from Philadelphia or New York and give items of intelligence. The arrival of the cars at the Junction begins to excite here the greatest interest, and both morning and evening I have had my room thronged."

And now at last the supreme moment had arrived. The line from Washington to Baltimore was completed, and on the 24th day of May, 1844, the company invited by the inventor assembled in the chamber of the United States Supreme Court to witness his triumph. True to his promise to Miss Annie Ellsworth, he had asked her to indite the first public message which should be flashed over the completed line, and she, in consultation with her good mother, chose the now historic words from the 23d verse of the 23d chapter of Numbers — "What hath God wrought!" The whole verse reads: "Surely there is no enchantment against Jacob, neither is there any divination against Israel: according to this time it shall be said of Jacob and of Israel, What hath God wrought!" To Morse, with his strong religious bent and his belief that he was but a chosen vessel, every word in this verse seemed singularly appropriate. Calmly he seated himself at the instrument and ticked off the inspired words in the dots and dashes of the Morse alphabet. Alfred Vail, at the other end of the line in Baltimore, received the message without an error, and immediately flashed it back again, and the Electro-Magnetic Telegraph was no longer the wild dream of a visionary, but an accomplished fact.

Mr. Prime's comments, after describing this historic occasion, are so excellent that I shall give them in full:—

"Again the triumph of the inventor was sublime. His confidence had been so unshaken that the surprise of his friends in the result was not shared by him. He knew what the instrument would do, and the fact accomplished was but the confirmation to others of what to him was a certainty on the packet-ship Sully in 1832. But the result was not the less gratifying and sufficient. Had his labors ceased at that moment, he would have cheerfully exclaimed in the words of Simeon: 'Lord, now lettest thou thy servant depart in peace, for mine eyes have seen thy salvation.'

"The congratulations of his friends followed. He

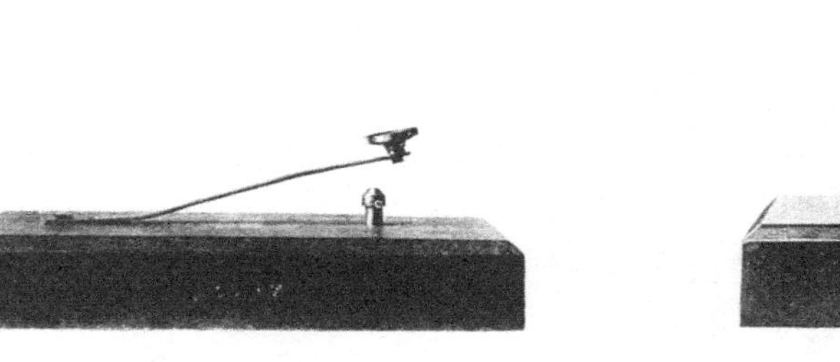

FIRST FORM OF KEY

IMPROVED FORM OF KEY

EARLY RELAY

The two keys and the relay are in the National Museum, Washington

FIRST WASHINGTON-BALTIMORE INSTRUMENT

The Washington-Baltimore instrument is owned by Cornell University

received them with modesty, in perfect harmony with the simplicity of his character. Neither then nor at any subsequent period of his life did his language or manner indicate exultation. He believed himself an instrument employed by Heaven to achieve a great result, and, having accomplished it, he claimed simply to be the original and only instrument by which that result had been reached. With the same steadiness of purpose, tenacity and perseverance, with which he had pursued the idea by which he was inspired in 1832, he adhered to his claim to the paternity of that idea, and to the merit of bringing it to a successful issue. Denied, he asserted it; assailed, he defended it. Through long years of controversy, discussion and litigation, he maintained his right. Equable alike in success and discouragement, calm in the midst of victories, and undismayed by the number, the violence, and the power of those who sought to deprive him of the honor and the reward of his work, he manfully maintained his ground, until, by the verdict of the highest courts of his country, and of academies of science, and the practical adoption and indorsement of his system by his own and foreign nations, those wires, which were now speaking only forty miles from Washington to Baltimore, were stretched over continents and under oceans making a network to encompass and unite, in instantaneous intercourse, for business and enjoyment, all parts of the civilized world."

It was with well-earned but modest satisfaction that he wrote to his brother Sidney on May 31: —

"You will see by the papers how great success has attended the first efforts of the Telegraph. That sentence of Annie Ellsworth's was divinely indited, for it is in my thoughts day and night. 'What hath God wrought!' It is his work, and He alone could have carried me thus far through all my trials and enabled me to triumph over the obstacles, physical and moral, which opposed me.

"'Not unto us, not unto us, but to thy name, O Lord, be all the praise.'

"I begin to fear now the effects of public favor, lest it should kindle that pride of heart and self-sufficiency which dwells in my own as well as in others' breasts, and which, alas! is so ready to be inflamed by the slightest spark of praise. I do indeed feel gratified, and it is right I should rejoice, but I rejoice with fear, and I desire that a sense of dependence upon and increased obligation to the Giver of every good and perfect gift may keep me humble and circumspect.

"The conventions at Baltimore happened most opportunely for the display of the powers of the Telegraph, especially as it was the means of correspondence, in one instance, between the Democratic Convention and the first candidate elect for the Vice-Presidency. The enthusiasm of the crowd before the window of the Telegraph Room in the Capitol was excited to the highest pitch at the announcement of the nomination of the Presidential candidate, and the whole of it afterwards seemed turned upon the Telegraph. They gave the Telegraph three cheers, and I was called to make my appearance at the window when three cheers were given to me by some hundreds present, composed mainly of members of Congress.

"Such is the feeling in Congress that many tell me they are ready to grant anything. Even the most inveterate opposers have changed to admirers, and one of them, Hon. Cave Johnson, who ridiculed my system last session by associating it with the tricks of animal magnetism, came to me and said: 'Sir, I give in. It is an astonishing invention.'

"When I see all this and such enthusiasm everywhere manifested, and contrast the present with the past season of darkness and almost despair, have I not occasion to exclaim 'What hath God wrought'? Surely none but He who has all hearts in His hands, and turns them as the rivers of waters are turned could so have brought light out of darkness. 'Sorrow may continue for a night, but joy cometh in the morning.' Pray for me then, my dear brother, that I may have a heart to praise the great Deliverer, and in future, when discouraged or despairing, be enabled to remember His past mercy, and in full faith rest all my cares on Him who careth for us.

"Mr. S. still embarrasses the progress of the invention by his stubbornness, but there are indications of giving way; mainly, I fear, because he sees his pecuniary interest in doing so, and not from any sense of the gross injury he has done me. I pray God for a right spirit in dealing with him."

The incident referred to in this letter with regard to the nomination for the Vice-Presidency by the Democratic Convention is worthy of more extended notice. The convention met in Baltimore on the 26th of May, and it was then that the two-thirds rule was first adopted. Van Buren had a majority of the votes, but could not secure the necessary two thirds, and finally James K. Polk was unanimously nominated. This news was instantly flashed to Washington by the telegraph and was received with mingled feelings of enthusiasm, disappointment, and wonder, and not believed by many until confirmed by the arrival of the mail.

The convention then nominated Van Buren's friend, Senator Silas Wright, of New York, for the Vice-Presidency. This news, too, was immediately sent by wire to Washington. Morse at once informed Mr. Wright, who was in the Capitol at the time, of his nomination, but he refused to accept it, and Morse wired his refusal to Vail in Baltimore, and it was read to the convention only a few moments after the nomination had been made. This was too much for the credulity of the assembly, and they adjourned till the following day and sent a committee to Washington to verify the dispatch. Upon the return of the committee, with the report that the telegraph had indeed performed this wonder, this new instrumentality received such an advertisement as could not fail to please the most exacting.

Then a scene was enacted new in the annals of civilization. In Baltimore the committee of conference surrounded Vail at his instrument, and in Washington Senator Wright sat beside Morse, all others being excluded. The committee urged Wright to accept the nomination, giving him good reasons for doing so. He replied, giving as good reasons for refusing. This first long-distance conversation was carried on until the committee was finally convinced that Wright was determined to refuse, and they so reported to the convention. Mr. Dallas was then nominated, and in November of that year Polk and Dallas were elected.

On June 3, Morse made his report to the Honorable McClintock Young, who was then Secretary of the Treasury *ad interim*. It was with great satisfaction that he was able to say: "Of the appropriation made there will remain in the Treasury, after the settlement of outstanding accounts, about $3500, which may be needed for contingent liabilities and for sustaining the line already constructed, until provision by law shall be made for such an organization of a telegraphic department or bureau as shall enable the Telegraph at least to support itself, if not to become a profitable source of revenue to the Government."

In the course of this report mention is also made of the following interesting incidents: —

"In regard to the *utility* of the Telegraph, time alone can determine and develop the whole capacity for good of so perfect a system. In the few days of its infancy it has already casually shown its usefulness in the relief, in various ways, of the anxieties of thousands; and, when such a sure means of relief is available to the public at large, the amount of its usefulness becomes incalculable. An instance or two will best illustrate this quality of the Telegraph.

"A family in Washington was thrown into great distress by a rumor that one of its members had met with a violent death in Baltimore the evening before. Several hours must have elapsed ere their state of suspense could be relieved by the ordinary means of conveyance. A note was dispatched to the telegraph rooms at the Capitol requesting to have inquiry made at Baltimore. The messenger had occasion to wait but *ten minutes* when the proper inquiry was made at Baltimore, and the answer returned that the rumor was without foundation. Thus was a worthy family relieved immediately from a state of distressing suspense.

"An inquiry from a person in Baltimore, holding the check of a gentleman in Washington upon the Bank of Washington, was sent by telegraph to ascertain if the gentleman in question had funds in that bank. A messenger was instantly dispatched from the Capitol who returned in a few minutes with an affirmative answer, which was returned to Baltimore instantly, thus establishing a confidence in a money arrangement which might have affected unfavorably (for many hours, at least) the business transactions of a man of good credit.

"Other cases might be given, but these are deemed sufficient to illustrate the point of utility, and to suggest to those who will reflect upon them thousands of cases in the public business, in commercial operations, and in private and social transactions, which establish beyond a doubt the immense advantages of such a speedy mode of conveying intelligence."

While such instances of the use of the telegraph are but the commonplaces of to-day, we can imagine with what wonder they were regarded in 1844.

Morse then addressed a memorial to Congress, on the same day, referring to the report just quoted from, and then saying: —

"The proprietors respectfully suggest that it is an engine of power, for good or for evil, which all opinions seem to concur in desiring to have subject to the control of the Government, rather than have it in the hands of private individuals and associations; and to this end the proprietors respectfully submit their willingness to transfer the exclusive use and control of it, from Washington City to the city of New York, to the United States, together with such improvements as shall be made by the proprietors, or either of them, if Congress shall proceed to cause its construction, and upon either of the following terms."

Here follow the details of the two plans: either outright purchase by the Government of the existing line and construction by the Government of the line from Baltimore to New York, or construction of the latter by the proprietors under contract to the Government; but no specific sum was mentioned in either case.

This offer was not accepted, as will appear further on, but $8000 was appropriated for the support of the line already built, and that was all that Congress would do. It was while this matter was pending that Morse wrote to his brother Sidney, on June 13: —

"I am in the crisis of matters, so far as this session of Congress is concerned, in relation to the Telegraph, which absorbs all my time. Perfect enthusiasm seems to pervade all classes in regard to it, but there is still the thorn in the flesh which is permitted by a wise Father to keep me humble, doubtless. May his strength be sufficient for me and I shall fear nothing, and will bear it till He sees fit to remove it. Pray for me, as I do for you, that, if prosperity is allotted to us, we may have hearts to use it to the glory of God."

CHAPTER XXXI

JUNE 23, 1844—OCTOBER 9, 1845

Fame and fortune now assured. — Government declines purchase of telegraph. — Accident to leg gives needed rest. — Reflections on ways of Providence. — Consideration of financial propositions. — F. O. J. Smith's fulsome praise. — Morse's reply. — Extension of telegraph proceeds slowly. — Letter to Russian Minister. — Letter to London "Mechanics' Magazine" claiming priority and first experiments in wireless telegraphy. — Hopes that Government may yet purchase. — Longing for a home. — Dinner at Russian Minister's. — Congress again fails him. — Amos Kendall chosen as business agent. — First telegraph company. — Fourth voyage to Europe. — London, Broek, Hamburg. — Letter of Charles T. Fleischmann. — Paris. — Nothing definite accomplished.

MORSE's fame was now secure, and fortune was soon to follow. Tried as he had been in the school of adversity, he was now destined to undergo new trials, trials incident to success, to prosperity, and to world-wide eminence. That he foresaw the new dangers which would beset him on every hand is clearly evidenced in the letters to his brother, but, heartened by the success which had at last crowned his efforts, he buckled on his armor ready to do battle to such foes, both within and without, as should in the future assail him. Fatalist as we must regard him, he believed in his star; or rather he went forward with sublime faith in that God who had thus far guarded him from evil, and in his own good time had given him the victory, and such a victory! For twelve years he had fought on through trials and privations, hampered by bodily ailments and the deep discouragements of those who should have aided him. Pitted against the trained minds and the wealth of other nations, he had gone forth a very David to battle, and, like David, the simplicity of his missile had given him the victory. Other telegraphs had been devised by other men; some had actually been put into operation, but it would seem as if all the nations had held their breath until his appeared, and, sweeping all the others from the field, demonstrated and maintained its supremacy.

From this time forward his life became more complex. Honors were showered upon him; fame carried his name to the uttermost parts of the earth; his counsel was sought by eminent scientists and by other inventors, both practical and visionary.

On the other hand, detractors innumerable arose; his rights to the invention were challenged, in all sincerity and in insincerity; infringements of his patent rights necessitated long and acrimonious lawsuits, and, like other men of mark, he was traduced and vilified. In addition to all this he took an active interest in the seething politics of the day and in religious questions which, to his mind and that of many others, affected the very foundations of the nation.

To follow him through all these labyrinthine ways would require volumes, and I shall content myself with selecting only such letters as may give a fair idea of how he bore himself in the face of these new and manifold trials, of how he sometimes erred in judgment and in action, but how through all he was sincere and firm in his faith, and how, at last, he was to find that home and that domestic bliss which he had all his life so earnestly desired, but which had until the evening of his days been denied to him.

Having won his great victory, retirement from the field of battle would have best suited him. He was now fifty-three years of age, and he felt that he had earned repose. To this end he sought to carry out his long-cherished idea that the telegraph should become the property of the Government, and he was willing to accept a very modest remuneration. As I have said before, he and the other proprietors joined in offering the telegraph to the Government for the paltry sum of $100,000. But the Administration of that day seems to have been stricken with unaccountable blindness, for the Postmaster-General, that same wise and sapient Cave Johnson who had sought to kill the telegraph bill by ridicule in the House, and in despite of his acknowledgment to Morse, reported: "That the operation of the Telegraph between Washington and Baltimore had not satisfied him that, under any rate of postage that could be adopted, its revenues could be made equal to its expenditures." Congress was equally lax, and so the Government lost its great opportunity, for when, in after years, the question of government ownership again came up, it was found that either to purchase outright or to parallel existing lines would cost many more millions than it would have taken thousands in 1844.

The failure of the Government to appreciate the value of what was offered to them was always a source of deep regret to Morse. For, while he himself gained much more by the operation of private companies, the evils which he had foretold were more than realized.

But to return to the days of '44, it would seem that in the spring of that year he met with a painful accident. Its exact nature is not specified, but it must have been severe, and yet we learn from the following letter to his brother Sidney, dated June 23, that he saw in it only another blessing: —

"I am still in bed, and from appearances I am likely to be held here for many days, perhaps weeks. The wound on the leg was worse than I at first supposed. It seems slow in healing and has been much inflamed, although now yielding to remedies. My hope was to have spent some weeks in New York, but it will now depend on the time of the healing of my leg.

"The ways of God are mysterious, and I find prayer answered in a way not at all anticipated. This accident, as we are apt to call it, I can plainly see is calculated to effect many salutary objects. I needed rest of body and mind after my intense anxieties and exertions, and I might have neglected it, and so, perhaps, brought on premature disease of both; but I am involuntarily laid up so that I must keep quiet, and, although the fall that caused my wound was painful at first, yet I have no severe pain with it now. But the principal effect is, doubtless, intended to be of a spiritual character, and I am afforded an opportunity of quiet reflection on the wonderful dealings of God with me.

"I cannot but constantly exclaim, 'What hath God wrought!' When I look back upon the darkness of last winter and reflect how, at one time everything seemed hopeless; when I remember that all my associates in the enterprise of the Telegraph had either deserted me or were discouraged, and one had even turned my enemy, reviler and accuser (and even Mr. Vail, who has held fast to me from the beginning, felt like giving up just in the deepest darkness of all); when I remember that, giving up all hope myself from any other source than his right arm which brings salvation, his salvation did come in answer to prayer, faith is strengthened, and did I not know by too sad experience the deceitfulness of the heart, I should say that it was impossible for me again to distrust or feel anxiety, undue anxiety, for the future. But He who knows the heart knows its disease, and, as the Good Physician, if we give ourselves unreservedly into his hands to be cured, He will give that medicine which his perfect knowledge of our case prescribes.

"I am well aware that just now my praises ring from one end of the country to the other. I cannot take up a paper in which I do not find something to flatter the natural pride of the heart. I have prayed, indeed, against it; I have asked for a right spirit under a trial of a new character, for prosperity is a trial, and our Saviour has denounced a woe on us 'when all men speak well of us.' May it not then be in answer to this prayer that He shuts me up, to strengthen me against the temptations which the praises of the world present, and so, by meditation on his dealings with me and reviewing the way in which He has led me, showing me my perfect helplessness without Him, He is preparing to bless me with stronger faith and more unreserved faith in Him?

"To Him, indeed, belongs all the glory. I have had evidence enough that without Christ I could do nothing. All my strength is there and I fervently desire to ascribe to Him all the praise. If I am to have influence, increased influence, I desire to have it for Christ, to use it for his cause; if wealth, for Christ; if more knowledge, for Christ. I speak sincerely when I say I fear prosperity lest I should be proud and forget whence it comes."

Having at length recovered from the accident which had given him, in spite of himself, the rest which he so much needed, Morse again devoted himself to his affairs with his accustomed vigor. The Government still delaying to take action, he was compelled, much to his regret, to consider the offers of private parties to extend the lines of the telegraph to important points in the Union. He had received propositions from various persons who were eager to push the enterprise, but in all negotiations he was hampered by the dilatoriness of Smith, who seemed bent on putting as many obstacles in the way of an amicable settlement as possible, and some of whose propositions had to be rejected for obvious reasons. Before Congress had finally put the quietus on his hopes in that direction, he considered the advisability of parting with his interest to some individual, and, on July 1, 1844, he wrote to Mr. David Burbank from Baltimore: —

"In reply to your query for what sum I would sell my share of the patent right in the Telegraph, which amounts to one half, I frankly say that, if *one hundred and ten thousand dollars* shall be secured to me in cash, current funds in the United States, or stocks at cash value, such as I may be disposed to accept if presented, so that in six months from this date I shall realize that sum, I will assign over all my rights and privileges in the Telegraph in the United States.

"I offer it at this price, not that I estimate the value of the invention so low, for it is perfectly demonstrable that the sum above mentioned is not half its value, but that I may have my own mind free to be occupied in perfecting the system, and in a general superintendence of it, unembarrassed by the business arrangements necessary to secure its utmost usefulness and value."

A Mr. Fry of Philadelphia had also made an offer, and, referring to this, he wrote to Smith from New York, on July 17: "A letter from Mr. Fry, of Philadelphia, in answer to the proposals which you sent, I have just received. I wish much to see you, as I cannot move in this matter until I know your views. I am here for about a fortnight and wish some arrangements made by which our business can be transacted without the necessity of so much waiting and so much writing."

All these negotiations seem to have come to nothing, and I have only mentioned them as showing Morse's willingness to part with his interest for much less than he knew it was worth, in order that he might not prove an obstacle in the expansion of the system by being too mercenary, and so that he might obtain some measure of freedom from care.

Mr. F. O. J. Smith, while still proving himself a thorn in the flesh to Morse in many ways, had compiled a Telegraph Dictionary which he called: "The Secret Corresponding Vocabulary, adapted for Use to Morse's Electro-Magnetic Telegraph, and also in conducting Written Correspondence transmitted by the Mails, or otherwise." The dedication reads as follows:

To Professor Samuel F. B. Morse, Inventor of the Electro-Magnetic Telegraph

SIR, — The homage of the world during the last half-century has been, and will ever continue to be, accorded to the name and genius of the illustrious American philosopher, Benjamin Franklin, for having first taught mankind that the wild and terrific ways and forces of the electric fluid, as it flies and flashes through the rent atmosphere, or descends to the surface of the earth, are guided by positive and fixed laws, as much as the movements of more sluggish matter in the physical creation, and that its terrible death-strokes may be rendered harmless by proper scientific precautions.

To another name of another generation, yet of the same proud national nativity, the glory has been reserved of having first taught mankind to reach even beyond the results of Franklin, and to subdue in a modified state, into the familiar and practical uses of a household servant who runs at his master's bidding, this same once frightful and tremendous element. Indeed the great work of science which Franklin commenced for the protection of man, you have most triumphantly subdued to his convenience. And it needs not the gift of prophecy to foresee, nor the spirit of personal flattery to declare, that the names of Franklin and Morse are destined to glide down the declivity of time together, the equals in the renown of inventive achievements, until the hand of History shall become palsied, and whatever pertains to humanity shall be lost in the general dissolution of matter.

Of one thus rich in the present applause of his countrymen, and in the prospect of their future gratitude, it affords the author of the following compilation, which is designed to contribute in a degree to the practical usefulness of your invention, a high gratification to speak in the presence of an enlightened public feeling.

That you may live to witness the full consummation of the vast revolution in the social and business relations of your countrymen, which your gonius has proved to be feasible, under the liberal encouragement of our national councils, and that you may, with this great gratification, also realize from it the substantial reward, which inventive merit too seldom acquires, in the shape of pecuniary independence, is the sincere wish of

Your most respectful and obedient servant
THE AUTHOR.

This florid and fulsome eulogy was written by that singular being who could thus flatter, and almost apotheosize, the inventor in public, while in secret he was doing everything to thwart him, and who never, as long as he lived, ceased to antagonize him, and later accused him of having claimed the credit of an invention all the essentials of which were invented by others. No wonder that Morse was embarrassed and at a loss how to reply to the letter of Smith's enclosing this eulogy and, at the same time, bringing up one of the subjects in dispute:—

NEW YORK, November 13, 1844.

DEAR SIR, — I have received yours of the 4th and 5th inst., and reply in relation to the several subjects you mention in their order.

I like very well the suggestion in regard to the presentation of a set of the Telegraph Dictionary you are publishing to each member of Congress, and, when I return to Washington, will see the Secretary of the Treasury and see if he will assent to it.

As to the dedication to me, since you have asked my opinion, I must say I should prefer to have it much curtailed and less laudatory. I must refer it entirely to you, however, as it is not for me to say what others should write and think of me.

In regard to the Bartlett claim against the Government and your plan for settling it, I cannot admit that, as proprietors of the Telegraph, we have anything to do with it. I regret that there has been any mention of it, and I had hoped that you yourself had come to the determination to leave the matter altogether, or at least until the Telegraph bill had been definitely settled in Congress. However much I may deprecate agitation of the subject in the Senate, to mar and probably to defeat all our prospects, it is a matter over which I have no control in the aspect that has been given to it, and therefore — "the suppression of details which had better not be pushed to a decision" — does not rest with me.

In regard, however, to such a division of the property of the Telegraph as shall enable each of us to labor for the general benefit without embarrassment from each other, I think it worthy of consideration, and the principle on which such a division is proposed to be made might be extended to embrace the entire property. The subject, however, requires mature deliberation, and I am not now prepared to present the plan, but will think it over and consult with Vail and Gale and arrange it, perhaps definitely, when I see you again in Washington.

I have letters from Vail at Washington and Rogers at Baltimore stating the fact that complete success has attended all the transmission of results by Telegraph, there not having been a failure in a single instance, and to the entire satisfaction of both political parties in the perfect impartiality of the directors of the Telegraph.

While the success of the Telegraph had now been fully demonstrated, and while congratulations and honors were showered on the inventor from all quarters, negotiations for its extension proceeded but slowly. Morse still kept hoping that the Government would eventually purchase all the rights, and it was not until well into 1845 that he was compelled to abandon this dream. In the mean time he was kept busy replying to enquiries from the representatives of Russia, France, and other European countries, and in repelling attacks which had already been launched against him in scientific circles. As an example of the former I shall quote from a letter to His Excellency Alexander de Bodisco, the Russian Minister, written in December, 1844: —

"In complying with your request to write you respecting my invention of the Electro-Magnetic Telegraph, I find there are but few points of interest not embraced in the printed documents already in your possession. The principle on which my whole invention rests is the power of the electro-magnet commanded at pleasure at any distance. The application of this power to the telegraph is original with me. If the electro-magnet is now used in Europe for telegraphic purposes, it has been subsequently introduced. All the systems of electric telegraphs in Europe from 1820 to 1840 are based on the *deflection of the magnetic needle*, while my system, invented in 1832, is based, as I have just observed, on the electro-magnet. . . .

"Should the Emperor be desirous of the superintendence of an experienced person to put the Telegraph in operation in Russia, I will either engage myself to visit Russia for that purpose; or, if my own or another government shall, previous to receiving an answer from Russia, engage my personal attendance, I will send an experienced person in my stead."

As a specimen of the vigorous style in which he repelled attacks on his merits as an inventor, I shall give the following: —

MESSRS. EDITORS, — The London "Mechanics' Magazine," for October, 1844, copies an article from the Baltimore "American" in which my discovery in relation to causing electricity to cross rivers without wires is announced, and then in a note to his readers the editor of the magazine makes the following assertion: "The English reader need scarcely be informed that Mr. Morse has in this, as in other matters relating to magneto telegraphs, only *re*discovered what was previously well known in this country."

More illiberality and deliberate injustice has been seldom condensed within so small a compass. From the experience, however, that I, in common with many American scientific gentlemen, have already had of the piratical conjoined with the abusive propensity of a certain class of English *savans* and writers, I can scarcely expect either liberality or justice from the quarter whence this falsehood has issued. But there is, fortunately, an appeal to my own countrymen, to the impartial and liberal-minded of Continental Europe, and the truly noble of England herself.

I claim to be the original inventor of the Electro-

Magnetic Telegraph; to be the first who planned and operated a really practicable Electric Telegraph. This is the broad claim I make in behalf of my country and myself before the world. If I cannot substantiate this claim, if any other, to whatever country he belongs, can make out a previous or better claim, I will cheerfully yield him the palm.

Although I had planned and completed my Telegraph unconscious, until after my Telegraph was in operation, that even the words "Electric Telegraph" had ever been combined until I had combined them, I have now made myself familiar with, I believe, all the plans, abortive and otherwise, which have been given to the world since the time of Franklin, who was the first to suggest the possibility of using electricity as a means of transmitting intelligence. With this knowledge, both of the various plans devised and the time when they were severally devised, I claim to be the first inventor of a really practicable telegraph on the electric principle. When this shall be seriously called in question by any responsible name, I have the proof in readiness.

As to English electric telegraphs, the telegraph of Wheatstone and Cooke, called the Magnetic Needle Telegraph, inefficient as it is, was invented five years after mine, and the printing telegraph, so-called (the title to the invention of which is litigated by Wheatstone and Bain) was invented seven years after mine.

So much for my *re*discovering what was previously known in England.

As to the discovery that electricity may be made to cross the water without wire conductors, above,through, or beneath the water, the very reference by the editor to another number of the magazine, and to the experiments of Cooke, or rather Steinheil, and of Bain, shows that the editor is wholly ignorant of the nature of my experiment. I have in detail the experiments of Bain and Wheatstone. They were merely in effect repetitions of the experiments of Steinheil. Their object was to show that the earth or water can be made one half of the circuit in conducting electricity, a fact proved by Franklin with ordinary electricity in the last century, and by Professor Steinheil, of Munich, with magnetic electricity in 1837. Mr. Bain, and after him Mr. Wheatstone, in England repeated, or (to use the English editor's phrase) rediscovered the same fact in 1841.

But what have these experiments, in which *one wire* is carried across the river, to do with mine *which dispenses with wires altogether* across the river? I challenge the proof that such an experiment has ever been tried in Europe, unless it be since the publication of my results.

The year 1844 was drawing to a close and Congress still was dilatory. Morse hated to abandon his cherished dream of government ownership, and, while carrying on negotiations with private parties in order to protect himself, he still hoped that Congress would at last see the light. He writes to his brother from Washington on December 30:—

"Telegraph matters look exceedingly encouraging, not only for the United States but for Europe. I have just got a letter from a special agent of the French Government, sent to Boston by the Minister of Foreign Affairs, in which he says that he has seen mine and 'is convinced of its superiority,' and wishes all information concerning it, adding: 'I consider it my duty to make a special report on your admirable invention.'"

And on January 18, 1845, he writes: —

"I am well, but anxiously waiting the action of Congress on the bill for extension of Telegraph. Texas drives everything else into a corner. I have not many fears if they will only get it up. I had to-day the Russian, Spanish, and Belgian Ministers to see the operation of the Telegraph; they were astonished and delighted. The Russian Minister particularly takes the deepest interest in it, and will write to his Government by next steamer. The French Minister also came day before yesterday, and will write in its favor to his Government. . . . Senator Woodbury gave a discourse before the Institute a few nights ago, in the Hall of the House of Representatives, in which he lauded the Telegraph in the highest terms, and thought I had gone a step beyond Franklin! The popularity of the Telegraph increases rather than declines."

The mention of Texas in this letter refers to the fact that Polk was elected to the Presidency on a platform which favored the annexation of that republic to the United States, and this question was, naturally, paramount in the halls of Congress. Texas was admitted to the Union in December, 1845.

Writing to his daughter, Mrs. Lind, in Porto Rico on February 8, he says: —

"The Telegraph operates to the perfect satisfaction of the public, as you perhaps see by the laudatory notices of the papers in all parts of the country. I am now in a state of unpleasant suspense waiting the passage of the bill for the extension of the Telegraph to New York.

I am in hopes they will take it up and pass it next week; if they should not, I shall at once enter into arrangements with private companies to take it and extend it.

"I do long for the time, if it shall be permitted, to have you with your husband and little Charles around me. I feel my loneliness more and more keenly every day. Fame and money are in themselves a poor substitute for domestic happiness; as means to that end I value them. Yesterday was the sad anniversary (the twentieth) of your dear mother's death, and I spent the most of it in thinking of her. . . ."

"*Thursday, February 12.* I dined at the Russian Ambassador's Tuesday. It was the most gorgeous dinner-party I ever attended in any country. Thirty-six sat down to table; there were eleven Senators, nearly half the Senate. . . . The table, some twenty or twenty-five feet long, was decorated with immense gilt vases of flowers on a splendid plateau of richly chased gilt ornaments, and candelabra with about a hundred and fifty lights. We were ushered into the house through eight liveried servants, who afterward waited on us at table.

"I go to-morrow evening to Mr. Wickliffe's, Postmaster General, and, probably, on Wednesday evening next to the President's. The new President, Polk, arrived this evening amid the roar of cannon. He will be inaugurated on the 4th of March, and I presume I shall be there.

"I am most anxiously waiting the action of Congress on the Telegraph. It is exceedingly tantalizing to suffer so much loss of precious time that cannot be recalled."

This time there was no eleventh-hour passage of the

bill, for Congress adjourned without reaching it, and while this, in the light of future events, was undoubtedly a tactical error on the part of the Government, it inured to the financial benefit of the inventor himself. The question now arose of the best means of extending the business of the telegraph through private companies, and Morse keenly felt the need of a better business head than he possessed to guide the enterprise through the shoals and quicksands of commerce. He was fortunate in choosing as his business and legal adviser the Honorable Amos Kendall.

Mr. James D. Reid, one of the early telegraphers and a staunch and faithful friend of Morse's, thus speaks of Mr. Kendall in his valuable book "The Telegraph in America": —

"Mr. Kendall is too well known in American history to require description. He was General Jackson's Postmaster General, incorruptible, able, an educated lawyer, clear-headed, methodical, and ingenious. But he was somewhat rigid in his manners and methods, and lacked the dash and *bonhomie* which would have carried him successfully into the business centres of the seaboard cities, and brought capital largely and cheerfully to his feet. Of personal magnetism, indeed, except in private intercourse, where he was eminently delightful, he had, at this period of his life, none. This made his work difficult, especially with railroad men. Yet the Telegraph could not have been entrusted to more genuinely honest and able hands. On the part of those he represented this confidence was so complete that their interests were committed to him without reserve."

Professor Gale and Alfred Vail joined with Morse in

entrusting their interests to Mr. Kendall's care, but F. O. J. Smith preferred to act for himself. This caused much trouble in the future, for it was a foregone conclusion that the honest, upright Kendall and the shifty Smith were bound to come into conflict with each other. The latter, as one of the original patentees, had to be consulted in every sale of patent rights, and Kendall soon found it almost impossible to deal with him.

At first Kendall had great difficulty in inducing capitalists to subscribe to what was still looked upon as a very risky venture. Mr. Corcoran, of Washington, was the first man wise in his generation, and others then followed his lead, so that a cash capital of $15,000 was raised. Mr. Reid says: "It was provided, in this original subscription, that the payment of $50 should entitle the subscriber to two shares of $50 each. A payment of $15,000, therefore, required an issue of $30,000 stock. To the patentees were issued an additional $30,000 stock, or half of the capital, as the consideration of the patent. The capital was thus $60,000 for the first link. W. W. Corcoran and B. B. French were made trustees to hold the patent rights and property until organization was effected. Meanwhile an act of incorporation was granted by the legislature of the State of Maryland, the first telegraphic charter issued in the United States."

The company was called "The Magnetic Telegraph Company," and was the first telegraph company in the United States.

Under the able, if conservative, management of Mr. Kendall the business of the telegraph progressed slowly but surely. Many difficulties were encountered, many obstacles had to be overcome, and the efforts of unprincipled men to pirate the invention, or to infringe on the patent, were the cause of numerous lawsuits. But it is not my purpose to write a history of the telegraph. Mr. Reid has accomplished this task much better than I possibly could, and, in following the personal history of Morse, the now famous inventor, I shall but touch incidentally on all these matters.

On the 18th of July, 1845, the following letter of introduction was sent to Morse from the Department of State:—

To the respective Diplomatic and Consular Agents of the United States in Europe.

SIR,—The bearer hereof, Professor Samuel F. B. Morse, of New York, Superintendent of Electro Magnetic Telegraphs for the United States, is about to visit Europe for the purpose of exhibiting to the various governments his own system, and its superiority over others now in use. From a personal knowledge of Professor Morse I can speak confidently of his amiability of disposition and high respectability. The merits of his discoveries and inventions in this particular branch of science are, I believe, universally conceded in this country. I take pleasure in introducing him to your acquaintance and in bespeaking for him, during his stay in your neighborhood, such attentions and good offices in aid of his object as you may find it convenient to extend to him.

I am, sir, with great respect,

Your obedient servant,

JAMES BUCHANAN,

Secretary of State.

S. F. B. MORSE

From a portrait by Daniel Huntington

With the assurance that he had left his business affairs in capable hands, Morse sailed from New York on August 6, 1845, and arrived in Liverpool on the 25th. For the fourth time he was crossing from America to Europe, but under what totally different circumstances. On previous occasions, practically unknown, he had voyaged forth to win his spurs in the field of art, or to achieve higher honors in this same field, or as a humble petitioner at the courts of Europe. Forced by circumstances to practise the most rigid economy, he had yet looked confidently to the future for his reward in material as well as spiritual gifts. Now, having abandoned his art, he had won such fame in a totally different realm that his name was becoming well-known in all the centres of civilization, and he was assured of a respectful hearing wherever he might present himself. Freed already from pecuniary embarrassment, he need no longer take heed for the morrow, but could with a light heart give himself up to the enjoyment of new scenes, and the business of proving to other nations the superiority of his system, secure in the knowledge that, whatever might betide him in Europe, he was assured of a competence at home.

His brother Sidney, with his family, had preceded him to Europe, and writing to Vail from London on September 1, Morse says: —

"I have just taken lodgings with my brother and his family preparatory to looking about for a week, when I shall continue my journey to Stockholm and St. Petersburg, by the way of Hamburg, direct from London.

"On my way from Liverpool I saw at Rugby the

telegraph wires of Wheatstone, which extend, I understood, as far as Northampton. I went into the office as the train stopped a moment, and had a glimpse of the instrument as we have seen it in the 'Illustrated Times.' The place was the ticket-office and the man very uncommunicative, but he told me it was not in operation and that they did not use it much. This is easily accounted for from the fact that the two termini are inconsiderable places, and Wheatstone's system clumsy and complicated. The advantage of recording is incalculable, and in this I have the undisputed superiority. As soon as I can visit the telegraph-office here I will give you the result of my observation. I shall probably do nothing until my return from the north."

Nothing definite was accomplished during his short stay in London, and on the 17th of September he left for the Continent with Mr. Henry Ellsworth and his wife. Mr. Ellsworth, the son of his old friend, had been appointed attaché to the American Legation at Stockholm. Morse's letters to his daughter give a detailed account of his journey, but I shall give only a few extracts from them: —

"*Hamburg, September 27, 1845.* Everything being ready on the morning of the 17th instant, we left Brompton Square in very rainy and stormy weather, and drove down to the Custom-house wharf and went on board our destined steamer, the William Joliffe, a dirty, black-looking, tub-like thing, about as large but not half so neat as a North River wood-sloop. The wind was full from the Southwest, blowing a gale with rain, and I confess I did not much fancy leaving land in so unpromising a craft and in such weather; yet our vessel proved an excellent seaboat, and, although all were sick on board but Mr. Ellsworth and myself, we had a safe but rough passage across the boisterous North Sea."

Stopping but a short time in Rotterdam, the party proceeded through the Hague and Haarlem to Amsterdam, and from the latter place they visited the village of Broek: —

"The inn at Broek was another example of the same neatness. Here we took a little refreshment before going into the village. We walked of course, for no carriage, not even a wheelbarrow, appeared to be allowed any more than in a gentleman's parlor. Everything about the exterior of the houses and gardens was as carefully cared for as the furniture and embellishments of the interior. The streets (or rather alleys, like those of a garden) were narrow and paved with small variously colored bricks forming every variety of ornamental figures. The houses, from the highest to the lowest class, exhibited not merely comfort but luxury, yet it was a selfish sort of luxury. The perpetually closed door and shut-up rooms of ceremony, the largest and most conspicuous of all in the house, gave an air of inhospitableness which, I should hope, was not indicative of the real character of the inhabitants. Yet it seemed to be a deserted village, a place of the dead rather than of the living, an ornamental graveyard. The liveliness of social beings was absent and was even inconsistent with the superlative neatness of all around us. It was a best parlor out-of-doors, where the gayety of frolicking children would derange the set order of the furniture, or an accidental touch of a sacrilegious foot might scratch the polish of a fresh-varnished fence, or flatten down the nap of the green carpet of grass, every blade of which is trained to grow exactly so.

"The grounds and gardens of a Mr. Vander Beck were, indeed, a curiosity from the strange mixture of the useful with the ridiculously ornamental. Here were the beautiful banks of a lake and Nature's embellishment of reeds and water plants, which, for a wonder, were left to grow in their native luxuriance, and in the midst a huge pasteboard or wooden swan, and a wooden mermaid of tasteless proportions blowing from a conch-shell. In another part was a cottage with puppets the size of life moving by clock-work; a peasant smoking and turning a reel to wind off the thread which his 'goed vrow' is spinning upon a wheel, while a most sheep-like dog is made to open his mouth and to bark — a dog which is, doubtless, the progenitor of all the barking, toy-shop dogs of the world. Directly in the vicinity is a beautiful grapery, with the richest clusters of grapes literally covering the top, sides and walls of the greenhouse, which stands in the midst of a garden, gay with dahlias and amaranths and every variety of flowers, with delicious fruits thickly studding the well-trained trees. Everything, however, was cut up into miniature landscapes; little bridges and little temples adorned little canals and little mounds, miniature representations of streams and hills.

"We visited the residence of the burgomaster. He was away and his servants permitted us to see the house. It was cleaning-day. Everything in the house was in keeping with the character of the village. But the kitchen! how shall I describe it? The polished marble floor, the dressers with glass doors like a bookcase, to keep the least particle of dust from the bright-polished utensils of brass and copper. The varnished mahogany handle of the brass spigot, lest the moisture of the hand in turning it should soil its polish, and, will you believe it, the very pothooks as well as the cranes (for there were two), in the fireplace were as bright as your scissors!

"Broek is certainly a curiosity. It is unique, but the impression left upon me is not, on the whole, agreeable. I should not be contented to live there. It is too ridiculously and uncomfortably nice. Fancy a lady always dressed throughout the day in her best evening-party dress, and say if she could move about with that ease which she would like. Such, however, must be the feeling of the inhabitants of Broek; they must be in perpetual fear, not only of soiling or deranging their clothes merely, but their very streets every step they take. But good-bye to Broek. I would not have missed seeing it but do not care to see it again."

Holland, which he had never visited before, interested him greatly, but he could not help saying: "One feels in Holland like being in a ship, constantly liable to spring a leak."

Hamburg he found more to his taste: —

"*September 26.* Hamburg, you may remember, was nearly destroyed by fire in 1842. It is now almost rebuilt and in a most splendid style of architecture. I am much prepossessed in its favor. We have taken up our quarters at the Victoria Hotel, one of the splendid new hotels of the city. I find the season so far advanced in these northern regions that I am thinking of giving up my journey farther north. My matters in London will demand all my spare time."

September 30. The windows of my hotel look out upon the Alster Basin, a beautiful sheet of water, three sides of which are surrounded with splendid houses. Boats and swans are gliding over the glassy surface, giving, with the well-dressed promenaders along the shores, an air of gayety and liveliness to the scene."

It will not be necessary to follow the traveller step by step during this visit to Europe. He did not go to Sweden and Russia, as he had at first planned, for he learned that the Emperor of Russia was in the South, and that nothing could be accomplished in his absence. He, therefore, returned to London from Hamburg. He was respectfully received everywhere and his invention was recognized as being one of great merit and simplicity, but it takes time for anything new to make its way. This is, perhaps, best summed up in the words of Charles T. Fleischmann, who at that time was agent of the United States Patent Office, and was travelling through Europe collecting information on agriculture, education, and the arts. He was a good friend of Morse's and an enthusiastic advocate of his invention. He carried with him a complete telegraphic outfit and lost no opportunity to bring it to the notice of the different governments visited by him, and his official position gave him the entrée everywhere. Writing from Vienna on October 7, he says: —

"There is no doubt Morse's telegraph is the best of that description I have yet seen, but the difficulty of introducing it is in this circumstance, that every scientific man invents a similar thing and, without having the practical experience and practical arrangement which make Morse's so preferable, they will experiment a few miles' distance only, and no doubt it works; but, when they come to put it up at a great distance, then they will find that their experience is not sufficient, and must come back ultimately to Morse's plan. The Austrian Government is much occupied selecting out of many plans (of telegraphs) one for her railroads. I have offered Morse's and proposed experiments. I am determined to stay for some time, to give them a chance of making up their minds."

Two other young Americans, Charles Robinson and Charles L. Chapin, were also travelling around Europe at this time for the purpose of introducing Morse's invention, but, while all these efforts resulted in the ultimate adoption by all the nations of Europe, and then of the world, of this system, the superiority of which all were compelled, sometimes reluctantly, to admit, no arrangement was made by which Morse and his co-proprietors benefited financially. The gain in fame was great, in money *nil*. It was, therefore, with mixed feelings that Morse wrote to his brother from Paris on November 1: —

"I am still gratified in verifying the fact that my Telegraph is ahead of all the other systems proposed. Wheatstone's is not adopted here. The line from Paris to Rouen is not on his plan, but is an experimental line of the Governmental Commission. I went to see it yesterday with my old friend the Administrator-in-Chief of the Telegraphs of France, Mr. Foy, who is one of the committee to decide on the best mode for France. The system on this line is his modification. . . . I have had a long interview with M. Arago. He is the same affable and polite man as in 1839. He is a warm friend of mine and contends for priority in my favor, and is also partial to my telegraphic system as the best. He is President of the Commission and is going to write the History of Electric Telegraphs. I shall give him the facts concerning mine. The day after to-morrow I exhibit my telegraphic system again to the Academy of Sciences, and am in the midst of preparations for a day important to me. I have strong hopes that mine will be the system adopted, but there may be obstacles I do not see. Wheatstone, at any rate, is not in favor here. . . .

"I like the French. Every nation has its defects and I could wish many changes here, but the French are a fine people. I receive a welcome here to which I was a perfect stranger in England. How deep this welcome may be I cannot say, but if one must be cheated I like to have it done in a civil and polite way."

He sums up the result of his European trip in a letter to his daughter, written from London on October 9, as he was on his way to Liverpool from where he sailed on November 19, 1845:—

"I know not what to say of my telegraphic matters here yet. There is nothing decided upon and I have many obstacles to contend against, particularly the opposition of the proprietors of existing telegraphs; but that mine is the best system I have now no doubt. All that I have seen, while they are ingenious, are more complicated, more expensive, less efficient and easier deranged. It may take some time to establish the superiority of mine over the others, for there is the usual array of prejudice and interest against a system which throws others out of use."

CHAPTER XXXII

DECEMBER 20, 1845 — APRIL 19, 1848

Return to America. — Telegraph affairs in bad shape. — Degree of LL.D. from Yale. — Letter from Cambridge Livingston. — Henry O'Reilly. — Grief at unfaithfulness of friends. — Estrangement from Professor Henry. — Morse's "Defense." — His regret at feeling compelled to publish it. — Hopes to resume his brush. — Capitol panel. — Again disappointed. — Another accident. — First money earned from telegraph devoted to religious purposes. — Letters to his brother Sidney. — Telegraph matters. — Mexican War. — Faith in the future. — Desire to be lenient to opponents. — Dr. Jackson. — Edward Warren. — Alfred Vail remains loyal. — Troubles in Virginia. — Henry J. Rogers. — Letter to J. D. Reid about O'Reilly. — F. O. J. Smith again. — Purchases a home at last. — "Locust Grove," on the Hudson, near Poughkeepsie. — Enthusiastic description. — More troubles without, but peace in his new home.

HAVING established to his satisfaction the fact that his system was better than any of the European plans, which was the main object of his trip abroad, Morse returned to his native land, but not to the rest and quiet which he had so long desired. Telegraph lines were being pushed forward in all directions, but the more the utility of this wonderful new agent was realized, the greater became the efforts to break down the lawful rights of the patentees, and competing lines were hurriedly built on the plea of fighting a baleful monopoly by the use of the inventions of others, said to be superior. Internal dissensions also arose in the ranks of the workers on the Morse lines, and some on whom he had relied proved faithless, or caused trouble in other ways. But, while these clouds arose to darken his sky, there was yet much sunshine to gladden his heart. His health was good, his children and the families of his brothers were well and prosperous. In the year 1846 his patent rights were extended for another period of years, and he was gradually accumulating a competence as the various lines in which he held stock began to declare dividends. In addition to all this his fame had so increased that he was often alluded to in the papers as "the idol of the nation," and honorary degrees were conferred on him by various institutions both at home and abroad. Of these the one that, perhaps, pleased him the most was the degree of LL.D. bestowed by his *alma mater*, Yale. He alludes to it with pride in many of his letters to his brother Sidney, and once playfully suggests that it must mean "Lightning Line Doctor."

One of the first letters which he received on his return to America was from Cambridge Livingston, dated December 20, 1845, and reads as follows: —

"The Trustees of the New York and Boston Magnetic Telegraph Association are getting up a certificate of stock, and are desirous of making it neat and appropriate. It has seemed to me very desirable that one of its decorations should be your coat of arms, and if you will do me the favor to transmit a copy, or a wax impression of the same, I shall be much obliged."

To this Morse replied: —

"I send you a sketch of the Morse coat of arms, according to your request, to do as you please with it. I am no advocate of heraldic devices, but the *motto* in this case sanctions it with me. I wish to live and die in its spirit: —

"'*Deo non armis fido.*'"

I have said that many on whom Morse relied proved faithless, and, while I do not intend to go into the details of all these troubles, it is only right that, in the interest of historical truth, some mention should be made of some of these men. The one who, next to F. O. J. Smith, caused the most trouble to Morse and his associates, was Henry O'Reilly. Mr. Reid, in his "Telegraph in America," thus describes him: —

"Henry O'Reilly was in many respects a wonderful man. His tastes were cultivated. His instincts were fine. He was intelligent and genial. His energy was untiring, his hopefulness shining. His mental activity and power of continuous labor were marvellous. He was liberal, generous, profuse, full of the best instincts of his nation. But he lacked prudence in money matters, was loose in the use of it, had little veneration for contracts, was more anxious for personal fame than wealth. He formed and broke friendships with equal rapidity, was bitter in his hates, was impatient of restraint. My personal attachment to him was great and sincere. We were friends for many years until he became the agent of F. O. J. Smith, and my duties threw me in collision with him."

It was not until some years after his first connection with the telegraph, in 1845, that O'Reilly turned against Morse and his associates. This will be referred to at the proper time, but I have introduced him now to give point to the following extract from a letter of his to Morse, dated December 23, 1845: —

"Do you recollect a person who, while under your hands for a daguerreotype in 1840–41, broke accidentally an eight-dollar lens? Tho' many tho't you 'visionary' in your ideas of telegraphic communication, that person, you may recollect, took a lively interest in the matter, and made some suggestions about the propriety of pressing the matter energetically upon Congress and upon public attention. You seemed then to feel pleased to find a person who took so lively an interest in your invention, and you will see by the enclosed circular that that person (your humble servant) has not lost any of his early confidence in its value. May you reap an adequate reward for the glorious thought!"

It was one of life's little ironies that the man who could thus call down good fortune on the head of the inventor should soon after become one of the chief instruments in the effort to rob him of his "adequate reward," and his good name as well. Morse had such bitter experiences with several persons, who turned from friends to enemies, that it is no wonder he wrote as follows to Vail some time after this date: —

"I am grieved to say that many things have lately come to my knowledge in regard to —— that show double-dealing. Be on your guard. I hope it is but *appearance*, and that his course may be cleared up by subsequent events.

"I declare to you that I have seen so much duplicity in those in whom I had confided as friends, that I feel in danger of entertaining suspicions of everybody. I have hitherto thought you were too much inclined to be suspicious of people, but I no longer think so.

"Keep this to yourself. It may be that appearances are deceptive, and I would not wrong one whom I had esteemed as a real friend without the clearest evidence of unfaithfulness. Yet when appearances are against, it is right to be cautious."

The name of the person referred to is left blank in the

copy of this letter which I have, so I do not know who it was, but the sentiments would apply to several of the early workers in the establishment of the telegraph.

I have said that Morse, being only human, was sometimes guilty of errors of judgment, but, in a careful study of the facts, the wonder is great that he committed so few. It is an ungracious task for a son to call attention to anything but the virtues of his father, especially when any lapses were the result of great provocation, and were made under the firm conviction that he was in the right. Yet in the interest of truth it is best to state the facts fairly and dispassionately, and let posterity judge whether the virtues do not far outweigh the faults. Such an error was committed, in my judgment, by Morse in the bitter controversy which arose between him and Professor Joseph Henry, and I shall briefly sketch the origin and progress of this regrettable incident.

In 1845, Alfred Vail compiled and published a "History of the American Electro-Magnetic Telegraph." In this work hardly any mention was made of the important discoveries of Professor Henry, and this caused that gentleman to take great offense, as he believed that Morse was the real author of the work, or had, at least, given Vail all the materials. As a matter of fact he had given Vail only his notes on European telegraphs and had not seen the proofs of the work, which was published while he was absent in Europe. As soon as Morse was made aware of Henry's feelings, he wrote to him regretting the omission and explaining his innocence in the matter, and he also draughted a letter, at Vail's request, which the latter copied and sent to Henry, stating that he, Vail, had been unable to obtain the particulars of Henry's discoveries, and that, if he had offended, he had done so innocently.

Henry was an extremely sensitive man and he paid no attention to Vail's letter, and sent only a curt acknowledgment of the receipt of Morse's. However, at a meeting somewhat later, the misunderstanding seemed to be smoothed over, on the assurance that, in a second edition of Vail's work, due credit should be given to Henry, and that whenever Morse had the opportunity he would gladly accord to that eminent man the discoveries which were his. There never was a true second edition of Vail's book, but in 1847 a few more copies were struck off from the old plates and the date was, unfortunately, changed from 1845 to 1847. Henry, naturally, looked upon this as a second edition and his resentment grew.

Morse's opportunity to do public honor to Henry came in 1848, when Professor Sears C. Walker, of the Coast Survey, published a report containing some remarks on the "Theory of Morse's Electro-Magnetic Telegraph." When Professor Walker submitted this report to Morse the latter said: "I have now the long-wished-for opportunity to do justice publicly to Henry's discovery bearing upon the telegraph. I should like to see him, however, previously, and learn definitely what he claims to have discovered. I will then prepare a paper to be appended and published as a note, if you see fit, to your Report."

This paper was written by Morse and sent to Professor Walker with the request that it be submitted to Professor Henry for his revision, which was done, but it was not included in Professor Walker's report, and this naturally nettled Morse, who also had sensitive nerves, and so the breach was widened. In this paper, after giving a brief history of electric discoveries bearing on the telegraph, and of his own inventions, Morse sums up: —

"While, therefore, I claim to be the first to propose the use of the *electro-magnet for telegraphic purposes*, and the *first* to *construct a telegraph on the basis of the electro-magnet*, yet to Professor Henry is unquestionably due the honor of the *discovery of a fact in science* which proves the practicability of exciting magnetism through a long coil or at a distance, either *to deflect a needle* or *to magnetize soft iron*."

I wish he had never revised this opinion, although he was sincere in thinking that a more careful study of the subject justified him in doing so.

A few years afterwards Morse and his associates became involved in a series of bitterly contested litigations with parties interested in breaking down the original patent rights, and Henry was called as a witness for the opponents of Morse.

He gave his testimony with great reluctance, but it was tinged with the bitterness caused by the failure of Vail to do him justice and his apparent conviction that Morse was disingenuous. He denied to the latter any scientific discoveries, and gave the impression (at least, to others) that Henry, and not Morse, was the real inventor of the telegraph. His testimony was used by the enemies of Morse, both at home and abroad, to invalidate the claims of the latter, and, stung by these aspersions on his character and attainments, and urged thereto by injudicious friends, Morse published a lengthy pamphlet entitled: "A Defense against the Injurious Deductions drawn from the Deposition of Professor Joseph Henry." In this pamphlet he not only attempted to prove that he owed nothing to the discoveries of Henry, but he called in question the truthfulness of that distinguished man.

The breach between these two honorable, highly sensitive men was now complete, and it was never healed.

The consensus of scientific opinion gives to Henry's discoveries great value in the invention of the telegraph. While they did not constitute a true telegraph in themselves; while they needed the inventions and discoveries, and, I might add, the sublime faith and indomitable perseverance of Morse to make the telegraph a commercial success; they were, in my opinion, essential to it, and Morse, I think, erred in denying this. But, from a thorough study of his character, we must give him the credit of being sincere in his denial. Henry, too, erred in ignoring the advances of Morse and Vail and in his proud sensitiveness. Professor Leonard D. Gale, the friend of both men, makes the following comment in a letter to Morse of February 9, 1852: "I fear Henry and I shall never again be on good terms. He is as cold as a polar berg, and, I am informed, very sensitive. It has been said by some busybody that his testimony was incompatible with mine, and so a sort of feeling is manifested as if it were so. I have said nothing about it yet." It would have been more dignified on the part of Morse to have disregarded the imputations contained in Henry's testimony, or to have replied much more briefly and dispassionately. On the other hand, the provocation was great and he was egged on by others, partly from motives of self-interest and partly from a sincere desire on the part of his friends that he should justify himself.

In a long letter to Vail, of January 15, 1851, in which he details the whole unfortunate affair, he says: "If there was a man in the world, not related to me, for whom I had conceived not merely admiration but affection, it was for Professor Joseph Henry. I think you will remember, and can bear me witness, that I often expressed the wish that I was able to put several thousand dollars at his service for scientific investigation. . . . The whole case has saddened me more than I can express. I have to fight hard against misanthropy, friend Vail, and I have found the best antidote to be, when the fit is coming on me, to seek out a case of suffering and to relieve it, that the act in the one case may neutralize the feeling in the other, and thus restore the balance in the heart."

In taking leave for the present of this unfortunate controversy I shall quote from the "Defense," to show that Morse sincerely believed it his duty to act as he did, but that he acted with reluctance: —

"That I have been slow to complain of the injurious character of his testimony; that I have so long allowed, almost entirely uncontradicted, its distortions to have all their legal weight against me in four separate trials, without public exposure and for a space of four years of time, will at least show, I humbly contend, my reluctance to appear opposed to him, even when self-defence is combined with the defence of the interests of a large body of assignees. . . . Painful, therefore, as is the task imposed upon me, I cannot shrink from it, but shall endeavor so to perform it as rather to parry the blows that have been aimed at me than to inflict any in return. If what I say shall wound, it shall be from the severity of the simple truth itself rather than from the manner of setting it forth."

In the year 1846 there still remained one panel in the rotunda of the Capitol at Washington to be filled by an historical painting. It had been assigned to Inman, but, that artist having recently died, Morse's friends, artists and others, sent a petition to Congress urging the appointment of Morse in his place. Referring to this in a letter to his brother Sidney, dated March 28, he says: —

"In regard to the rotunda picture I learn that my friends are quite zealous, and it is not improbable that it may be given me to execute. If so, what should you say to seeing me in Paris?

"However, this is but castle-building. I am quite indifferent as to the result except that, in case it is given me, I shall be restored to my position as an artist by the same power that prostrated me, and then shall I not more than ever have cause to exclaim: 'Surely Thou hast led me in a way which I knew not'? I have already, in looking back, seen enough of the dealings of Providence with me to excite my wonder and gratitude. How singularly has my way been hedged up in my profession at the very moment when, to human appearance, everything seemed prosperously tending to the accomplishment of my desire in painting a national picture. The language of Providence in all his dealings with me has been almost like that to Abraham: 'Take now thy son, thine only son Isaac whom thou lovest, and offer him for a burnt offering,' etc.

"It has always seemed a mystery to me how I should have been led on to the acquirement of the knowledge I possess of painting, with so much sacrifice of time and money, and through so many anxieties and perplexities, and then suddenly be stopped as if a wall were built across my path, so that I could pursue my profession no longer. But, I believe, I had grace to trust in God in the darkest hour of trial, persuaded that He could and would clear up in his own time and manner all the mystery that surrounded me.

"And now, if not greatly deceived, I have a glimpse of his wonderful, truly wonderful, mercy towards me. He has chosen thus to order events that my mind might be concentrated upon that invention which He has permitted to be born for the blessing, I trust, of the world. And He has chosen me as the instrument, and given me the honor, and at the moment when all has been accomplished which is essential to its success, He so orders events as again to turn my thoughts to my almost sacrificed Isaac."

In this, however, he did not read the fates aright, for a letter from his friend, Reverend E. Goodrich Smith, dated March 2, 1847, conveys the following intelligence: "I have just learned to-day that, with their usual discrimination and justice, Congress have voted $6000 to have the panel filled by young Powell. He enlisted all Ohio, and they all electioneered with all their might, and no one knew that the question would come up. New York, I understand, went for you. I hope, however, you may yet yourself resume the pencil, and furnish the public the most striking commentary on their utter disregard of justice, by placing somewhere 'The Germ of the Republic' in such colors that shall make them blush and hang their heads to think themselves such men."

But, while he was to be blessed in the fulfilment of a long cherished dream, it was not the dream of painting a great historic picture. He never seriously touched a brush again, for all his energies were needed in the defence of himself and his invention from defamation and attack.

In the summer of 1846 he met with another accident giving him a slight period of rest which he would not otherwise have taken. He writes of it to his brother on July 30: "On Monday last I had the misfortune to fall into one of those mantraps on Broadway, set principally to break people's legs and maim them, and *incidentally* for the deposit of the coal of the household."

Vail refers jestingly to this mishap in a letter of August 21: "I trust your unfortunate and unsuccessful attempt to get down cellar has not been a serious affair."

And Morse replies in the same vein: "My *cellar experiment* was not so unsuccessful as you imagine. I succeeded to my entire satisfaction in taking three inches of skin, a little of the flesh and a trifle of bone from the front of my left leg, and, as the result, got one week's entire leisure with my leg in a chair. The experiment was so satisfactory that I deem it needless to try it again, having established beyond a doubt that skin, flesh and bone are no match against wood, iron and stone. I am entirely well of it and enjoyed my visit to the western lines very much."

It was characteristic of Morse that the first money which he received from the actual sale of his patent rights ($45 for the right to use his patent on a short line from the Post-Office to the National Observatory in Washington) was devoted by him to a religious purpose. From a letter of October 20, 1846, we learn that, adding $5 to this sum, he presented $25 to a Sunday School, and $25 to the fund for repairs.

The attachment of the three Morse brothers to each other was intense, and lasted to the end of their lives. The letters of Finley Morse to his brother Sidney, in particular, would alone fill a volume and are of great interest. Most of them have never before been published and I shall quote from them freely in following Morse's career.

Sidney and his family were still in Europe, and the two following extracts are from letters to him: —

"*October 29, 1846.* I don't know where this will find you, but, as the steamer Caledonia goes in a day or two, and as I did not write you by the last steamer, I thought I would occupy a few moments (not exactly of leisure) to write you. . . . Charles has little to do, but does all he can. He is desirous of a farm and I have made up my mind to indulge him. . . . I shall go up the river in a day or two and look in the vicinity of Po'keepsie. . . .

"Telegraph matters are every day assuming a more and more interesting aspect. All physical and scientific difficulties are vanquished. If conductors are well put up there is nothing more to wish for in the facilities of intercourse. My operators can easily talk with each other as fast as persons usually write, and faster than this would be faster than is necessary. The Canadians are alive on the subject, and lines are projected from Toronto to Montreal, from Montreal to Quebec and to Halifax. Lines are also in contemplation from Toronto to Detroit, on the Canada side, and from Buffalo to Chicago on this side, so that it may not be visionary to say that our first news from England may reach New York via Halifax, Detroit, Buffalo and Albany. . . .

"The papers will inform you of the events of the war. Our people are united on this point so far as to pursue it with vigor to a speedy termination. However John Bull may sneer and endeavor to detract from the valor of our troops, his own annals do not furnish proofs of greater skill and more fearless daring and successful result. The Mexican race is a worn-out race, and God in his Providence is taking this mode to regenerate them. Whatever may be the opinions of some in relation to the justness or unjustness of our quarrel, there ought to be but one opinion among all good men, and that should be that the moment should be improved to throw a light into that darkened nation, and to raise a standard there which, whatever may become of the Stars and Stripes, or Eagle and Prickly Pear, shall be never taken down till all nations have flocked to it. Our Bible and Tract Societies and missionaries ought to be in the wake of our armies."

"*January 28, 1847.* Telegraph matters are becoming more and more interesting. The people of the country everywhere are desirous of availing themselves of its facilities, and the lines are being extended in all directions. As might be expected then, I have my plans interfered with by mercenary speculators who threaten to put up rival telegraphs and contest my patent. *I am ready for them.* We have had to apply for an injunction on the Philadelphia and Pittsburg line. The case is an aggravated one and will be decided on Monday or Tuesday at Philadelphia in Circuit Court of United States. I have no uneasiness as to the result. [It was decided against him, however, but this proved only a temporary check.]

"There are more F. O. Js. than one, yet not one quite so bad. I think amid all the scramble I shall probably have enough come to my share, and it does not matter by what means our Heavenly Father chooses to curtail my receipts, for I shall have just what he pleases, none can hinder it, and more I do not want. . . . House and his associates are making most strenuous efforts to interfere and embarrass me by playing on the ignorance of the public and the natural timidity of capitalists. I shall probably have to lay the law on him and make an example before my patent is confirmed in the minds of the public. It is the course, I am told, of every substantial patent. It has to undergo the ordeal of one trial in the courts. . . .

"Although I thus write, you need have no fears that my operations will be seriously affected by any schemes of common letter printing telegraphs. I have just filed a caveat for one which I have invented, which as far transcends in simplicity and efficiency any previous plan for the purpose, as my telegraph system is superior to the old visual telegraphs. I will have it in operation by the time you return."

Apropos of the attacks made upon him by would-be

infringers, the following from a letter of his legal counsel, Daniel Lord, Esq., dated January 12, 1847, may not come amiss: "It ought to be a source of great satisfaction to you to have your invention stolen and counterfeited. Think what an acknowledgment it is, and what a tribute to its merits."

Referring to this in a letter to Mr. Lord of a later date, Morse answers: "The plot thickens all around me; I think a *dénouement* not far off. I remember your consoling me under these attacks with bidding me think that I had invented something worth contending for. Alas! my dear sir, what encouragement is there to an inventor if, after years of toil and anxiety, he has only purchased for himself the pleasure of being a target for every vile fellow to shoot at, and, in proportion as his invention is of public utility, so much the greater effort is to be made to defame, that the robbery may excite the less sympathy? I know, however, that beyond all this is a clear sky, but the clouds may not break away until I am no longer personally interested whether it be foul or fair. I wish not to complain, but I have feelings and cannot play the stoic if I would."

It was a new experience for Morse to become involved in the intricacies of the law, and, in a letter to a friend, Henry I. Williams, Esq., dated February 22, 1847, he naïvely remarks: "A student all my life, mostly in a profession which is adverse in its habits and tastes from those of the business world, and never before engaged in a lawsuit, I confess to great ignorance even of the ordinary, commonplace details of a court."

His desire to be both just and merciful is shown in a letter to Mr. Kendall, written on February 16, just before the decision was rendered against him: "I have been in court all day, and have been much pleased with the clearness and, I think, conclusiveness of Mr. Miles's argument. I think he has produced an evident change in the views of the judge. Yet it is best to be prepared for the worst, and, even if we succeed in getting the injunction, I wish as much leniency as possible to be shown to the opposing parties. Indeed, in this I know my views are seconded by you. However we may have 'spoken daggers,' let us use none, and let us make every allowance for honest mistake, even where appearances are at first against such a supposition. O'Reilly may have acted hastily, under excitement, under bad advisement, and in that mood have taken wrong steps. Yet I still believe he may be recovered, and, while I would use every precaution to protect our just rights, I wish not to take a single step that can be misconstrued into vindictiveness or triumph."

It was well that it was his invariable rule to be prepared for the worst, for, writing to his brother Sidney on February 24, he says: "We have just had a lawsuit in Philadelphia before Judge Kane. We applied for an injunction to stay irregular and injurious proceedings on the part of Western (Pittsburg and Cincinnati) Company, and our application has been *refused* on technical grounds. I know not what will be the issue. I am trying to have matters compromised, but do not know if it can be done, and we may have to contest it in *law.* Our application was in court of equity. A movement of Smith was the cause of all."

Another sidelight is thrown on Morse's character by the following extract from a letter to one of his lieutenants, T. S. Faxton, written on March 15: "We must raise the salaries of our operators or they will all be taken from us, that is, all that are good for anything. You will recollect that, at the first meeting of the Board of Directors, I took the ground that 'it was our policy to make the office of operator desirable, to pay operators well and make their situation so agreeable that intelligent men and men of character will seek the place and dread to lose it.' I still think so, and, depend upon it, it is the soundest economy to act on this principle."

Just about this time, to add to Morse's other perplexities, Doctor Charles T. Jackson began to renew his claims to the invention of the telegraph, while also disputing with Morton the discovery of ether as an anæsthetic, then called "Letheon," and claiming the invention of gun-cotton and the discovery of the circulation of the blood. Morse found a willing and able champion in Edward Warren, Esq., of Boston, and many letters passed between them. As Jackson's wild claims were effectually disposed of, I shall not dwell upon this source of annoyance, but shall content myself with one extract from a letter to Mr. Warren of March 23: "I wish not to attack Dr. Jackson nor even to defend myself in *public* from his *private* attacks. If in any of his publications he renews his claim, which I consider as long since settled by default, then it will be time and proper for me to notice him. . . . The most charitable construction of the Dr's. conduct is to attribute it to a monomania induced by excessive vanity."

While many of those upon whom he had looked as friends turned against him in the mad scramble for power and wealth engendered by the extension of the telegraph lines, it is gratifying to turn to those who remained true to him through all, and among these none was more loyal than Alfred Vail. Their correspondence, which was voluminous, is always characterized by the deepest confidence and affection. In a long letter of March 24, Vail shows his solicitude for Morse's peace of mind: "I think I would not be bothered with a directorship in the New York and Buffalo line, nor in any other. I should wish to keep clear of them. It will only tend to harass and vex when you should be left quiet and undisturbed to pursue your improvements and the enjoyment of what is most gratifying to you."

And Morse, writing to Vail somewhat later in this same year, exclaims: "You say you hope I shall not forget that we have spent many hours together. You might have added 'happy hours.' I have tried you, dear Vail, as a friend, and think I know you as a zealous and honest one."

Still earlier, on March 18, 1845, in one of his reports to the Postmaster-General, Cave Johnson, he adds: "In regard to the salary of the 'one clerk at Washington — $1200,' Mr. Vail, who would from the necessity of the case take that post, is my right-hand man in the whole enterprise. He has been with me from the year 1837, and is as familiar with all the mechanism and scientific arrangements of the Telegraph as I am myself. . . . His time and talent are more essential to the success of the Telegraph than [those of] any two persons that could be named."

Returning now to the letters to his brother Sidney, I shall give the following extracts: —

"*March 29, 1847.* I am now in New York permanently; that is I have no longer any official connection with Washington, and am thinking of *fixing* somewhere so soon as I can get my telegraphic matters into such a state as to warrant it; but my patience is still much tried. Although the enterprise looks well and is prospering, yet somehow I do not command the *cash* as some business men would if they were in the same situation. The property is doubtless good and is increasing, but I cannot use it as I could the money, for, while everybody seems to think I have the wealth of John Jacob, the only sum I have actually realized is my first dividend on one line, about fourteen hundred dollars, and with this I cannot purchase a house. But time will, perhaps, enable me to do so, if it is well that I should have one. . . . I have had some pretty threatening obstacles, but they as yet are summer clouds which seem to be dissipating through the smiles of our Heavenly Father. House's affair I think is dead. I believe it has been held up by speculators to drive a better bargain with me, thinking to scare me; but they don't find me so easily frightened. In Virginia I had to oppose a most bigoted, narrow, illiberal clique in a railroad company, which had the address to get a bill through the House of Delegates giving them actually the monopoly of telegraphs, and ventured to halloo before they were out of the woods. Mr. Kendall went post-haste to Richmond, met the bill and its supporters before the Committee of the Senate, and, after a sharp contest, procured its rejection in the Senate, and the adoption, by a vote of 13 to 7, of a substitute granting me *right of way* and *corporate powers*, which bill, after violent opposition in the House, was finally passed, 44 to 27.

So a mean intrigue was defeated most signally, and I came off triumphant."

"*April 27.* This you will recognize by the date is my birthday; 36 years old. Only think, I shall never be 26 again. Don't you wish you were as young as I am? Well, if *feelings* determined age I should be in reality what I have above stated, but that leaf in the family Bible, those boys and that daughter, those nieces and nephews of younger brothers, and especially that *grandson*, they all concur in putting twenty years more to those 36. I cannot get them off; there they are 56! . . .

"There is an underhand intrigue against my telegraph interests in Virginia, fostered by a friend turned enemy in the hope to better his own interests, a man whom I have ever treated as a friend while I had the governmental patronage to bestow, and gave him office in Baltimore. Having no more of patronage to give I have no more friendship from him. Mr. R. has proved himself false, notwithstanding his naming his son after me as a proof of friendship."

The Mr. R. referred to was Henry J. Rogers, and, writing of him to Vail on April 26, Morse says: "I am truly grieved at Rogers's conduct. He must be conscious of doing great injustice; for a man that has wronged another is sure to invent some cause for his act if there has been none given. In this case he endeavors to excuse his selfish and injurious acts by the false assertion that 'I had cast him overboard.' Why, what does he mean? Was I not overboard myself? Does he or anyone else suppose I have nothing else to do than to find them places, and not only intercede for them, which in Rogers's case and Zantziger's I have constantly and perseveringly done to the present hour, but I am bound to force the companies, over which I have no control, to take them at any rate, on the penalty of being traduced and injured by them if they do not get the office they seek? As to Rogers, you know my feelings towards him and his. I had received him as a *friend*, not as a mere employee, and let no opportunity pass without urging forward his interests. I recollected his naming his son for me, and had determined, if the wealth actually came which has been predicted to me, that that child should be remembered."

Always desirous of being just and merciful, Morse writes to Vail on May 1: "Rogers is here. I have had a good deal of conversation with him, and the result is that I think that some circumstances which seemed to inculpate him are explicable on other grounds than intention to injure us."

But he was finally forced to give him up, for on August 7 he writes: "You cannot tell how pained I am at being compelled to change my opinion of R. Your feelings correspond entirely with my own. I was hoping to do something gratifying to him and his family, and soon should have done it if he would permit it; but no! The mask of friendship covered a deep selfishness that scrupled not to sacrifice a real friendship to a short-sighted and overreaching ambition. Let him go. I wished to befriend him and his, and would have done so from the heart, but as he cannot trust me I have enough who can and do."

The case of Rogers was typical, and I have, therefore, given it in some detail. It was always a source of grief to Morse when men, whom in his large-hearted way he had admitted to his intimacy, turned against him; and he was called upon to suffer many such blows. He has been accused of having quarrelled with all his associates. This, of course, is not true, for we have only to name Vail, and Gale, and Kendall, and Reid, and a host of others to prove the contrary. But, like all men who have achieved great things, he made bitter enemies, some of whom at first professed sincere friendship for him and were implicitly trusted by him. However, a dispassionate study of all the circumstances leading up to the rupture of these friendly ties will prove that, in practically every case he was sinned against, not sinning.

A letter to James D. Reid, written on December 21, will show that the quality of his mercy was not strained: "You may recollect when I met you in Philadelphia, on the unpleasant business of attending in a court to witness the contest of two parties for their rights, you informed me of the destitute condition of O'Reilly's family. At that moment I was led to believe, from consultation with the counsel for the Patentees, that the case would undoubtedly go in their (the Patentees') favor. Your statement touched me, and I could not bear to think that an innocent wife and inoffensive children should suffer, even from the wrong-doing of their proper protector, should this prove to be the case. You remember I authorized you to draw on me for twenty dollars to be remitted to Mr. O'Reilly's family, and to keep the source from whence it was derived secret. My object in writing is to ask if this was done, and, in case it was, to request you to draw on me for that amount."

In an earlier letter to his brother he remarks philosophically: "Smith is Smith yet and so likely to be, but I have become used to him and you would be surprised to find how well oil and water appear to agree. There must be crosses and the aim should be rather to bear them gracefully, graciously, and patiently, than to have them removed."

While thus harassed on all sides by those who would filch from him his good name as well as his purse, his reward was coming to him for the patience and equanimity with which he was bearing his crosses. The longing for a home of his own had been intense all through his life and now, in the evening of his years, this dream was to be realized. He thus announces to his brother the glorious news: —

POUGHKEEPSIE, NORTH RIVER,
July 30, 1847.

In my last I wrote you that I had been looking out for a farm in this region, and gave you a diagram of a place which I fancied. Since then I was informed of a place for sale south of this village 2 miles, on the bank of the river, part of the old Livingston Manor, and far superior. *I have this day concluded a bargain for it.* There are about one hundred acres. I pay for it $17,500.

I am almost afraid to tell you of its beauties and advantages. It is just such a place as in England could not be purchased for double the number of pounds sterling. Its "capabilities," as the landscape gardeners would say, are unequalled. There is every variety of surface, plain, hill, dale, glens, running streams and fine forest, and every variety of different prospect; the Fishkill Mountains towards the south and the Catskills towards the

HOUSE AT LOCUST GROVE, POUGHKEEPSIE, N.Y.

north; the Hudson with its varieties of river craft, steamboats of all kinds, sloops, etc., constantly showing a varied scene.

I will not enlarge. I am congratulated by all in having made an excellent purchase, and I find a most delightful neighborhood. Within a few miles around, approached by excellent roads, are Mr. Lenox, General Talmadge, Philip Van Rensselaer, etc., on one side; on the other, Harry Livingston, Mrs. Smith Thomson (Judge Thomson's widow, and sister to the first Mrs. Arthur Breese), Mr. Crosby, Mr. Boorman, etc., etc. The new railroad will run at the foot of the grounds (probably) on the river, and bring New York within two hours of us. There is every facility for residence — good markets, churches, schools. Take it all in all I think it just the place *for us all.* If you should fancy a spot on it for building, I can accommodate you, and Richard wants twenty acres reserved for him. Singularly enough this was the very spot where Uncle Arthur found his wife. The old trees are pointed out where he and she used to ramble during their courtship.

On September 12, after again expatiating on the beauties and advantages of his home, he adds: "I have some clouds and mutterings of thunder on the horizon (the necessary attendants, I suppose, of a lightning project) which I trust will give no more of storm than will suffice, under Him who directs the elements, to clear the air and make a serener and calmer sunset."

On October 12, he announces the name which he has given to his country place, and a singular coincidence: —

"*Locust Grove.* You see by the date where I am.

Locust Grove, it seems, was the original name given to this place by Judge Livingston, and, without knowing this fact, I had given the same name to it, so that there is a natural appropriateness in the designation of my home. The wind is howling mournfully this evening, a second edition, I fear, of the late destructive equinoctial, but, dreary as it is out-of-doors, I have comfortable quarters within."

In the world of affairs the wind was howling, too, and the storm was gathering which culminated in the series of lawsuits brought by Morse and his associates against the infringers on his patents. The letters to his brother are full of the details of these piratical attacks, but throughout all the turmoil he maintained his poise and his faith in the triumph of justice and truth. In the letter just quoted from he says: "These matters do not annoy me as formerly. I have seen so many dark storms which threatened, and particularly in relation to the Telegraph, and I have seen them so often hushed at the 'Peace, be still' of our covenant God, that now the fears and anxieties on any fresh gathering soon subside into perfect calm."

And on November 27, he writes: "The most annoying part of the matter to me is that, notwithstanding my matters are all in the hands of agents and I have nothing to do with any of the arrangements, I am held up by name to the odium of the public. Lawsuits are commenced against them at Cincinnati and will be in Indiana and Illinois as well as here, and so, notwithstanding all my efforts to get along peaceably, I find the fate of Whitney before me. I think I may be able to secure my farm, and so have a place to retire to for the evening of my days, but even this may be denied me. A few months will decide. . . . You have before you the fate of an inventor, and, take as much pains as you will to secure to yourself your valuable invention, make up your mind from my experience now, in addition to others, that you will be robbed of it and abused into the bargain. This is the lot of a successful inventor or discoverer, and no precaution, I believe, will save him from it. He will meet with a mixed estimate; the enlightened, the liberal, the good, will applaud him and respect him; the sordid, the unprincipled will hate him and detract from his reputation to compass their own contemptible and selfish ends."

While events in the business world were rapidly converging towards the great lawsuits which should either confirm the inventor's rights to the offspring of his brain, or deprive him of all the benefits to which he was justly and morally entitled, he continued to find solace from all his cares and anxieties in his new home, with his children and friends around him. He touches on the lights and shadows in a letter to his brother, who was still in England, dated New York, April 19, 1848: —

"I snatch a moment by the Washington, which goes to-morrow, to redeem my character in not having written of late so often as I could wish. I have been so constantly under the necessity of watching the movements of the most unprincipled set of pirates I have ever known, that all my time has been occupied in defense, in putting evidence into something like legal shape that I am the inventor of the Electro-Magnetic Telegraph!! Would you have believed it ten years ago that a question could be raised on that subject? Yet this very morning in the 'Journal of Commerce' is an article from a New Orleans paper giving an account of a public meeting convened by O'Reilly, at which he boldly stated that I had '*pirated my invention from a German invention*' a great deal better than mine. And the 'Journal of Commerce' has a sort of halfway defense of me which implies there is some doubt on the subject. I have written a note which may appear in to-morrow's 'Journal,' quite short, but which I think, will stop that game here.

"A trial in court is the only event now which will put public opinion right, so indefatigable have these unprincipled men been in manufacturing a spurious public opinion.

"Although these events embarrass me, and I do not receive, and may not receive, my rightful dues, yet I have been so favored by a kind Providence as to have sufficient collected to free my farm from mortgage on the 1st of May, and so find a home, a beautiful home, for me and mine, unencumbered, and sufficient over to make some improvements. . . .

"I do not wish to raise too many expectations, but every day I am more and more charmed with my purchase. I can truly say I have never before so completely realized my wishes in regard to situation, never before found so many pleasant circumstances associated together to make a home agreeable, and, so far as earth is concerned, I only wish now to have you and the rest of the family participate in the advantages with which a kind God has been pleased to indulge me.

"Strange, indeed, would it be if clouds were not in the sky, but the Sun of Righteousness will dissipate as many and as much of them as shall be right and good, and this is all that should be required. I look not for freedom from trials; they must needs be; but the number, the kind, the form, the degree of them, I can safely leave to Him who has ordered and will still order all things well."

CHAPTER XXXIII

JANUARY 9, 1848—DECEMBER 19, 1849

Preparation for lawsuits. — Letter from Colonel Shaffner. — Morse's reply deprecating bloodshed. — Shaffner allays his fears. — Morse attends his son's wedding at Utica. — His own second marriage. — First of great lawsuits. — Almost all suits in Morse's favor. — Decision of Supreme Court of United States. — Extract from an earlier opinion. — Alfred Vail leaves the telegraph business. — Remarks on this by James D. Reid. — Morse receives decoration from Sultan of Turkey. — Letter to organizers of Printers' Festival. — Letter concerning aviation. — Optimistic letter from Mr. Kendall. — Humorous letter from George Wood. — Thomas R. Walker. — Letter to Fenimore Cooper. — Dr. Jackson again. — Unfairness of the press. — Letter from Charles C. Ingham on art matters. — Letter from George Vail. — F. O. J. Smith continues to embarrass. — Letter from Morse to Smith.

THE year 1848 was a momentous one to Morse in more ways than one. The first of the historic lawsuits was to be begun at Frankfort, Kentucky,—lawsuits which were not only to establish this inventor's claims, but were to be used as a precedent in all future patent litigation. In his peaceful retreat on the banks of the Hudson he carefully and systematically prepared the evidence which should confound his enemies, and calmly awaited the verdict, firm in his faith that, however lowering the clouds, the sun would yet break through. Finding relaxation from his cares and worries in the problems of his farm, he devoted every spare moment to the life out-of-doors, and drank in new strength and inspiration with every breath of the pure country air. Although soon to pass the fifty-seventh milestone, his sane, temperate habits had kept him young in heart and vigorous in body, and in this same year he was to be rewarded for his long and lonely vigil during the dark decades of his middle life, and to enter upon an Indian Summer of happy family life.

While spending as much time as possible at his beloved Locust Grove, he was yet compelled, in the interests of his approaching legal contests, to consult with his lawyers in New York and Washington, and it was while in the latter city that he received a letter from Colonel Tal. P. Shaffner, one of the most energetic of the telegraph pioneers, and a devoted, if sometimes injudicious, friend It was he who, more than any one else, was responsible for the publication of Morse's "Defense" against Professor Henry.

The letter was written from Louisville on January 9, 1848, and contains the following sentences: "We are going ahead with the line to New Orleans. I have twenty-five hands on the road to Nashville, and will put on more next week. I have ten on the road to Frankfort, and my associate has gangs at other parts. O'Reilly has fifteen hands on the Nashville route and I confidently expect a few fights. My men are well armed and I think they can do their duty. I shall be with them when the parties get together, and, if anything does occur, the use of Dupont's best will be appreciated by me. This is to be lamented, but, if it comes, we shall not back out."

Deeply exercised, Morse answers him post-haste: "It gives me real pain to learn that there is any prospect of physical collision between the O'Reilly party and ours, and I trust that this may arrive in time to prevent any movement of those friendly to me which shall provoke so sad a result. I emphatically say that, if *the law* cannot protect me and my rights in your region, I shall never sanction the appeal to force to sustain myself, however conscious of being in the right. I infinitely prefer to suffer still more from the gross injustice of unprincipled men than to gain my rights by a single illegal step. . . . I hope you will do all in your power to prevent collision. If the parties meet in putting up posts or wires, let our opponents have their way unmolested. I have no patent for putting up posts or wires. They as well as we have a right to put them up. It is the use made of them afterwards which may require legal adjustment. The men employed by each party are not to blame. Let no ill-feeling be fomented between the two, no rivalry but that of doing their work the best; let friendly feeling as between them be cherished, and teach them to refer all disputes to the principals. I wish no one to fight for me physically. He may 'speak daggers but use none.' However much I might appreciate his friendship and his motive, it would give me the deepest sorrow if I should learn that a single individual, friend or foe, has been injured in life or limb by any professing friendship for me."

He was reassured by the following from Colonel Shaffner: —

"*January 27*. Your favor of the 21st was received yesterday. I was sorry that you allowed your feelings to be so much aroused in the case of contemplated difficulties between our hands and those of O'Reilly. They held out the threats that we should not pass them, and we were determined to do it. I had them notified that we were prepared to meet them under any circumstances. We were prepared to have a real 'hug,' but, when our hands overtook them, they only 'yelled' a little and mine followed, and for fifteen miles they were side by side, and when a man finished his hole, he ran with all his might to get ahead. But finally, on the 24th, we passed them about eighty miles from here, and now we are about twenty-five miles ahead of them without the loss of a drop of blood, and we shall be able to beat them to Nashville, if we can get the wire in time, which is doubtful."

There were many such stirring incidents in the early history of the telegraph, and the half of them has not been told, thus leaving much material for the future historian.

But, while so much that was exciting was taking place in the outside world, the cause of it all was turning his thoughts towards matters more domestic. On June 13, he writes to his brother: "Charles left me for Utica last evening, and Finley and I go this evening to be present at his marriage on Thursday the 15th."

It was at his son's wedding that he was again strongly attracted to his young second cousin (or, to be more exact, his first cousin once removed), the first cousin of his son's bride, and the result is announced to his brother in a letter of August 7: "Before your return I shall be again married. I leave to-morrow for Utica where cousin (second cousin) Sarah Elizabeth Griswold now is. On Thursday morning the 10th we shall (God willing) be married, and I shall immediately proceed to Louisville and Frankfort in Kentucky to be present at my first suit against O'Reilly, the pirate of my invention. It comes off on the 23d inst. So far as the justice of the case is concerned I am confident of final success, but there are so many crooks in the law that I ought to be prepared for disappointment."

Continuing, he tells his brother that he has been secretly in love with his future wife for some years: "But, reflecting on it, I found I was in no situation to indulge in any plans of marrying. She had nothing, I had nothing, and the more I loved her the more I was determined to stifle my feelings without hinting to her anything of the matter, or letting her know that I was at all interested in her."

But now, with increasing wealth, the conditions were changed, and so they were married, and in their case it can with perfect truth be said, "They lived happy ever after," and failed by but a year of being able to celebrate their silver wedding. Soon a young family grew up around him, to whom he was always a patient and loving father. We his children undoubtedly gave him many an anxious moment, as children have a habit of doing, but through all his trials, domestic as well as extraneous, he was calm, wise, and judicious.

But now the first of the great lawsuits, which were to confirm Morse's patent rights or to throw his invention open to the world, was begun, and, with his young bride, he hastened to Frankfort to be present at the trial. To follow these suits through all their legal intricacies would make dry reading and consume reams of paper. Mr. Prime in a footnote remarks: "Mr. Henry O'Reilly has deposited in the Library of the New York Historical Society more than one hundred volumes containing a complete history of telegraphic litigation in the United States. These records are at all times accessible to any persons who wish to investigate

SARAH ELIZABETH GRISWOLD

Second wife of S. F. B. Morse

the claims and rights of individuals or companies. The *testimony* alone in the various suits fills several volumes, each as large as this."

It will, therefore, only be necessary to say that almost all of these suits, including the final one before the Supreme Court of the United States, were decided in Morse's favor. Every legal device was used against him; his claims and those of others were sifted to the uttermost, and then as now expert opinion was found to uphold both sides of the case. To quote Mr. Prime:

"The decision of the Supreme Court was unanimous on all the points involving the right of Professor Morse to the claim of being the original inventor of the Electro-Magnetic Recording Telegraph. A minority of the court went still further, and gave him the right to the motive power of magnetism as a means of operating machinery to imprint signals or to produce sounds for telegraphic purposes. The testimony of experts in science and art is not introduced because it was thoroughly weighed and sifted by intelligent and impartial men, whose judgment must be accepted as final and sufficient. The justice of the decision has never been impugned. Each succeeding year has confirmed it with accumulating evidence.

"One point was decided against the Morse patent, and it is worthy of being noticed that this decision, which denied to Morse the *exclusive* use of electro-magnetism for recording telegraphs, has never been of injury to his instrument, because no other inventor has devised an instrument to supersede his.

"The court decided that the Electro-Magnetic Telegraph was the sole and exclusive invention of Samuel

F. B. Morse. If others could make better instruments for the same purpose, they were at liberty to use electro-magnetism. Twenty years have elapsed since this decision was rendered; the Morse patent has expired by limitation of time, but it is still without a rival in any part of the world."

This was written in 1873, but I think that I am safe in saying that the same is true now after the lapse of forty more years. While, of course, there have been both elaboration and simplification, the basic principle of the universal telegraph of to-day is embodied in the drawings of the sketch-book of 1832, and it was the invention of Morse, and was entirely different from any form of telegraph devised by others.

I shall make but one quotation from the long opinion handed down by the Supreme Court and delivered by Chief Justice Taney: —

"Neither can the inquiries he made, nor the information or advice he received from men of science, in the course of his researches, impair his right to the character of an inventor. No invention can possibly be made, consisting of a combination of different elements of power, without a thorough knowledge of the properties of each of them, and the mode in which they operate on each other. And it can make no difference in this respect whether he derives his information from books, or from men skilled in the science. If it were otherwise, no patent in which a combination of different elements is used could ever be obtained. For no man ever made such an invention without having first obtained this information, unless it was discovered by some fortunate accident. And it is evident that such an invention as the Electro-Magnetic Telegraph could never have been brought into action without it. For a very high degree of scientific knowledge, and the nicest skill in the mechanic arts, are combined in it, and were both necessary to bring it into successful operation. *And the fact that Morse sought and obtained the necessary information and counsel from the best sources, and acted upon it, neither impairs his rights as an inventor, nor detracts from his merits.*"

The italics are mine, for it has over and over been claimed for everybody who had a part in the early history of the telegraph, either by hint, help, or discovery, that more credit should be given to him than to Morse himself — to Henry, to Gale, to Vail, to Doctor Page, and even to F. O. J. Smith. In fact Morse used often to say that some people thought he had no right to claim his invention because he had not discovered electricity, nor the copper from which his wires were made, nor the brass of his instruments, nor the glass of his insulators.

I shall make one other quotation from the opinion of Judge Kane and Judge Grier at one of the earlier trials, in Philadelphia, in 1851: —

"That he, Mr. Morse, was the first to devise and practise the art of recording language, at telegraphic distances, by the dynamic force of the electro-magnet, or, indeed, by any agency whatever, is, to our minds, plain upon all the evidence. It is unnecessary to review the testimony for the purpose of showing this. His application for a patent, in April, 1838, was preceded by a series of experiments, results, illustrations and proofs of final success, which leave no doubt whatever but that his great invention was consummated before the early spring of 1837. There is no one person, whose invention has been spoken of by any witness, or referred to in any book as involving the principle of Mr. Morse's discovery, but must yield precedence of date to this. Neither Steinheil, nor Cooke and Wheatstone, nor Davy, nor Dyar, nor Henry, had at this time made a recording telegraph of any sort. The devices then known were merely *semaphores*, that spoke to the eye for a moment — bearing about the same relation to the great discovery before us as the Abbé Sicard's invention of a visual alphabet for the purposes of conversation bore to the art of printing with movable types. Mr. Dyar's had no recording apparatus, as he expressly tells us, and Professor Henry had contented himself with the abundant honors of his laboratory and lecture-rooms."

One case was decided against him, but this decision was afterwards overruled by the Supreme Court, so that it caused no lasting injury to his claims.

As decision after decision was rendered in his favor he received the news calmly, always attributing to Divine Providence every favor bestowed upon him. Letters of congratulation poured in on him from his friends, and, among others, the following from Alfred Vail must have aroused mingled feelings of pleasure and regret. It is dated September 21, 1848: —

I congratulate you in your success at Frankfort in arresting thus far that pirate O'Reilly. I have received many a hearty shake from our friends, congratulating me upon the glorious issue of the application for an injunction. The pirate dies hard, and well he may. It is his privilege to kick awhile in this last death struggle.

These pirates must be followed up and each in his turn nailed to the wall.

The Wash. & N.O. Co. is at last organized, and for the last three weeks we have received daily communications from N.O. Our prospects are flattering. And what do you think they have done with me? Superintendent of Washington & N.O. line all the way from Washington to Columbia at $900!!!!!

This game will not be played long. I have made up my mind to leave the Telegraph to take care of itself, since it cannot take care of me. I shall, in a few months, leave Washington for New Jersey, family, kit and all, and bid adieu to the subject of the Telegraph for some more profitable business. . . .

I have just finished a most beautiful register with a *pen lever key* and an expanding reel. Have orders for six of the same kind to be made at once; three for the south and three for the west.

I regret you could not, on your return from the west, have made us at least a flying visit with your charming lady. I am happy to learn that your cup of happiness is so full in the society of one who, I learn from Mr. K., is well calculated to cheer you and relieve the otherwise solitude of your life. . . . My kindest wishes for yourself and Mrs. Morse, and believe me to be, now as ever,

Yours, etc.,

ALFRED VAIL.

Mr. James D. Reid in an article in the "Electrical World," October 12, 1895, after quoting from this letter, adds: —

"The truth is Mr. Vail had no natural aptitude for executive work, and he had a temper somewhat variable and unhappy. He and I got along very well together until I determined to order my own instruments, his being too heavy and too difficult, as I thought, for an operator to handle while receiving. We had our instruments made by the same maker — Clark & Co., Philadelphia. Yet even that did not greatly separate us, and we were always friends. About some things his notions were very crude. It was under his guidance that David Brooks, Henry C. Hepburn and I, in 1845, undertook to insulate the line from Lancaster to Harrisburg, Pennsylvania, by saturating bits of cotton cloth in beeswax and wrapping them round projecting arms. The bees enjoyed it greatly, but it spoiled our work.

"But I have no desire to criticize him. He seemed to me to have great opportunities which he did not use. He might have had, I thought, the register work of the country and secured a large business. But it went from him to others, and so he left the field."

This eventful year of 1848 closed with the great telegraph suits in full swing, but with the inventor calm under all his trials. In a letter, of December 18, to his brother Sidney, who had now returned to America, he says: "My affairs (Telegraphically) are only under a slight mist, hardly a cloud; I see through the mist already."

And in another part of this letter he says: "I may see you at the end of the week. If I can bring Sarah down with me, I will, to spend Christmas, but the weather may change and prevent. What weather! I am working on the lawn as if it were spring. You have no idea how lovely this spot is. Not a day passes that

I do not feel it. If I have trouble abroad, I have peace, and love, and happiness at home. My sweet wife I find, indeed, a rich treasure. Uniformly cheerful and most affectionate, she makes sunshine all the day. God's gifts are worthy of the giver."

It was in the early days of 1849 that a gift of another kind was received by him which could not fail to gratify him. This was a decoration, the "Nichan Iftikar" or "Order of Glory," presented to him by the Sultan of Turkey, the first and only decoration which the Sultan of the Ottoman Empire had conferred upon a citizen of the United States. It was a beautiful specimen of the jeweller's art, the monogram of the Sultan in gold, surrounded by 130 diamonds in a graceful design. It was accompanied by a diploma (or *berait*) in Turkish, which being translated reads: —

In the Name of Him
Sultan Abdul Hamid Khan

Son of Mahmoud Khan, son of Abdul Hamid Khan — may he ever be victorious!

The object of the present sovereign decoration of Noble Exalted Glory, of Elevated Place, and of this Illustrious World Conquering Monogram is as follows:

The bearer of this Imperial Monogram of exalted character, Mr. Morse, an American, a man of science and of talents, and who is a model of the Chiefs of the nation of the Messiah — may his grade be increased — having invented an Electrical Telegraph, a specimen of which has been exhibited in my Imperial presence; and it being proper to patronize knowledge and to express my sense of the value of the attainments of the

Inventor, as well as to distinguish those persons who are the Inventors of such objects as serve to extend and facilitate the relations of mankind, I have conferred upon him, on my exalted part, an honorable decoration in diamonds, and issued also this present diploma, as a token of my benevolence for him.

Written in the middle of the moon Sefer, the fortunate, the year of the Flight one thousand two hundred and sixty-four, in Constantinople the well-guarded.

The person who was instrumental in gaining for the inventor this mark of recognition from the Sultan was Dr. James Lawrence Smith, a young geologist at that time in the employ of the Sultan. He, aided by the Reverend C. Hamlin, of the Armenian Seminary at Bebek, gave an exhibition of the working of the telegraph before the Sultan and all the officers of his Government, and when it was proposed to decorate him for his trouble and lucid explanation, he modestly and generously disclaimed any honor, and begged that any such recognition should be given to the inventor himself. Other decorations and degrees were bestowed upon the inventor from time to time, but these will be summarized in a future chapter. I have enlarged upon this one as being the first to be received from a foreign monarch.

As his fame increased, requests of all sorts poured in on him, and it is amazing to find how courteously he answered even the most fantastic, overwhelmed as he was by his duties in connection with the attacks on his purse and his reputation. Two of his answers to correspondents are here given as examples: —

January 17, 1849.

Gentlemen, — I have received your polite invitation to the Printers' Festival in honor of Franklin, on his birthday the 17th of the present month, and regret that my engagements in the city put it out of my power to be present.

I thank you kindly for the flattering notice you are pleased to take of me in connection with the telegraph, and made peculiarly grateful at the present time as coming from a class of society with whom are my earliest pleasurable associations. I may be allowed, perhaps, to say that in my boyhood it was my delight, during my vacations, to seek my pastime in the operations of the printing-office. I solicited of my father to take the corrected proofs of his Geography to the printing-office, and there, through the day for weeks, I made myself practically acquainted with all the operations of the printer. At 9 years of age I compiled a small volume of stories, called it the 'Youth's Friend,' and then set it up, locked the matter in its form, prepared the paper and worked it off; going through the entire process till it was ready for the binder. I think I have some claim, therefore, to belong to the fraternity.

The other letter was in answer to one from a certain Solomon Andrews, President of the Inventors' Institute of Perth Amboy, who was making experiments in aviation, and I shall give but a few extracts: —

"I know by experience the language of the world in regard to an untried invention. He who will accomplish anything useful and new must steel himself against the sneers of the ignorant, and often against the unimaginative sophistries of the learned....

"In regard to the subject on which you desire an opinion, I will say that the *idea* of navigating the air has been a favorite one with the inventive in all ages; it is naturally suggested by the flight of a bird. I have watched for hours together in early life, in my walks across the bridge from Boston to Charlestown, the motions of the sea-gulls.... Often have I attempted to unravel the mystery of their motion so as to bring the principle of it to bear upon this very subject, but I never experimented upon it. Many ingenious men, however, have experimented on air navigation, and have so far succeeded as to travel in the air many miles, but always with the current of wind in their favor. By *navigating* the atmosphere is meant something more than dropping down with the tide in a boat, without sails, or oars or other means of propulsion.... Birds not only rise in the air, but they can also propel themselves against the ordinary currents. A study, then, of the conditions that enable a bird thus to defy the ordinary currents of the atmosphere seems to furnish the most likely mode of solving the problem. Whilst a bird flies, whilst I see a mass of matter overcoming, by its structure and a power within it, the natural forces of gravitation and a current of air, I dare not say that air navigation is absurd or impossible.

"I consider the difficulties to be overcome are the combining of strength with lightness in the machine sufficient to allow of the exercise of a force without the machine from a source of power within. A difficulty will occur in the right adaptation of propellers, and, should this difficulty be overcome, the risks of derangement of the machinery from the necessary lightness of its parts would be great, and consequently the risks to life would be greater than in any other mode of travelling. From a wreck at sea or on shore a man may be rescued with his life, and so by the running off the track by the railroad car, the majority of passengers will be saved; but from a fall some thousands, or only hundreds, of feet through the air, not one would escape death. . . .

"I have no time to add more than my best wishes for the success of those who are struggling with these difficulties."

These observations, made nearly sixty-five years ago, are most pertinent to present-day conditions, when the conquest of the air has been accomplished, and along the very lines suggested by Morse, but at what a terrible cost in human life.

That the inventor, harassed on all sides by pirates, unscrupulous men, and false friends, should, in spite of his Christian philosophy, have suffered from occasional fits of despondency, is but natural, and he must have given vent to his feelings in a letter to his true friend and able business agent, Mr. Kendall, for the latter thus strives to hearten him in a letter of April 20, 1849: —

"You say, 'Mrs. Morse and Elizabeth are both sitting by me.' How is it possible, in the midst of so much that is charming and lovely, that you *could* sink into the gloomy spirit which your letter indicates? Can there be a Paradise without Devils in it — Blue Devils, I mean? And how is it that now, instead of addressing themselves first to the woman, they march boldly up to the man?

"Faith in our Maker is a most important Christian virtue, but man has no right to rely on Faith alone until he has exhausted his own power. When we have done all we can with pure hands and honest hearts, then may we rely with confidence on the aid of Him who governs worlds and atoms, controls, when He chooses, the will of man, restrains his passions and makes his bad designs subservient to the best of ends.

"Now for a short application of a short sermon. We must do our best to have the Depositions and Affidavits prepared and forwarded in due time. This done we may have *Faith* that we will gain our cause. Or, if with our utmost exertions, we fail in our preparations, we shall be warranted in having Faith that no harm will come of it.

"But if, like the Jews in the Maccabees, we rely upon the Lord to fight our battles, without lifting a weapon in our defence, or, like the wagoner in the fable, we content ourselves with calling on Hercules, we shall find in the end that 'Faith without Works is dead.' . . . The world, as you say, is '*the world*' — a quarrelling, vicious, fighting, plundering world — yet it is a very good world for good men. Why should man torment himself about that which he cannot help? If we but enjoy the good things of earth and endure the evil things with a cheerful resignation, bad spirits — blue devils and all — will fly from our bosoms to their appropriate abode."

Another true and loyal friend was George Wood, associated with Mr. Kendall in Washington, from whom are many affectionate and witty letters which it would be a pleasure to reproduce, but for the present

I shall content myself with extracts from one dated May 4, 1849: —

"It does seem to me that Satan has, from the jump, been at war with this invention of yours. At first he strove to cover you up with a F.O.G. of Egyptian hue; then he ran your wires through leaden pipe, constructed by his 'pipe-laying' agents, into the ground and 'all aground.' And when these were hoisted up, like the Brazen Serpent, on poles for all to gaze at and admire, then who so devout a worshipper as the Devil in the person of one of his children of darkness, who came forward at once to contract for a line reaching to St. Louis — *and round the world* — upon that principle of the true construction of *constitutions*, and such like *contracts*, first promulgated by that 'Old Roman' the 'Hero of two Wars,' and approved by the 'whole hog' Democracy of the 'first republic of the world,' and which, like the moral law is summarily comprehended in a few words — 'The constitution (or contract) is what I understand it to be.'

"Now without stopping to show you that O'Reilly was a true disciple of O'Hickory, I think you will not question his being a son of Satan, whose brazen instruments (one of whom gave his first born the name of Morse) instigated by the Gent in Black, not content with inflicting us with the Irish Potato Rot, has recently brought over the Scotch Itch, if, perhaps, by plagues Job was never called upon to suffer (for there were no Courts of Equity and Chancery in those early days) the American inventor might be tempted to curse God and die. But, Ah! you have such a sweet wife, and Job's was such a vinegar cruet."

It is, perhaps, hardly necessary to explain that F. O. J. Smith was nicknamed "Fog" Smith, and that the "Scotch Itch" referred to the telegraph of Alexander Bain, which, for a time, was used by the enemies of Morse in the effort to break down his patent rights. The other allusions were to the politics of the day.

Another good friend and business associate was Thomas R. Walker, who in 1849 was mayor of Utica, New York. Mr. Walker's wife was the half-sister of Mrs. Griswold, Morse's mother-in-law, so there were ties of relationship as well as of friendship between the two men, and Morse thought so highly of Mr. Walker that he made him one of the executors of his will.

In a letter of July 11, 1849, Mr. Walker says: "The course pursued by the press is simply mercenary. Were it otherwise you would receive justice at their hands, and your fame and merits would be vindicated instead of being tarnished by the editorials of selfish and ungenerous men. But — '*magna est veritas et prevalebit.*' There is comfort in that at any rate."

It would seem that not only was the inventor forced to uphold his rights through a long series of lawsuits, but a great part of the press of the country was hostile to him on the specious plea that they were attempting to overthrow a baleful monopoly. In this connection the following extract from a letter to J. Fenimore Cooper, written about this time, is peculiarly apt: —

"It is not because I have not thought of you and your excellent family that I have not long since written to you to know your personal welfare. I hear of you often, it is true, through the papers. They praise you, as usual, for it is praise to have the abuse of such as abuse you. In all your libel suits against these degraded wretches I sympathize entirely with you, and there are thousands who now thank you in their hearts for the moral courage you display in bringing these licentious scamps to a knowledge of their duty. Be assured the good sense, the intelligence, the right feeling of the community at large are with you. The licentiousness of the press needed the rebuke which you have given it, and it feels it too despite its awkward attempts to brave it out.

"I will say nothing of your 'Home as Found.' I will use the frankness to say that I wish you had not written it. . . . When in Paris last I several times passed 59 Rue St. Dominique. The gate stood invitingly open and I looked in, but did not see my old friends although everything else was present. I felt as one might suppose another to feel on rising from his grave after a lapse of a century."

An attack from another and an old quarter is referred to in a letter to his brother Sidney of July 10, also another instance of the unfairness of the press: —

"Dr. Jackson had the audacity to appear at Louisville by *affidavit* against me. My *counter-affidavit*, with his original letters, contradicting *in toto* his statement, put him *hors de combat*. Mr. Kendall says he was 'completely used up.' . . . I have got a copy of Jackson's affidavit which I should like to show you. There never was a more finished specimen of wholesale lying than is contained in it. He is certainly a monomaniac; no other conclusion could save him from an indictment for perjury.

"By the Frankfort paper sent you last week, and the

extract I now send you, you can give a very effective shot to the 'Tribune.' It is, perhaps, worthy of remark that, while all the papers in New York were so forward in publishing a *false* account of O'Reilly's success in the Frankfort case, not one that I have seen has noticed the decision just given at Louisville *against* him in every particular. This shows the animus of the press towards me. Nor have they taken any pains to correct the false account given of the previous decision."

Although no longer President of the National Academy of Design, having refused reëlection in 1845 in order to devote his whole time to the telegraph, Morse still took a deep interest in its welfare, and his counsel was sought by its active members. On October 13, 1849, Mr. Charles C. Ingham sent him a long letter detailing the trials and triumphs of the institution, from which I shall quote a few sentences: "'Lang syne,' when you fought the good fight for the cause of Art, your prospects in life were not brighter than they are now, and in bodily and mental vigor you are just the same, therefore do not, at this most critical moment, desert the cause. It is the same and our enemies are the same old insolent quacks and impostors, who wish to make a footstool of the profession on which to stand and show themselves to the public. . . . Now, with this prospect before you, rouse up a little of your old enthusiasm, put your shoulder to the wheel, and place the only school of Art on all this side of the world on a firm foundation."

Unfortunately the answer to this letter is not in my possession, but we may be sure that it came from the heart, while it must have expressed the writer's deep regret that the multiplicity of his other cares would prevent him from undertaking what would have been to him a labor of love.

Although Alfred Vail had severed his active connection with the telegraph, he and his brother George still owned stock in the various lines, and Morse did all in his power to safeguard and further their interests. They, on their part, were always zealous in championing the rights of the inventor, as the following letter from George Vail, dated December 19, 1849, will show: —

"Enclosed I hand you a paragraph cut from the 'Newark Daily' of 17th inst. It was evidently drawn out by a letter which I addressed to the editor some months ago, stating that I could not see what consistency there was in his course; that, while he was assuming the championship of American manufactures, ingenuity, enterprise, etc., etc., he was at the same time holding up an English inventor to praise, while he held all the better claims of Morse in the dark, — alluding to his bespattering Mr. Bain and O'Reilly with compliments at our expense, etc.

"I would now suggest that, if you are willing, we give *Mr. Daily* a temperate article on the rise and progress of telegraphs, asserting claims for yourself, and, as I must father the article, give the Vails and New Jersey all the 'sodder' they are entitled to, and a little more, if you can spare it.

"Will you write something adapted to the case and forward it to me as early as possible, that it may go in on the heels of this paragraph enclosed?"

F. O. J. Smith continued to embarrass and thwart the other proprietors by his various wild schemes for self-aggrandizement. As Mr. Kendall said in a letter of August 4: "There is much *Fog* in Smith's letter, but it is nothing else."

And on December 4, he writes in a more serious vein: "Mr. Smith peremptorily refuses an arbitration which shall embrace a separation of all our interests, and I think it inexpedient to have any other. He is so utterly unprincipled and selfish that we can expect nothing but renewed impositions as long as we have any connection with him. He asks me to make a proposition to buy or sell, which I have delayed doing, because I know that nothing good can come of it; but I have informed him that I will consider any proposition he may make, if not too absurd to deserve it. I do not expect any that we can accede to without sacrifices to this worse than patent pirate which I am not prepared to make."

Mr. Kendall then concludes that the only recourse will be to the law, but Morse, always averse to war, and preferring to exhaust every effort to bring about an amicable adjustment of difficulties, sent the following courteous letter to Smith on December 8, which, however, failed of the desired result: —

"I deeply regret to learn from my agent, Mr. Kendall, that an unpleasant collision is likely to take place between your interest in the Telegraph and the rest of your coproprietors in the patent. I had hoped that an amicable arbitrament might arrange all our mutual interests to our mutual advantage and satisfaction; but I learn that his proposition to that effect has been rejected by you.

"You must be aware that the rest of your coproprietors have been great sufferers in their property, for some time past, from the frequent disagreements between their agent and yourself, and that, for the sake of peace, they have endured much and long. It is impossible for me to say where the fault lies, for, from the very fact that I put my affairs into the hands of an agent to manage for me, it is evident I cannot have that minute, full and clear view of the matters at issue between him and yourself that he has, or, under other circumstances, that I might have. But this I can see, that mutual disadvantage must be the consequence of litigation between us, and this we both ought to be desirous to avoid.

"Between fair-minded men I cannot see why there should be a difference, or at least such a difference as cannot be adjusted by uninterested parties chosen to settle it by each of the disagreeing parties.

"I write this in the hope that, on second thought, you will meet my agent Mr. Kendall in the mode of arbitration proposed. I have repeatedly advised my agent to refrain from extreme measures until none others are left us; and if such are now deemed by him necessary to secure a large amount of our property, hazarded by perpetual delays, while I shall most sincerely regret the necessity, there are interests which I am bound to protect, connected with the secure possession of what is rightfully mine, which will compel me to oppose no further obstacle to his proceeding to obtain my due, in such manner as, in his judgment, he may deem best."

CHAPTER XXXIV

MARCH 5, 1850—NOVEMBER 10, 1854

Precarious financial condition. — Regret at not being able to make loan. — False impression of great wealth. — Fears he may have to sell home. — F. O. J. Smith continues to give trouble. — Morse system extending throughout the world. — Death of Fenimore Cooper. — Subscriptions to charities, etc. — First use of word "Telegram." — Mysterious fire in Supreme Court clerk's room. — Letter of Commodore Perry. — Disinclination to antagonize Henry. — Temporary triumph of F. O. J. Smith. — Order gradually emerging. — Expenses of the law. — Triumph in Australia. — Gift to Yale College. — Supreme Court decision and extension of patent. — Social diversions in Washington. — Letters of George Wood and P. H. Watson on extension of patent. — Loyalty to Mr. Kendall; also to Alfred Vail. — Decides to publish "Defense." — Controversy with Bishop Spaulding. — Creed on Slavery. — Political views. — Defeated for Congress.

WHILE I have anticipated in giving the results of the various lawsuits, it must be borne in mind that these dragged along for years, and that the final decision of the Supreme Court was not handed down until January 30, 1854. During all this time the inventor was kept in suspense as to the final outcome, and often the future looked very dark indeed, and he was hard pressed to provide for the present.

On March 5, 1850, he writes to a friend who had requested a loan of a few hundred dollars: —

"It truly pains me to be obliged to tell you of my inability to make you a loan, however small in amount or amply secured. In the present embarrassed state of my affairs, consequent upon these never-ending and vexatious suits, I know not how soon all my property may be taken from me. The newspapers, among their other innumerable falsehoods, circulate one in regard to my 'enormous wealth.' The object is obvious. It is to destroy any feeling of sympathy in the public mind from the gross robberies committed upon me. 'He is rich enough; he can afford to give something to the public from his extortionate monopoly,' etc., etc.

"Now no man likes to proclaim his poverty, for there is a sort of satisfaction to some minds in being esteemed rich, even if they are not. The evil of this is that from a rich man more is expected in the way of pecuniary favors (and justly too), and consequently applications of all kinds are daily, I might say for the last few months almost hourly, made to me, and the fabled wealth attributed to me, or to Crœsus, would not suffice to satisfy the requests made."

And, after stating that, of the 11,607 miles of telegraph at that time in operation, only one company of 509 miles was then paying a dividend, he adds: "If this fails I have nothing. On this I solely depend, for I have now no profession, and at my age, with impaired eyesight, I cannot resume it.

"I have indeed a farm out of which a farmer might obtain his living, but to me it is a source of expense, and I have not actually, though you may think it strange, the means to make my family comfortable."

In a letter to Mr. Kendall of January 4, 1851, he enlarges on this subject: —

"I have been taking in sail for some time past to prepare for the storm which has so long continued and still threatens destruction, but with every economy my family must suffer for the want of many comforts which the low state of my means prevents me from procuring. I contrived to get through the last month without incurring debt, but I see no prospect now of being able to do so the present month. . . . I wish much to know, and, indeed, it is indispensably necessary I should be informed of the precise condition of things; for, if my property is but *nominal* in the stocks of the companies, and is to be soon rendered valueless from the operations of pirates, I desire to know it, that I may sell my home and seek another of less pretension, one of humbler character and suited to my change of circumstances. It will, indeed, be like cutting off a right hand to leave my country home, but, if I cannot retain it without incurring debt, it must go, and before debt is incurred and not after. I have made it a rule from my childhood to live always within my means, to have no debts; for if there is a terror which would unman me more than any other in this world, it is the sight of a man to whom I owed money, however inconsiderable in amount, without my being in a condition to pay him. On this point I am nervously sensitive, to a degree which some might think ridiculous. But so it is and I cannot help it. . . .

"Please tell me how matters stand in relation to F. O. G. I wish nothing short of entire separation from that unprincipled man if it can possibly be accomplished. . . . I can suffer his frauds upon myself with comparative forbearance, but my indignation boils when I am made, *nolens volens*, a *particeps criminis* in his frauds on others. I will not endure it if I must suffer the loss of all the property I hold in the world."

The beloved country place was not sacrificed, and a way out of all his difficulties was found, but his faith and Christian forbearance were severely tested before his path was smoothed. Among all his trials none was so hard to bear as the conduct of F. O. J. Smith, whose strange tergiversations were almost inconceivable. Like the old man of the sea, he could not be shaken off, much as Morse and his partners desired to part company with him forever. The propositions made by him were so absurd that they could not for a moment be seriously considered, and the reasonable terms submitted by Mr. Kendall were unconditionally rejected by him. It will be necessary to refer to him and his strange conduct from time to time, but to go into the matter in detail would consume too much valuable space. It seems only right, however, to emphasize the fact that his animosity and unscrupulous self-seeking constituted the greatest cross which Morse was called upon to bear, even to the end of his life, and that many of the aspersions which have been cast upon the inventor's fame and good name, before and after his death, can be traced to the fertile brain of this same F. O. J. Smith.

While the inventor was fighting for his rights in his own country, his invention, by the sheer force of its superiority, was gradually displacing all other systems abroad. Even in England it was superseding the Cooke and Wheatstone needle telegraph, and on the Continent it had been adopted by Prussia, Austria, Bavaria, Hanover, and Turkey. It is worthy of note that that broad-minded scientist, Professor Steinheil, of Bavaria, who had himself invented an ingenious plan of telegraph when he was made acquainted with the Morse system, at once acknowledged its superiority and urged its adoption by the Bavarian Government. In France, too, it was making its way, and Morse, in answer to a letter of inquiry as to terms, etc., by M. Brequet, thus characteristically avows his motives, after finishing the business part of the letter, which is dated April 21, 1851: —

"To be frank with you, my dear sir (and I feel that I can be frank with you), while I am not indifferent to the pecuniary rewards of my invention (which will be amply satisfactory if my own countrymen will but do me justice), yet as these were not the stimulus to my efforts in perfecting and establishing my invention, so they now hold but a subordinate position when I attempt to comprehend the full results of the Telegraph upon the welfare of my fellow men. I am more solicitous to see its benefits extended world-wide during my lifetime than to turn the stream of wealth, which it is generating to millions of persons, into my own pocket. A few drops from the sea, which may not be missed, will suffice for me."

In the early days of 1852 death took from him one of his dearest friends, and the following letter, written in February, 1852, to Rufus Griswold, Esq., expresses his sentiments: —

"I sincerely regret that circumstances over which I have no control prevent my participation in the services commemorative of the character, literary and moral, of my lamented friend the late James Fenimore Cooper, Esq.

"I can scarcely yet realize that he is no longer with us, for the announcement of his death came upon me most unexpectedly. The pleasure of years of close intimacy with Mr. Cooper was never for a moment clouded by the slightest coolness. We were in daily, I can truly say, almost hourly, intercourse in the year

1831 in Paris. I never met with a more sincere, warm-hearted, constant friend. No man came nearer to the ideal I had formed of a truly high-minded man. If he was at times severe or caustic in his remarks on others, it was when excited by the exhibition of the little arts of little minds. His own frank, open, generous nature instinctively recoiled from contact with them. His liberalities, obedient to his generous sympathies, were scarcely bounded by prudence; he was always ready to help a friend, and many such there are who will learn of his departure with the most poignant sorrow. Although unable to be with you, I trust the Committee will not overlook me when they are collecting the funds for the monument to his genius."

It might have been said of Morse, too, that "his liberalities were scarcely bounded by prudence," for he gave away or lost through investments, urged upon him by men whom he regarded as friends but who were actuated by selfish motives, much more than he retained. He gave largely to the various religious organizations and charities in which he was interested, and it was characteristic of him that he could not wait until he had the actual cash in hand, but, even while his own future was uncertain, he made donations of large blocks of stocks, which, while of problematical value while the litigation was proceeding, eventually rose to much above par.

While he strove to keep his charities secret, they were bruited abroad, much to his sorrow, for, although at the time he was hard pressed to make both ends meet, they created a false impression of great wealth, and the importunities increased in volume.

It is always interesting to note the genesis of familiar words, and the following is written in pencil by Morse on a little slip of paper: —

"*Telegram* was first proposed by the Albany 'Evening Journal,' April 6, 1852, and has been universally adopted as a legitimate word into the English language."

On April 21, 1852, Mr. Kendall reports a mysterious occurrence: —

"Our case in the Supreme Court will very certainly be reached by the middle of next week. A most singular incident has occurred. The papers brought up from the court below, not entered in the records, were on a table in the clerk's room. There was no fire in the room. One of the clerks after dark lighted a lamp, looked up some papers, blew out the lamp and locked the door. Some time afterwards, wishing to obtain a book, he entered the room without a light and got the book in the dark. In the morning our papers were burnt up, and *nothing else.*

"The papers burnt are all the drawings, all the books filed, Dana's lectures, Chester's pamphlet, your sketch-book (if the original was there), your bag of type, etc., etc. But we shall replace them as far as possible and go on with the case. *Was* your original sketch-book there? If so, has any copy been taken?"

The original sketch-book was in this collection of papers so mysteriously destroyed, but most fortunately a certified copy had been made, and this is now in the National Museum in Washington. Also, most fortunately, this effort on the part of some enemy to undermine the foundations of the case proved abortive, if, indeed, it was not a boomerang, for, as we have seen, the decision of the Supreme Court was in Morse's favor.

In the year 1852, Commodore Perry sailed on his memorable trip to Japan, which, as is well known, opened that wonderful country to the outside world and started it on its upward path towards its present powerful position among the nations. The following letter from Commodore Perry, dated July 22, 1852, will, therefore, be found of unusual interest: —

I shall take with me, on my cruise to the East Indias, specimens of the most remarkable inventions of the age, among which stands preëminent your telegraph, and I write a line by Lieutenant Budd, United States Navy, not only to introduce him to your acquaintance, but to ask as a particular favour that you would give him some information and instruction as to the most practicable means of exhibiting the Telegraph, as well as a daguerreotype apparatus, which I am also authorized to purchase, also other articles connected with drawing.

I have directed Lieutenant Budd to visit Poughkeepsie in order to confer with you. He will have lists, furnished by Mr. Norton and a daguerreotype artist, which I shall not act upon until I learn the result of his consultation with you.

I hope you will pardon this intrusion upon your time. I feel almost assured, however, that you will take a lively interest in having your wonderful invention exhibited to a people so little known to the world, and there is no one better qualified than yourself to instruct Lieutenant Budd in the duties I have entrusted to his charge, and who will fully explain to you the object I have in view.

I leave this evening for Washington and should be much obliged if you would address me a line to that place.

Most truly and respectfully yours

M. C. PERRY.

It was about this time that the testimony of Professor Joseph Henry was being increasingly used by Morse's opponents to discredit him in the scientific world and to injure his cause in the courts. I shall, therefore, revert for a moment to the matter for the purpose of emphasizing Morse's reluctance to do or say anything against his erstwhile friend.

In a letter to H. J. Raymond, editor of the New York "Times," he requests space in that journal for a fair exposition of his side of the controversy in reply to an article attacking him. To this Mr. Raymond courteously replies on November 22, 1852: "The columns of the 'Times' are entirely at your service for the purpose you mention, or, indeed, for almost any other. The writer of the article you allude to was Dr. Bettner, of Philadelphia."

Morse answers on November 30: —

"I regret finding you absent; I wished to have had a few moments' conversation with you in relation to the allusion I made to Professor Henry. If possible I wish to avoid any course which might weaken the influence for good of such a man as Henry. I will forbear exposure to the last moment, and, in view of my duty as a Christian at least, I will give him an opportunity to explain to me in private. If he refuses, then I shall feel it my duty to show how unfairly he has conducted himself in allowing his testimony to be used to my detriment.

"I write in haste, and will merely add that, to consummate these views, I shall for the present delay the article I had requested you to insert in your columns, and allow the various misrepresentations to remain yet a little longer unexposed, at the same time thanking you cordially for your courteous accordance of my request."

A slight set-back was encountered by Morse and his associates at this time by the denial of an injunction against F. O. J. Smith, and, in a letter to Mr. Kendall of December 4, the long-suffering inventor exclaims: —

"F. O. J. crows at the top of his voice, and I learned that he and his man Friday, Foss, had a regular spree in consequence, and that the latter was noticed in Broadway drunk and boisterously huzzaing for F. O. J. and cursing me and my telegraph.

"I read in my Bible: 'The triumph of the wicked is short.' This may have a practical application, in this case at any rate. I have full confidence in that Power that, for wise purposes, allows wickedness temporarily to triumph that His own designs of bringing good out of evil may be the more apparent."

Another of Morse's fixed principles in life is referred to in a letter to Judge E. Fitch Smith of February 4, 1853: "Yours of the 31st ulto. is this moment received. Your request has given me some trouble of spirit on this account, to wit: My father lost a large property, the earnings of his whole life of literary labor, by simply endorsing. My mother was ever after so affected by this fact that it was the constant theme of her disapprobation, and on her deathbed I gave her my promise, in accordance with her request, that *I never would endorse a note.* I have never done such a thing, and, of course, have never requested the endorsement of another. I cannot, therefore, in that mode accommodate you, but I can probably aid you as effectually in another way."

It will not be necessary to dwell at length on further happenings in the year 1853. Order was gradually emerging from chaos in the various lines of telegraph, which, under the wise guidance of Amos Kendall, were tending towards a consolidation into one great company. The decision of the Supreme Court had not yet been given, causing temporary embarrassment to the patentees by allowing the pirates to continue their depredations unchecked. F. O. J. Smith continued to give trouble. To quote from a letter of Morse's to Mr. Kendall of January 10, 1853: "The Good Book says that 'one sinner destroyeth much good,' and F. O. J. being (as will be admitted by all, perhaps, except himself) a sinner of that class bent upon destroying as much good as he can, I am desirous, even at much sacrifice (a desire, of course, *inter nos*) to get rid of controversy with him."

Further on in this letter, referring to another cause for anxiety, he says: "Law is expensive, and we must look it in the face and expect to pay roundly for it. . . . It is a delicate task to dispute a professional man's charges, and, though it may be an evil to find ourselves bled so freely by lawyers, it is, perhaps, the least of evils to submit to it as gracefully as we can."

But, while he could not escape the common lot of

man in having to bear many and severe trials, there were compensatory blessings which he appreciated to the full. His home life was happy and, in the main, serene; his farm was a source of never-ending pleasure to him; he was honored at home and abroad by those whose opinion he most valued; and he was almost daily in receipt of the news of the extension of the "Morse system" throughout the world. Even from far-off Australia came the news of his triumph. A letter was sent to him, written from Melbourne on December 3, 1853, by a Mr. Samuel McGowan to a friend in New York, which contains the following gratifying intelligence: —

"Since the date of my last to you matters with me have undergone a material change. I have come off conqueror in my hard fought battle. The contract has been awarded to me in the faces of the representatives of Messrs. Wheatstone and Cooke, Brett and other telegraphic luminaries, much to their chagrin, as I afterwards ascertained; several of them, it appears, having been leagued together in order, as they stated, to thwart a speculating Yankee. However, matters were not so ordained, and I am as well satisfied. I hope they will all live to be the same."

In spite of his financial difficulties, caused by bad management of some of the lines in which he was interested, he could not resist the temptation to give liberally where his heart inclined him, and in a letter of January 9, 1854, to President Woolsey of Yale, he says: —

"Enclosed, therefore, you have my check for one thousand dollars, which please hand to the Treasurer of the College as my subscription towards the fund which is being raised for the benefit of my dearly loved *Alma Mater*.

"I wish I could make it a larger sum, and, without promising what I may do at some future time, yet I will say that the prosperity of Yale College is so near my heart that, should my affairs (now embarrassed by litigations in self-defence yet undecided) assume a more prosperous aspect, I have it in mind to add something more to the sum now sent."

The year 1854 was memorable in the history of the telegraph because of two important events — the decision of the Supreme Court in Morse's favor, already referred to, and the extension of his patent for another period of seven years. The first established for all time his legal right to be called the "Inventor of the Telegraph," and the second enabled him to reap some adequate reward for his years of privation, of struggle, and of heroic faith. It was for a long time doubtful whether his application for an extension of his patent would be granted, and much of his time in the early part of 1854 was consumed in putting in proper form all the data necessary to substantiate his claim, and in visiting Washington to urge the justice of an extension. From that city he wrote often to his wife in Poughkeepsie, and I shall quote from some of these letters.

"*February 17*. I am at the National Hotel, which is now quite crowded, but I have an endurable room with furniture hardly endurable, for it is hard to find, in this hotel at least, a table or a bureau that can stand on its four proper legs, rocking and tetering like a gold-digger's washing-pan, unless the lame leg is propped up with an old shoe, or a stray newspaper fifty times folded, or a magazine of due thickness (I am using 'Harper's Magazine' at this moment, which is somewhat a desecration, as it is too good to be trampled under foot, even the foot of a table), or a coal cinder, or a towel. Well, it is but for a moment and so let it pass.

"Where do you think I was last evening? Read the invitation on the enclosed card, which, although forbidden to be *transferable*, may without breach of honor be transferred to my other and better half. I felt no inclination to go, but, as no refusal would be accepted, I put on my best and at nine o'clock, in company with Mr. and Mrs. Shaffner (the latter of whom, by the by, is quite a pleasant and pretty woman, with a boy one year older than Arthur and about as mischievous) and Mr. and Mrs. John Kendall.

"I went to the ladies' parlor and was presented to the ladies, six in number, who did the honors (if that is the expression) of the evening. There was a great crowd, I think not less than three hundred people, and from all parts of the country — Senators and their wives, members of the House and their wives and daughters, and there was a great number of fine looking men and women. I was constantly introduced to a great many, who uniformly showered their compliments on your *modest* husband."

The card of invitation has been lost, but it was, perhaps, to a President's Reception, and the "great" crowd of three hundred would not tax the energies of the President's aides at the present day.

The next letter is written in a more serious vein: —

"*February 26.* I am very busily engaged in the preparation of my papers for an extension of my patents. This object is of vital importance to me; it is, in fact, the moment to reap the harvest of so many years of labor, and expense, and toil, and neglected would lose me the fruits of all. . . . F. O. J. Smith is here, the same ugly, fiendlike, dog-in-the-manger being he has ever been, the 'thorn in the flesh' which I pray to be able to support by the sufficient grace promised. It is difficult to know how to feel and act towards such a man, so unprincipled, so vengeful, so bent on injury, yet the command to bless those that curse, to pray for those who despitefully use us and persecute us, to love our enemies, to forgive our enemies, is in full force, and I feel more anxious to comply with this injunction of our blessed Saviour than to have the thorn removed, however strongly this latter must be desired."

"*March 4.* You have little idea of the trouble and expense to which I am put in this 'extension' matter. . . . I shall have to pay *hundreds of dollars* more before I get through here, besides being harassed in all sorts of ways from now till the 20th of June next. If I get my extension then I may expect some respite, or, at least, opposition in another shape. I hope eventually to derive some benefit from the late decision, but the reckless and desperate character of my opponents may defeat all the good I expect from it. Such is the reward I have purchased for myself by my invention. . . .

"Mr. Wood is here also. He is the same firm, consistent and indefatigable friend as ever. I know not what I should do in the present crisis without him. I could not possibly put my accounts into proper shape without his aid, and he exerts himself for me as strongly as if I were his brother. . . . Mr. Kendall has been ill almost all the time that I have been here, which has caused me much delay and consumption of time."

It was not until the latter part of June that the extension of his patents was granted, and his good friend, alluded to in the preceding letter, Mr. George Wood, tells, in a letter of June 21st, something of the narrow escape it had: —

"Your Patent Extension is another instance of God's wonder working Providence towards you as expressed in the history of this great discovery. Of that history, of all the various shapes and incidents you may never know, not having been on the spot to watch all its moments of peril, and the way in which, like many a good Christian, it was 'scarcely saved.'

"In this you must see God's hand in giving you a man of remarkable skill, energy, talent, and power as your agent. I refer to P. H. Watson, to whom mainly and mostly, I think, this extension is due. God works by means, and, though he designed to do this for you, he selected the proper person and gave him the skill, perseverance and power to accomplish this result. I hope now you have got it you will make it do for you all it can accomplish pecuniarily. But as for the money, I don't think so much as I do the effect of this upon your reputation. This is the apex of the pyramid."

And Mr. Watson, in a letter of June 20, says: "We had many difficulties to contend with, even to-day, for at one time the Commissioner intended to withhold his decision for reasons which I shall explain at length when we meet. It seemed to give the Commissioner much pleasure to think that, in extending the patent, he was doing an act of justice to you as a great public benefactor, and a somewhat unfortunate man of genius. Dr. Gale and myself had to assure him that the extension would legally inure to your benefit, and not to that of your agents and associates before he could reconcile it with his duty to the public to grant the extension."

Morse himself, in a letter to Mr. Kendall, also of June 20, thus characteristically expresses himself: —

"A memorable day. I never had my anxieties so tried as in this case of extension, and after weeks of suspense, this suspense was prolonged to the last moment of endurance. I have just returned with the intelligence from the telegraph office from Mr. Watson — 'Patent extended. All right.'

"Well, what is now to be done? I am for taking time by the forelock and placing ourselves above the contingencies of the next expiration of the patent. While keeping our vantage ground with the pirates I wish to meet them in a spirit of compromise and of magnanimity. I hope we may now be able to consolidate on advantageous terms."

It appears that at this time he was advised by many of his friends, including Dr. Gale, to sever his business connection with Mr. Kendall, both on account of the increasing feebleness of that gentleman, and because, while admittedly the soul of honor, Mr. Kendall had kept their joint accounts in a very careless and slipshod manner, thereby causing considerable financial loss to the inventor. But, true to his friends, as he always was, he replies to Dr. Gale on June 30: —

"Let me thank you specially personally for your solicitude for my interests. This I may say without disparagement to Mr. Kendall, that, were the contract with an agent to be made anew, I might desire to have a younger and more healthy man, and better acquainted with regular book-keeping, but I could not desire a more upright and more honorable man. If he has committed errors, (as who has not?) they have been of the head and not of the heart. I have had many years experience of his conduct, think I have seen him under strong temptation to do injustice with prospects of personal benefit, and with little chance of detection, and yet firmly resisting."

Among the calumnies which were spread broadcast, both during the life of the inventor and after his death, even down to the present day, was the accusation of great ingratitude towards those who had helped him in his early struggles, and especially towards Alfred Vail. The more the true history of his connection with his associates is studied, the more baseless do these accusations appear, and in this connection the following extracts from letters to Alfred Vail and to his brother George are most illuminating. The first letter is dated July 15, 1854: —

"The legal title to my Patent for the American Electro-Magnetic Telegraph of June 20th, 1840, is, by the late extension of said patent for seven years from the said date, now vested in me alone; but I have intended that the pecuniary interest which was guaranteed to you in my invention as it existed in 1838, and in my patent of 1840, should still inure to your benefit (yet in a different shape) under the second patent and the late extension of the first.

"For the simplification of my business transactions

I prefer to let the Articles of Agreement, which expired on the 20th June, 1854, remain cancelled and not to renew them, retaining in my sole possession the *legal title;* but I hereby guarantee to you two sixteenths of such sums as may be paid over to me in the sale of patent rights, after the proportionate deductions of such necessary expenses as may be required in the business of the agency for conducting the sales of said patent rights, subject also to the terms of your agreement with Mr. Kendall.

"Mr. Kendall informs me that no assignment of an interest in my second patent (the patent of 1846) was ever made to you. This was news to me. I presumed it was done and that the assignment was duly recorded at the Patent Office. The examination of the records in the progress of obtaining my extension has, doubtless, led to the discovery of the omission."

After going over much the same ground in the letter to George Vail, also of July 15th, he gives as one of the reasons why the new arrangement is better: "The annoyances of Smith are at an end, so far as the necessity of consulting him is concerned."

And then he adds: —

"I presume it can be no matter of regret with Alfred that, by the position he now takes, strengthening our defensive position against the annoyances of Smith, he can receive *more pecuniarily* than he could before. Please consult with Mr. Kendall on the form of any agreement by which you and Alfred may be properly secured in the pecuniary benefits which you would have were he to stand in the same legal relation to the patent that he did before the expiration of its original term, so as to give me the position in regard to Smith that I must take in self-defense, and I shall cheerfully accede to it.

"Poor Alfred, I regret to know, torments himself needlessly. I had hoped that I was sufficiently known to him to have his confidence. I have never had other than kind feelings towards him, and, while planning for his benefit and guarding his interests at great and almost ruinous expense to myself, I have had to contend with difficulties which his imprudence, arising from morbid suspicions, has often created. My wish has ever been to act towards him not merely justly but generously."

In a letter to Mr. Kendall of July 17, 1854, Morse declares his intention of publishing that "Defense" which he had held in reserve for several years, hoping that the necessity for its publication might be avoided by a personal understanding with Professor Henry, which, however, that gentleman refused: —

"You will perceive what injury I have suffered from the machinations of the sordid pirates against whom I have had to contend, and it will also be noticed how history has been falsified in order to detract from me, and how the conduct of Henry, in his deposition, has tended to strengthen the ready prejudice of the English against the American claim to priority. An increasing necessity, on this account, arises for my 'Defense,' and so soon as I can get it into proper shape by revision, I intend to publish it.

"This I consider a duty I owe the country more than myself, for, so far as I am personally concerned, I am conscious of a position that History will give me when the facts now suppressed by interested pirates and their abettors shall be known, which the verdict of posterity, no less than that of the judicial tribunals already given, is sure to award."

While involved in apparently endless litigation which necessitated much correspondence, and while the compilation and revision of his "Defense" must have consumed not only days but weeks and months, he yet found time to write a prodigious number of letters and newspaper articles on other subjects, especially on those relating to religion and politics. Although more tolerant as he grew older, he was still bitterly opposed to the methods of the Roman Catholic Church, and to the Jesuits in particular. He, in common with many other prominent men of his day, was fearful lest the Church of Rome, through her emissaries the Jesuits, should gain political ascendancy in this country and overthrow the liberty of the people. He took part in a long and heated newspaper controversy with Bishop Spaulding of Kentucky concerning the authenticity of a saying attributed to Lafayette — "If ever the liberty of the United States is destroyed it will be by Romish priests."

It was claimed by the Roman Catholics that this statement of Lafayette's was ingeniously extracted from a sentence in a letter of his to a friend in which he assures this friend that such a fear is groundless. Morse followed the matter up with the patience and keenness of a detective, and proved that no such letter had ever been written by Lafayette, that it was a clumsy forgery, but that he really had made use of the sentiment quoted above, not only to Morse himself, but to others of the greatest credibility who were still living.

In the field of politics he came near playing a more active part than that of a mere looker-on and humble voter, for in the fall of 1854 he was nominated for Congress on the Democratic ticket. It would be difficult and, perhaps, invidious to attempt to state exactly his political faith in those heated years which preceded the Civil War. In the light of future events he and his brothers and many other prominent men of the day were on the wrong side. He deprecated the war and did his best to prevent it.

"Sectional division" was abhorrent to him, but on the question of slavery his sympathies were rather with the South, for I find among his papers the following: —

"My creed on the subject of slavery is short. Slavery *per se* is not sin. It is a social condition ordained from the beginning of the world for the wisest purposes, benevolent and disciplinary, by Divine Wisdom. The mere holding of slaves, therefore, is a condition having *per se* nothing of moral character in it, any more than the being a parent, or employer, or ruler, but is moral or immoral as the duties of the relation of master, parent, employer or ruler are rightly used or abused. The subject in a national view belongs not, therefore, to the department of Morals, and is transferred to that of Politics to be politically regulated.

"The accidents of the relation of master and slave, like the accidents of other social relations, are to be praised or condemned as such individually and in accordance with the circumstances of every case, and, whether adjudged good or bad, do not affect the character of the relation itself."

On the subject of foreign immigration he was most

outspoken, and replying to an enquiry of one of his political friends concerning his attitude towards the so-called "Know Nothings," he says: —

"So far as I can gather from the public papers, the object of this society would seem to be to resist the aggression of foreign influence and its insidious and dangerous assaults upon all that Americans hold dear, politically and religiously. It appears to be to prevent injury to the Republic from the ill-timed and, I may say, unbecoming tamperings with the laws, and habits, and deeply sacred sentiments of Americans by those whose position, alike dictated by modesty and safety, to them as well as to us, is that of minors in training for American, not European, liberty.

"I have not, at this late day, to make up an opinion on this subject. My sentiments 'On the dangers to the free institutions of the United States from foreign immigration' are the same now that I have ever entertained, and these same have been promulgated from Maine to Louisiana for more than twenty years.

"This subject involves questions which, in my estimation, make all others insignificant in the comparison, for they affect all others. To the disturbing influence of foreign action in our midst upon the political and religious questions of the day may be attributed in a great degree the present disorganization in all parts of the land.

"So far as the Society you speak of is acting against this great evil it, of course, meets with my hearty concurrence. I am content to stand on the platform, in this regard, occupied by Washington in his warnings against foreign influence, by Lafayette, in his personal conversation and instructions to me, and by Jefferson in his condemnation of the encouragement given, even in his day, to foreign immigration. If this Society has ulterior objects of which I know nothing, of these I can be expected to speak only when I know something."

As his opinions on important matters, political and religious, appear in the course of his correspondence, I shall make note of them. It is more than probable that, as he differed radically from his father and the other Federalists on the question of men and measures during the War of 1812, so I should have taken other ground than his had I been born and old enough to have opinions in the stirring *ante-bellum* days of the fifties. And yet, as hindsight makes our vision clearer than foresight, it is impossible to say definitely what our opinions would have been under other conditions, and there can, at any rate, be no question of the absolute sincerity of the man who, from his youth up, had placed the welfare of his beloved country above every other consideration except his duty to his God.

It would take a keen student of the political history of this country to determine how far the opinions and activities of those who were in opposition on questions of such prime importance as slavery, secession, and unrestricted immigration, served as a wholesome check on the radical views of those who finally gained the ascendancy. The aftermath of two of these questions is still with us, for the negro question is by no means a problem solved, and the subject of proper restrictions on foreign immigration is just now occupying the attention of our Solons.

That Morse should make enemies on account of the

outspoken stand he took on all these questions was to be expected, but I shall not attempt to sit in judgment, but shall simply give his views as they appear in his correspondence. At any rate he was not called upon to state and maintain his opinions in the halls of Congress, for, in a letter of November 10, 1854, to a friend, he says at the end: "I came near being in Congress at the late election, but had *not quite votes enough*, which is the usual cause of failure on such occasions."

CHAPTER XXXV

JANUARY 8, 1855—AUGUST 14, 1856

Payment of dividends delayed. — Concern for welfare of his country. — Indignation at corrupt proposal from California. — Kendall hampered by the Vails. — Proposition by capitalists to purchase patent rights. — Cyrus W. Field. — Newfoundland Electric Telegraph Company. — Suggestion of Atlantic Cable. — Hopes thereby to eliminate war. — Trip to Newfoundland. — Temporary failure. — F. O. J. Smith continues to give trouble. — Financial conditions improve. — Morse and his wife sail for Europe. — Fêted in London. — Experiments with Dr. Whitehouse. — Mr. Brett. — Dr. O'Shaughnessy and the telegraph in India. — Mr. Cooke. — Charles R. Leslie. — Paris. — Hamburg. — Copenhagen. — Presentation to king. — Thorwaldsen Museum. — Oersted's daughter. — St. Petersburg. — Presentation to Czar at Peterhoff.

I HAVE said in the preceding chapter that order was gradually emerging from chaos in telegraphic matters, but the progress towards that goal was indeed gradual, and a perusal of the voluminous correspondence between Morse and Kendall, and others connected with the different lines, leaves the reader in a state of confused bewilderment and wonder that all the conflicting interests, and plots and counterplots, could ever have been brought into even seeming harmony. Too much praise cannot be given to Mr. Kendall for the patience and skill with which he disentangled this apparently hopeless snarl, while at the same time battling against physical ills which would have caused most men to give up in despair. That Morse fully appreciated the sterling qualities of this faithful friend is evidenced by the letter to Dr. Gale in the preceding chapter, and by many others. He always refused to consider for a moment the substitution of a younger man on the plea of Mr. Kendall's failing health, and his carelessness in the keeping of their personal accounts. It is true that, because of this laxity on Mr. Kendall's part, Morse was for a long time deprived of the full income to which he was entitled, but he never held this up against his friend, always making excuses for him.

Affairs seem to have been going from bad to worse in the matter of dividends, for, while in 1850 he had said that only 509 miles out of 1150 were paying him personally anything, he says in a letter to Mr. Kendall of January 8, 1855: —

"I perceive the Magnetic Telegraph Company meet in Washington on Thursday the 11th. Please inform me by telegraph the amount of dividend they declare and the time payable. This is the only source on which I can calculate for the means of subsistence from day to day with any degree of certainty.

"It is a singular reflection that occurs frequently to my mind that out of 40,000 miles of telegraph, all of which should pay me something, only 225 miles is all that I can depend upon with certainty; and the case is a little aggravated when I think that throughout all Europe, which is now meshed with telegraph wires from the southern point of Corsica to St. Petersburg, on which my telegraph is universally used, not a mile contributes to my support or has paid me a farthing.

"Well, it is all well. I am not in absolute want, for I have some credit, and painful as is the state of debt to me from the apprehension that creditors may suffer from my delay in paying them, yet I hope on."

Mr. Kendall was not so sensitive on the subject of debt as was Morse, and he was also much more optimistic and often rebuked his friend for his gloomy anticipations, assuring him that the clouds were not nearly so dark as they appeared.

Always imbued with a spirit of lofty patriotism, Morse never failed, even in the midst of overwhelming cares, to give voice to warnings which he considered necessary. Replying to an invitation to be present at a public dinner he writes: —

Gentlemen, — I have received your polite invitation to join with you in the celebration of the birthday of Washington. Although unable to be present in person, I shall still be with you in heart.

Every year, indeed every day, is demonstrating the necessity of our being wide awake to the insidious sapping of our institutions by foreign emissaries in the guise of friends, who, taking advantage of the very liberality and unparalleled national generosity which we have extended to them, are undermining the foundations of our political fabric, substituting (as far as they are able to effect their purpose) on the one hand a dark, cold and heartless atheism, or, on the other, a disgusting, puerile, degrading superstition in place of the God of our fathers and the glorious elevating religion of love preached by his Son.

The American mind, I trust, is now in earnest waking up, and no one more rejoices at the signs of the times than myself. Twenty years ago I hoped to have seen it awake, but, alas! it proved to be but a spasmodic yawn preparatory to another nap. If it shall now have waked in earnest, and with renewed strength shall gird itself to the battle which is assuredly before it, I shall feel not a little in the spirit of good old Simeon — "Now let thy servant depart in peace, for mine eyes have seen thy salvation."

Go forward, my friends, in your patriotic work, and may God bless you in your labors with eminent success.

It has been shown, I think, in the course of this work, that Morse, while long-suffering and patient under trials and afflictions, was by no means poor-spirited, but could fight and use forceful language when roused by acts of injustice towards himself, his country, or his sense of right. Nothing made him more righteously angry than dishonesty in whatever form it was manifested, and the following incident is characteristic.

On June 26, 1855, Mr. Kendall forwarded a letter which he had received from a certain Milton S. Latham, member of Congress from California, making a proposition to purchase the Morse patent rights for lines in California. In this letter occur the following sentences: "For the use of Professor Morse's patent for the State of California in perpetuity, with the reservations named in yours of the 3d March, 1855, addressed to me, they are willing to give you $30,000 in their stock. This is all they will do. It is proper I should state that the capital stock of the California State Telegraph in cash was $75,000, which they raised to $150-000, and subsequently to $300,000. The surplus stock over the cash stock was used among members of the Legislature to procure the passage of the act incorporating the company, and securing for it certain privileges."

Mr. Kendall in his letter enclosing this naïve business proposition, remarks: "It is an impressive commentary on the principles which govern business in California that this company doubled their stock to bribe members of the State Legislature, and are now willing to add but ten per cent to be relieved from the position of patent pirates and placed henceforth on an honest footing."

Morse more impulsively exclaims in his reply: —

"Is it possible that there are men who hold up their heads in civilized society who can unblushingly take the position which the so-called California State Telegraph Company has deliberately taken?

"Accept the proposition? Yes, I will accept it when I can consent to the housebreaker who has entered my house, packed up my silver and plated ware, and then coolly says to me — 'Allow me to take what I have packed up and I will select out that which is worthless and give it to you, after I have used it for a few years, provided any of it remain!'

"A more unprincipled set of swindlers never existed. Who is this Mr. Latham that he could recommend our accepting such terms?"

In addition to the opposition of open enemies and unprincipled pirates, Morse and Kendall were sometimes hampered by the unjust suspicions of some of those whose interests they were striving to safeguard. Referring to one such case in a letter of June 15, 1855, Mr. Kendall says: —

"If there should be opposition I count on the Vails against me. Alfred has for some time been hostile because I could not if I would, and would not if I could, find him a snug sinecure in some of the companies. I fear George has in some degree given way to the same spirit. I have heard of his complaining of me, and when, before my departure for the West, I tendered my services to negotiate a connection of himself and brother with the lessees of the N. O. & O. line, he declined my offer, protesting against the entire arrangements touching that line.

"Having done all I could and much more than I was bound to do for the benefit of those gentlemen, I shall not permit their jealousy to disturb me, but I am anxious to have them understand the exact position I am to occupy in relation to them. I understood your purpose to be that they should share in the benefits of the extension, whether legally entitled to them or not, yet nothing has been paid over to them for sales since made. All the receipts, except a portion of my commissions, have been paid out on account of expenses, and to secure an interest for you in the N. O. & O. line."

It is easy to understand that the Vails should have been somewhat suspicious when little or nothing in the way of cash was coming in to them, but they seem not to have realized that Morse and Kendall were in the same boat, and living more on hope than cash. Mr. Kendall enlarges somewhat on this point in a letter of June 22, 1855: —

"Most heartily will I concur in a sale of all my interests in the Telegraph at any reasonable rate to such a company as you describe. I fully appreciate your reasons for desiring such a consummation, and, in addition to them, have others peculiar to my own position. Any one who has a valuable patent can profit by it only by a constant fight with some of the most profligate and, at the same time, most shrewd members of society.

I have found myself not only the agent of yourself and the Messrs. Vail to sell your patent rights, but the soldier to fight your battles, as well in the country as in the courts of justice. Almost single-handed, with the deadly enmity of one of the patentees, and the annoying jealousies of another, I have encountered surrounding hosts, and, I trust, been instrumental in saving something for the Proprietors of this great invention, and done something to maintain the rights and vindicate the fame of its true author. Nothing but your generous confidence has rendered my position tolerable, and enabled me to meet the countless difficulties with which my path has been beset with any degree of success. And now, at the end of a ten years' war, I am prepared to retire from the field and leave the future to other hands, if I can but see your interests secured beyond contingency, and a moderate competency provided for my family and myself."

The company referred to in this letter was one proposed by Cyrus W. Field and other capitalists of New York. The plan was to purchase the patent rights of Morse, Kendall, Vail, and F. O. J. Smith, and, by means of the large capital which would be at their command, fight the pirates who had infringed on the patent, and gradually unite the different warring companies into one harmonious concern. A monopoly, if you will, but a monopoly which had for its object better, cheaper, and quicker service to the people. This object was achieved in time, but, unfortunately for the peace of mind of Morse and Kendall, not just then.

The name of Cyrus Field naturally suggests the Atlantic Cable, and it was just at this time that steps were being seriously taken to realize the prophecy made by Morse in 1843 in his letter to the Secretary of the Treasury: "The practical inference from this law is that a telegraphic communication on the electro-magnetic plan may with certainty be established across the Atlantic Ocean! Startling as this may now seem I am confident the time will come when this project will be realized."

In 1852 a company had been formed and incorporated by the Legislature of Newfoundland, called the "Newfoundland Electric Telegraph Company." The object of this company was to connect the island by means of a cable with the mainland, but this was not accomplished at that time, and no suggestion was made of the possibility of crossing the ocean. One of the officers of that company, however, Mr. F. N. Gisborne, came to New York in 1854 and tried to revive the interest of capitalists and engineers in the scheme. Among others he consulted Matthew D. Field, and through him met his brother Cyrus W. Field, and the question of a through line from Newfoundland to New York was seriously discussed. Cyrus Field, a man of great energy and already interested financially and otherwise in the terrestrial telegraph, was fascinated by the idea of stretching long lines under the waters also. He examined a globe, which was in his study at home and, suddenly realizing that Newfoundland and Ireland were comparatively near neighbors, he said to himself: "Why not cross the ocean and connect the New World with the Old?" He had heard that Morse long ago had prophesied that this link would some day be welded, and he became possessed with the idea that he was the person to accomplish this marvel, just as Morse had received the inspiration of the telegraph in 1832.

A letter to Morse, who was just then in Washington, received an enthusiastic and encouraging reply, coupled with the information that Lieutenant Maury of the Navy had, by a series of careful soundings, established the existence of a plateau between Ireland and Newfoundland, at no very great depth, which seemed expressly designed by nature to receive and carefully guard a telegraphic cable. Mr. Field lost no time in organizing a company composed originally of himself, his brother the Honorable David Dudley Field, Peter Cooper, Moses Taylor, Marshall O. Roberts, and Chandler White. After a liberal charter had been secured from the legislature of Newfoundland the following names were added to the list of incorporators: S. F. B. Morse, Robert W. Lowber, Wilson G. Hunt, and John W. Brett. Mr. Field then went to England and with characteristic energy soon enlisted the interest and capital of influential men, and the Atlantic Telegraph Company was organized to coöperate with the American company, and liberal pledges of assistance from the British Government were secured. Similar pledges were obtained from the Congress of the United States, but, quite in line with former precedents, by a majority of only *one* in the Senate. Morse was appointed electrician of the American company and Faraday of the English company, and much technical correspondence followed between these two eminent scientists.

In the spring of 1855, Morse, in a letter to his friend and relative by marriage, Thomas R. Walker, of Utica, writes enthusiastically of the future: "Our *Atlantic line* is in a fair way. We have the governments and capitalists of Europe zealously and warmly engaged to carry it through. *Three years* will not pass before a *submarine telegraph communication will be had with Europe,* and I do not despair of sitting in my office and, by a touch of the telegraph-key, asking a question simultaneously to persons in London, Paris, Cairo, Calcutta, and Canton, and getting the answer from all of them in *five minutes* after the question is asked. Does this seem strange? I presume if I had even suggested the thought some twenty years ago, I might have had a quiet residence in a big building in your vicinity."

The first part of this prophecy was actually realized, for in 1858, just three years after the date of this letter, communication was established between the two continents and was maintained for twenty days. Then it suddenly and mysteriously ceased, and not till 1866 was the indomitable perseverance of Cyrus Field crowned with permanent success.

More of the details of this stupendous undertaking will be told in the proper chronological order, but before leaving the letter to Mr. Walker, just quoted from, I wish to note that when Morse speaks of sitting in his office and communicating by a touch of the key with the outside world, he refers to the fact that the telegraph companies with which he was connected had obligingly run a short line from the main line (which at that time was erected along the highway from New York to Albany) into his office at Locust Grove, Poughkeepsie, so that he was literally in touch with every place of any importance in the United States.

Always solicitous for the welfare of mankind in general, he says in a letter to Norvin Green, in July, 1855, after discussing the proposed cable: "The effects of the Telegraph on the interests of the world, political, social and commercial have, as yet, scarcely begun to be apprehended, even by the most speculative minds. I trust that one of its effects will be to bind man to his fellow-man in such bonds of amity as to put an end to war. I think I can predict this effect as in a not distant future."

Alas! in this he did not prove himself a true prophet, although it must be conceded that many wars have been averted or shortened by means of the telegraph, and there are some who hope that a warless age is even now being conceived in the womb of time.

On July 18, 1855, he writes to his good friend Dr. Gale: "I have no time to add, as every moment is needed to prepare for my Newfoundland expedition, to be present at laying down the first submarine cable *of any considerable length* on this side the water, although the first for telegraph purposes, you well remember, we laid between Castle Garden and Governor's Island in 1842."

On the 7th of August, Morse, with his wife and their eldest son, a lad of six, joined a large company of friends on board the steamer James Adger which sailed for Newfoundland. There they were to meet the Sarah L. Bryant, from England, with the cable which was to be laid across the Gulf of St. Lawrence. The main object of the trip was a failure, like so many of the first attempts in telegraphic communication, for a terrific storm compelled them to cut the cable and postpone the attempt, which, however, was successfully accomplished the next year.

The party seems to have had a delightful time otherwise, for they were fêted wherever they stopped, notably at Halifax, Nova Scotia, and St. Johns, Newfoundland. At the latter place a return banquet was given on board the James Adger, and the toastmaster, in calling on Morse for a speech, recited the following lines:—

> "The steed called Lightning (say the Fates)
> Was tamed in the United States.
> 'T was Franklin's hand that caught the horse,
> 'T was harnessed by Professor Morse."

To turn again for a moment to the darker side of the picture of those days, it must be kept in mind that annoying litigation was almost constant, and in the latter part of 1855 a decision had been rendered in favor of F. O. J. Smith, who insisted on sharing in the benefits of the extension of the patent, although, instead of doing anything to deserve it, he had done all in his power to thwart the other patentees. Commenting on this in a letter to Mr. Kendall of November 22, 1855, Morse, pathetically and yet philosophically, says:—

"Is there any mode of arrangement with Smith by which matters in partnership can be conducted with any degree of harmony? I wish him to have his legal rights in full, however unjustly awarded to him. I must suffer for my ignorance of legal technicalities. Mortifying as this is it is better, perhaps, to suffer it with a good grace and even with cheerfulness, if possible, rather than endure the wear and tear of the spirits which a brooding over the gross fraud occasions. An opportunity of setting ourselves right in regard to him may be not far off in the future. Till then let us stifle at least all outward expressions of disgust or indignation at the legal swindle."

And, with the keen sense of justice which always actuated him, he adds in a postscript: "By the by, if Judge Curtis's decision holds good in regard to Smith's *inchoate* right, does it not equally hold good in regard to Vail, and is he not entitled to a proportionate right in the extension?"

During the early months of 1856 the financial affairs of the inventor had so far been straightened out that he felt at liberty to leave the country for a few months' visit to Europe. The objects of this trip were threefold. He wished, as electrician of the Cable Company, to try some experiments over long lines with certain English scientists, with a view to determining beyond peradventure the practicability of an ocean telegraph. He also wished to visit the different countries on the continent where his telegraph was being used, to see whether their governments could not be induced to make him some pecuniary return for the use of his invention. Last, but not least, he felt that he had earned a short vacation from the hard work and the many trials to which he had been subjected for so many years, and a trip abroad with his wife, who had never been out of her own country, offered the best means of relaxation and enjoyment. On the 7th of June, 1856, he sailed from New York on the Baltic, accompanied by his wife and his niece Louisa, daughter of his brother Richard.

The trip proved a delightful one in every way; he was acclaimed as one of the most noted men of his day wherever he went, and emperors, kings, and scientists vied with each other in showering attentions upon him.

His letters contain minute descriptions of many of his experiences and I shall quote liberally from them.

To Cyrus Field he writes, on July 5, of the results of some of his experiments with Dr. Whitehouse: —

"I intended to have written you long before this and have you receive my letter previous to your departure from home, but every moment of my time has been occupied, as you can well conceive, since my arrival. I have especially been occupied in experiments with Dr. Whitehouse of the utmost importance. Their results, except in a general way, I am not at present at liberty to divulge; besides they are not, as yet, by any means completed so as to assure commercial men that they may enter upon the great project of uniting Europe to America with a certainty of success."

And then, after dwelling upon the importance of Dr. Whitehouse's services, and expressing the wish that he should be liberally rewarded for his labors, he continues: —

"I can say on this subject generally that the experiments Dr. Whitehouse has made favorably affect the project so far as its *practicability* is concerned, but to certainly assure its *practicality* further experiments are essential. To enable Dr. Whitehouse to make these, and that he may derive the benefit of them, I conceive it to be a wise outlay to furnish him with adequate means for his purpose.

"I wish I had time to give you in detail the kind receptions I have everywhere met with. To Mr. Statham and his family in a special manner are we indebted for the most indefatigable and constant attentions. Were we relatives they could not have been more assiduous in doing everything to make our stay in London agreeable. To Mr. Brett also I am under great obligations. He has manifested (as have, indeed, all the gentlemen connected with the Telegraph here) the utmost liberality and the most ample concession to the excellence of my telegraphic system. I have been assured now from the *highest sources* that my system is not only the most practical for general use, but that it is fast becoming the *world's telegraph.*"

His brother Sidney was at this time also in Europe with his wife and some other members of his family, and the brothers occasionally met in their wanderings to and fro. Finley writes to Sidney from Fenton's Hotel, London, on July 1: —

"Yours from Edinburgh of the 28th ulto. is just received. I regret we did not see you when you called the evening before you left London. We all wished to see you and all yours before we separated so widely apart, but you know in what a whirl one is kept on a first arrival in London and can make allowances for any seeming neglect. From morning till night we have been overwhelmed with calls and the kindest and most flattering attentions.

"On the day before you called I dined at Greenwich with a party invited by Mr. Brett, representing the great telegraph interests of Europe and India. I was most flatteringly received, and Mr. Brett, in the only toast given, gave my name as the Inventor of the Telegraph and of the system which has spread over the whole world and is superseding all others. Dr. O'Shaughnessy, who sat opposite to me, made some remarks warmly seconding Mr. Brett, and stating that he had come from India where he had constructed more than four thousand miles of telegraph; that he had tried many systems upon his lines, and that a few days before I arrived he had reported, in his official capacity as the Director of the East India lines, to the East India Company that my system was the best, and recommended to them its adoption, which I am told will undoubtedly be the case.

"This was an unexpected triumph to me, since I had heard from one of our passengers in the Baltic that in the East Indies they were reluctant to give any credit to America for the Telegraph, claiming it exclusively for Wheatstone. It was, therefore, a surprise to me to hear from the gentleman who controls all the Eastern lines so warm, and even enthusiastic, acknowledgment of the superiority of mine.

"But I have an additional cause for gratitude for an acknowledgment from a quarter whence I least expected any favor to my system. Mr. Cooke, formerly associated with Wheatstone, told one of the gentlemen, who informed me of it, that he had just recommended to the British Government the substitution of my system for their present system, and had no doubt his recommendation would be entertained. He also said that he had heard I was about to visit Europe, and that he should take the earliest opportunity to pay his respects to me. Under these circumstances I called and left my card on Mr. Cooke, and I have now a note from him stating he shall call on me on Thursday. Thus the way seems to be made for the adoption of my Telegraph throughout *the whole world*.

"I visited one of the offices with Dr. Whitehouse and

Mr. Brett where (in the city) I found my instruments in full activity, sending and receiving messages from and to Paris and Vienna and other places on the Continent. I asked if all the lines on the Continent were now using my system, that I had understood that some of the lines in France were still worked by another system. The answer was — 'No, *all the lines on the Continent* are now *Morse lines*.' You will undoubtedly be pleased to learn these facts."

While he was thus being wined, and dined, and praised by those who were interested in his scientific achievements, he harked back for a few hours to memories of his student days in London, for his old friend and room-mate, Charles R. Leslie, now a prosperous and successful painter, gave him a cordial invitation to visit him at Petworth, near London. Morse joyfully accepted, and several happy hours were spent by the two old friends as they wandered through the beautiful grounds of the Earl of Egremont, where Leslie was then making studies for the background of a picture.

The next letter to his brother Sidney is dated Copenhagen, July 19: —

"Here we are in Copenhagen where we arrived yesterday morning, having travelled from Hamburg to Kiel, and thence by steamboat to Corsoer all night, and thence by railroad here, much fatigued owing to the miserable *dis*commodations on board the boat. I have delivered my letters here and am awaiting their effect, expecting calls, and I therefore improve a few moments to apprise you of our whereabouts. . . . In Paris I was most courteously received by the Count de Vouchy, now at the head of the Telegraphs of France, who, with many compliments, told me that my system was the one in universal use, the simplest and the best, and desired me to visit the rooms in the great building where I should find my instruments at work. Sure enough, I went into the Telegraph rooms where some twenty of my own children (beautifully made) were chatting and chattering as in American offices. I could not but think of the contrast in that same building, even as late as 1845, when the clumsy semaphore was still in use, and but a single line of electric wire, an experimental one to Rouen, was in existence in France. . . . When we left Paris we took a courier, William Carter, an Englishman, whom thus far we find to be everything we could wish, active, vigilant, intelligent, honest and obliging. As soon as he learned who I was he made diligent use of his information, and wherever I travelled it was along the lines of the Telegraph. The telegraph posts seemed to be posted to present arms (shall I say?) as I passed, and the lines of conductors were constantly stooping and curtsying to me. At all the stations the officials received me with marked respect; everywhere the same remark met me — 'Your system, Sir, is the only one recognized here. It is the best; we have tried others but have settled down upon yours as the best.' But yesterday, in travelling from Corsoer to Copenhagen, the Chief Director of the Railroads told me, upon my asking if the Telegraph was yet in operation in Denmark, that it was and was in process of construction along this road. 'At first,' said he, 'in using the needle system we found it so difficult to have employees skilled in its operation that we were about to abandon the idea, but now, having adopted yours, we find no difficulty and are constructing telegraphs on all our roads.'

"At all the custom-houses and in all the railroad depots I found my name a passport. My luggage was passed with only the form of an examination, and although I had taken second-class tickets for my party of four, yet the inspectors put us into first-class carriages and gave orders to the conductors to put no one in with us without our permission. I cannot enumerate all the attentions we have received.

"At Hamburg we were delighted, not only with its splendor and cleanliness, but having made known to Mrs. Lind (widow of Edward's brother Henry) that we were in Hamburg, we received the most hearty welcome, passed the day at her house and rode out in the environs. At dinner a few friends were invited to meet us. Mr. Overman, a distant connection of the Linds, was very anxious for me to stay a few days, hinting that, if I would consent, the authorities and dignitaries of Hamburg would show me some mark of respect, for my name was well known to them. I was obliged to decline as I am anxious to be in St. Petersburg before the Emperor is engaged in his coronation preparations."

While in Denmark Morse was granted a private interview with the king at his castle of Frederiksborg, whither he was accompanied by Captain Raasloff: —

"After a few minutes the captain was called into the presence of the king, and in a few minutes more I was requested to go into the audience-chamber and was introduced by the captain to Frederick VII, King of Denmark. The king received me standing and very courteously. He is a man of middle stature, thick-set, and resembles more in the features of his face the busts and pictures of Christian IV than those of any of his predecessors, judging as I did from the numerous busts and portraits of the Kings of Denmark which adorn the city palace and the Castle of Frederiksborg. The king expressed his pleasure at seeing the inventor of the Telegraph, and regretted he could not speak English as he wished to ask me many questions. He thanked me, he said, for the beautiful instrument I had sent him; told me that a telegraph line was now in progress from the castle to his royal residence in Copenhagen; that when it was completed he had decided on using my instrument, which I had given him, in his own private apartments. He then spoke of the invention as a most wonderful achievement, and wished me to inform him how I came to invent it. I accordingly in a few words gave him the early history of it, to which he listened most attentively and thanked me, expressing himself highly gratified. After a few minutes more of conversation of the same character, the king shook me warmly by the hand and we took our leave. . . .

"We arrived in the afternoon at Copenhagen. Mrs. F. called in her carriage. We drove to the Thorwaldsen Museum or Depository where are all the works of this great man. This collection of the greatest sculptor since the best period of Greek art is attractive enough in itself to call travellers of taste to Copenhagen. After spending some hours in Thorwaldsen's Museum I went to see the study of Oersted, where his most important discovery of the *deflection of the needle* by a galvanic current was made, which laid the foundation of the science of electro-magnetism, and without which my invention could not have been made. It is now a drawing school. I sat at the table where he made his discovery.

"We went to the Porcelain Manufactory, and, singularly enough, met there the daughter of Oersted, to whom I had the pleasure of an introduction. Oersted was a most amiable man and universally beloved. The daughter is said to resemble her father in her features, and I traced a resemblance to him in the small porcelain bust which I came to the manufactory to purchase."

"*St. Petersburg, August 8, 1856.* Up to this date we have been in one constant round of visits to the truly wonderful objects of curiosity in this magnificent city. I have seen, as you know, most of the great and marvellous cities of Europe, but I can truly say none of them can at all compare in splendor and beauty to St. Petersburg. It is a city of palaces, and palaces of the most gorgeous character. The display of wealth in the palaces and churches is so great that the simple truth told about them would incur to the narrator the suspicion of romancing. England boasts of her regalia in the Tower, her crown jewels, her Kohinoor diamond, etc. I can assure you that they fade into insignificance, as a rushlight before the sun, when brought before the wealth in jewels and gold seen here in such profusion. What think you of nosegays, as large as those our young ladies take to parties, composed entirely of diamonds, rubies, emeralds, sapphires and other precious stones, chosen to represent accurately the colors of various flowers? — The imperial crown, globular in shape, composed of diamonds, and containing in the centre of the Greek cross which surmounts it an unwrought ruby at least two inches in diameter? The sceptre has a diamond very nearly as large as the Kohinoor. At the Arsenal at Tsarskoye Selo we saw the trappings of a horse, bridle, saddle and all the harness, with an immense saddle-cloth, set with tens of thousands of diamonds. On those parts of the harness where we have rosettes, or knobs, or buckles, were rosettes of diamonds an inch and a half to two inches in diameter, with a diamond in the centre as large as the first joint of your thumb, or say three quarters of an inch in diameter. Other trappings were as rich. Indeed there seemed to be no end to the diamonds. All the churches are decorated in the most costly manner with diamonds and pearls and precious stones."

The following account of his reception by the czar is written in pencil: "On the paper found in my room in Peterhoff." It differs somewhat from the letter written to his children and introduced by Mr. Prime in his book, but is, to my mind, rather more interesting.

"*August 14, 1856.* This day is one to be remembered by me. Yesterday I received notice from the Russian Minister of Foreign Affairs, through our Minister Mr. Seymour, that his Imperial Majesty, the Emperor Alexander II, had appointed the hour of 1.30 this day to see me at his palace at Peterhoff. I accordingly waited upon our minister to know the etiquette to be observed on such an occasion. It was necessary, he said, to be at the boat by eight o'clock in the morning, which would arrive at Peterhoff about 9.30. I must dress in black coat, vest and pantaloons and white cravat, and appear with my Turkish nishan [or decoration]. So this morning I was up early and, upon taking the boat, found our Minister Mr. Seymour, Colonel Colt and Mr. Jarvis, attachés to the Legation, with Mrs. Colt and Miss Jarvis coming on board. I learned also that there were to be many presentations of various nations' attachés to the various special deputations sent to represent their different courts at the approaching coronation at Moscow.

"The day is most beautiful, rendered doubly so by its contrast with so many previous disagreeable ones. On our arrival at the quay at Peterhoff we found, somewhat to my surprise, the imperial carriages in waiting for us, with coachmen and footmen in the imperial livery, which, as in England and France, is scarlet, and splendid black horses, ready to take us to our quarters in the portion of the palace buildings assigned to the Americans. We were attended by four or five servants in livery loaded with gold lace, and shown to our apartments upon the doors of which we found our names already written.

"After throwing off our coats the servants inquired if we would have breakfast, to which, of course, we had no objection, and an excellent breakfast of coffee and sandwiches was set upon the table, served up in silver with the imperial arms upon the silver waiter and tea set. Everything about our rooms, which consisted of parlor and bedroom, was plain but exceedingly clean and neat. After seeing us well housed our attendant chamberlain left us to prepare ourselves for the presentation, saying he would call for us at the proper time. As there were two or three hours to spare I took occasion to improve the time by commencing this brief notice of the events of the day.

"About two o'clock our attendant, an officer named Thörner, under the principal chamberlain who is, I believe, Count Borsch, called to say our carriages were ready. We found three carriages in waiting with three servants each, the coachman and two footmen, in splendid liveries; some in the imperial red and gold lace, and others in blue and broad gold lace emblazoned throughout with the double headed eagle. We seated ourselves in the carriages which were then driven at a rapid rate to the great palace, the entrance to which directly overlooked the numerous and celebrated grand fountains. Hundreds of well-dressed people thronged on each side of the carriageway as we drove up to the door. After alighting we were ushered through a long hall and through a double row of servants of various grades, loaded with gold lace and with *chapeaux bras*. Ascending the broad staircase, on each side of which we found more liveried servants, we entered an anteroom between two Africans dressed in the costume of Turkey, and servants of a higher grade, and then onward into a large and magnificent room where were assembled those who were to be presented. Here we found ourselves among princes and nobles and distinguished persons of all nations. Among the English ladies were Lady Granville and Lady Emily Peel, the wife of Sir Robert Peel, the latter a beautiful woman and dressed with great taste, having on her head a Diana coronet of diamonds. . . . Among the gentlemen were officers attached to the various deputations from England, Austria, France and Sardinia. Several princes were among them, and conspicuous for splendor of dress was Prince Esterhazy; parts of his dress and the handle and scabbard of his sword blazed with diamonds.

"Here we remained for some time. From the windows of the hall we looked out upon the magnificent fountains and the terrace crowned with gorgeous vases of blue and gold and gilded statues. At length the master of ceremonies appeared and led the way to the southern veranda that overlooked the garden, ranging us in line and reading our names from a list, to see if we were truly mustered, after which a side door opened and the Emperor Alexander entered. His majesty was dressed in military costume, a blue sash was across his breast passing over the right shoulder; on his left breast were stars and orders. He commenced at the head of the column, which consisted of some fourteen or fifteen persons, and, on the mention of the name by the master of ceremonies, he addressed a few words to each. To Mr. Colt he said: 'Ah! I have seen you before. When did you arrive? I am glad to see you.' When he came to me the master of ceremonies miscalled my name as Mr. More. I instantly corrected him and said, 'No, Mr. Morse.' The emperor at once said: 'Ah! that name is well known here; your system of Telegraph is in use in Russia. How long have you been in St. Petersburg? I hope you have enjoyed yourself.' To which I appropriately replied. After a few more unimportant questions and answers the emperor addressed himself to the other gentlemen and retired.

"After remaining a few moments, the master of ceremonies, who, by the by, apologized to me for miscalling my name, opened the door from the veranda into the empress' drawing-room, where we were again put in line to await the appearance of the empress. The doors of an adjoining room were suddenly thrown open and the empress, gorgeously but appropriately attired, advanced towards us. She was dressed in a beautiful blue silk terminating in a long flowing train of many flounces of the richest lace; upon her head a crown of diamonds, upon her neck a superb necklace of diamonds, some twenty of which were as large as the first joint of the finger. The upper part of her dress was embroidered with diamonds in a broad band, and the dress in front buttoned to the floor with rosettes of diamonds, the central diamond of each button being at least a half inch in diameter. A splendid bouquet of diamonds and precious stones of every variety of color, arranged to imitate flowers, was upon her bosom. She addressed a few words gracefully to each, necessarily commonplace, for what could she say to strangers but the common words of enquiry — when we came and whether we had been pleased with St. Petersburg.

"Gratifying as it was to us to see her, I could not but think it was hardly possible for her to have any other gratification in seeing us than that which I have no doubt she felt, that she was giving pleasure to others. To me she appeared to be amiable and truly feminine. Her manner was timid yet dignified without the least particle of hauteur. The impression left on my mind by both the emperor and empress is that they are most truly amiable and kind.

"After speaking to each of us she gracefully bowed to us, we, of course, returning the salutation, and she retired followed by her maids of honor, her long train sweeping the floor for a distance of several yards behind her. We were then accompanied by the master of ceremonies back to the large reception-room, and soon after we left the palace, descending the staircase through the same lines of liveried servants to the royal carriages drawn up at the door, and returned to our rooms. On descending to our parlor we found a beautiful collation with tropical fruits and confectionery provided for us. Our polite attendant, who partook with us, said that the carriages were at our service and waiting for us to take a drive in the gardens previous to dinner, which was to be served at five o'clock in the English Palace and to which we were invited.

"Two carriages called charabancs, somewhat like the Irish vehicle of the same name, with four servants in the imperial livery to each, we found at the door, and we drove for several miles through the splendid gardens and grounds laid out with all the taste of the most beautiful English grounds, with lakes, and islands, and villas, and statues, and fountains, and the most perfect neatness marked every step of our way.

"The most attractive object in our ride was the Italian villa, a favorite resort of the emperor, a perfect gem of its kind. We alighted here and visited all the apartments and the grounds around it. No description could do it justice; a series of pictures alone could give an idea of its beauties. While here several other royal carriages with the various deputations to the coronation ceremonies, soon to occur at Moscow, arrived, and the cortège of carriages with the gorgeous costumes of the visitors alone furnished an exciting scene, heightened by the proud bearing of the richly caparisoned horses, chiefly black, and the showy trappings of the liveried attendants.

"On our return to our rooms we dressed for dinner and proceeded in the same manner to the palace in the gardens called the English Palace. Here we found assembled in the great reception hall the distinguished company, in number forty-seven, of many nations, who were to sit down to the table together. When dinner was announced we entered the grand dining-hall and found a table most gorgeously prepared with gold and silver service and flowers. At table I found myself opposite three princes, an Austrian, a Hungarian, and one from some other German state, and near me on my left Lord Ward, one of the most wealthy nobles of England, with whom I had a good deal of conversation. Opposite and farther to my right was Prince Esterhazy, seated between Lady Granville and the beautiful Lady Emily Peel. On the other side of Lady Peel was Lord Granville and near him Sir Robert Peel. Among the guests, a list of whom I regret I did not obtain, was the young Earl of Lincoln and several other noblemen in the suite of Lord Granville. . . . Some twenty servants in the imperial livery served the table which was furnished with truly royal profusion and costliness. The rarest dishes and the costliest wines in every variety were put before us. I need not say that in such a party everything was conducted with the highest decorum. No noise, no boisterous mirth, no loud talking, but a quiet cheerfulness and perfect ease characterized the whole entertainment.

"After dinner all arose, both ladies and gentlemen, and left the room together, not after the English fashion of the gentlemen allowing the ladies to retire and then seating themselves again by themselves to drink, etc. We retired for a moment to the great reception-hall for coffee, but, being fearful that we should be too late for the last steamer from Peterhoff to St. Petersburg, we were hurrying to get through and to leave, but the moment our fears had come to the knowledge of Lord Granville, he most kindly came to us and told us to feel at ease as his steam-yacht was lying off the quay to take them up to the city, and he was but too proud to have the opportunity of offering us a place on board; an offer which we, of course, accepted with thanks.

"Having thus been entertained with truly imperial hospitality for the entire day, ending with this sumptuous entertainment, we descended once more to the carriages and drove to the quay, where a large barge belonging to the Jean d'Acre, English man-of-war (which is the ship put in commission for the service of Lord Granville), manned by stalwart man-of-war's-men, was waiting to take the English party of nobles, etc., on board the steam-yacht. When all were collected we left Peterhoff and were soon on board. The weather was fine and the moon soon rose over the palace of Peterhoff, looking for a moment like one of the splendid gilded domes of the palace.

"On board the yacht I had much conversation with Lord Granville, who brought the various members of his suite and introduced them to me, — Sir Robert Peel; the young Earl of Lincoln, the son of the Duke of Newcastle, who, when himself the Earl of Lincoln in 1839, showed me such courtesy and kindness in London; Mr. Acton, a nephew of Lord Granville, with whom I had some conversation in which, while I was speaking of the Greek religion as compared with the Romish, he informed me he was a Roman Catholic. I wished much to have had more conversation with him, but the time was not suitable, and the steamer was now near the end of the voyage.

"We landed at the quay in St. Petersburg about eleven o'clock, and I reached my lodgings in the Hotel de Russie about twelve, thus ending a day of incidents which I shall long remember with great gratification, having only one unpleasant reflection connected with it, to wit that my dear wife, my niece and our friend Miss L. were not with me to participate in the pleasure and novelty of the scenes."

CHAPTER XXXVI

AUGUST 23, 1856 — SEPTEMBER 15, 1858

Berlin. — Baron von Humboldt. — London, successful cable experiments with Whitehouse and Bright. — Banquet at Albion Tavern. — Flattering speech of W. F. Cooke. — Returns to America. — Troubles multiply. — Letter to the Honorable John Y. Mason on political matters. — Kendall urges severing of connection with cable company. — Morse, nevertheless, decides to continue. — Appointed electrician of company. — Sails on U.S.S. Niagara. — Letter from Paris on the crinoline. — Expedition sails from Liverpool. — Queenstown harbor. — Accident to his leg. — Valencia. — Laying of cable begun. — Anxieties. — Three successful days. — Cable breaks. — Failure. — Returns to America. — Retires from cable enterprise. — Predicts in 1858 failure of apparently successful laying of cable. — Sidney E. Morse. — The Hare and the Tortoise. — European testimonial: considered niggardly by Kendall. — Decorations, medals, etc., from European nations. — Letter of thanks to Count Walewski.

His good democratic eyes a trifle dazzled by all this imperial magnificence, Morse left St. Petersburg and, with his party, journeyed to Berlin. What was to him the most interesting incident of his visit to that city is thus described: —

"*August 23.* To-day I went to Potsdam to see Baron Humboldt, and had a delightful interview with this wonderful man. Although I had met with him at the soirées of Baron Gerard, the distinguished painter, in Paris in 1832, and afterward at the Academy of Sciences, when my Telegraph was exhibited to the assembled academicians in 1838, I took letters of introduction to him from Baron Gerolt, the Prussian Minister. But they were unnecessary, for the moment I entered his room, which is in the Royal Palace, he called me by name and greeted me most kindly, saying, as I presented my letters: 'Oh! sir, you need no letters, your name is a sufficient introduction'; and so, seating myself, he rapidly touched upon various topics relating to America."

On the margin of a photograph of himself, presented to Morse by the baron, is an inscription in French of which the following is a translation: —

To Mr. S. F. B. Morse, whose philosophic and useful labors have rendered his name illustrious in two worlds, the homage of the high and affectionate esteem of Alexander Humboldt.

POTSDAM, August 1856.

The next thirty days were spent in showing the beauties of Cologne, Aix-la-Chapelle, Brussels and Paris to his wife and niece, and in the latter part of September the little party returned to London. Here Morse resumed his experiments with Dr. Whitehouse and Mr. Bright, and on October 3, he reports to Mr. Field: —

"As the electrician of the New York, Newfoundland and London Telegraph Company, it is with the highest gratification that I have to apprise you of the result of our experiments of this morning upon a single continuous conductor of more than two thousand miles in extent, a distance, you will perceive, sufficient to cross the Atlantic Ocean from Newfoundland to Ireland.

"The admirable arrangements made at the Magnetic Telegraph office in Old Broad Street for connecting ten subterranean gutta-percha insulated conductors of over two hundred miles each, so as to give one continuous length of more than two thousand miles, during the hours of the night when the Telegraph is not commercially employed, furnished us the means of conclusively settling by actual experiment the question of the practicability as well as the practicality of telegraphing through our proposed Atlantic cable. . . . I am most happy to inform you that, as a crowning result of a long series of experimental investigation and inductive reasoning upon this subject, the experiments under the direction of Dr. Whitehouse and Mr. Bright which I witnessed this morning — in which the induction-coils and receiving-magnets, as modified by these gentlemen, were made to actuate one of my recording instruments — have most satisfactorily resolved all doubts of the practicability as well as practicality of operating the Telegraph from Newfoundland to Ireland."

In 1838, Morse had been curtly and almost insultingly refused a patent for his invention in England, a humiliation for which he never quite forgave the English. Now, eighteen years after this mortifying experience, the most eminent scientists of this same England vied with each other in doing him honor. Thus was his scientific fame vindicated, but, let it be remarked parenthetically, this kind of honor was all that he ever received from the land of his ancestors. While other nations of Europe united, two years later, in granting him a pecuniary gratuity, and while some of their sovereigns bestowed upon him decorations or medals, England did neither. However, it was always a source of the keenest gratification that two of those who had invented rival telegraphs proved themselves broad-minded and liberal enough to acknowledge the superiority of his system, and to urge its adoption by their respective Governments. The first of these was Dr. Steinheil, of Munich, to whom I have already referred, and to whom is due the valuable discovery that the earth can be used as a return circuit.

The second was the Englishman, W. F. Cooke, who, with Wheatstone, devised the needle telegraph.

On October 9, a banquet was tendered to Morse by the telegraph companies of England. It was given at the Albion Tavern. Mr. Cooke presided and introduced the guest of the evening in the following charming speech:—

"I was consulted only a few months ago on the subject of a telegraph for a country in which no telegraph at present exists. I recommended the system of Professor Morse. I believe that system to be one of the simplest in the world, and in that lies its permanency and certainty. [Cheers.] There are others which may be as good in other circumstances, but for a wide country I hesitate not to say Professor Morse's is the best adapted. It is a great thing to say, and I do so after twenty years' experience, that Professor Morse's system is one of the simplest that ever has been and, I think, ever will be conceived. [Cheers.]

"It was a great thing for me, after having been so long connected with the electric telegraph, to be invited to preside at this interesting meeting, and I have travelled upward of one hundred miles in order to be present to-day, having, when asked to preside, replied by electric telegraph 'I will.' [Cheers.] But I may lower your idea of the sacrifice I made in so doing when I tell you that I knew the talents of Professor Morse, and was only too glad to accept an invitation to do honor to a man I really honored in my heart. [Cheers.]

"I have been thinking during the last few days on what Professor Morse has done. He stands alone in America as the originator and carrier out of a grand conception. We know that America is an enormous country, and we know the value of the telegraph, but I think we have a right to quarrel with Professor Morse for not being content with giving the benefit of it to his own country, but that he extended it to Canada and Newfoundland, and, even beyond that, his system has been adopted all over Europe [cheers] — and the nuisance is that we in England are obliged to communicate by means of his system. [Cheers and laughter.]

"I as a director of an electric telegraph company, however, should be ashamed of myself if I did not acknowledge what we owe him. But he threatens to go further still, and promises that, if we do not, he will carry out a communication between England and Newfoundland across the Atlantic. I am nearly pledged to pay him a visit on the other side of the Atlantic to see what he is about, and, if he perseveres in his obstinate attempt to reach England, I believe I must join him in his endeavors. [Cheers.]

"To think that he has united all the stripes and stars of America, which are increasing day by day — and I hope they will increase until they are too numerous to mention — that he has extended his system to Canada and is about to unite those portions of the world to Europe, is a glorious thing for any man; and, although I have done something in the same cause myself, I confess I almost envy Professor Morse for having forced from an unwilling rival a willing acknowledgment of his services. [Cheers.]

"I am proud to see Professor Morse this side of the water. I beg to give you 'The health of Professor Morse,' and may he long live to enjoy the high reputation he has attained throughout the world!"

Soon after this, with these flattering words still ringing in his ears, he and his party sailed for New York and, once arrived at home, the truth of the trite saying that "A prophet is not without honor save in his own country" was soon to be brought to his attention. While he had been fêted and honored abroad, while he had every reason to believe that his petition to the European governments for some pecuniary compensation would, in time, be granted, he returned to be plunged anew into vexatious litigation, intrigues and attacks upon his purse, his fame, and his good name. On November 27, 1856, he refers to his greatest cross in a letter to Mr. Kendall: —

"I have just returned from Boston, having accomplished the important duty for which I alone went there, to wit, to say 'yes' before a gentleman having U.S. Commissioner after his name, instead of 'yes' before one who had only S. Commissioner after his name; and this at a cost of exactly twenty dollars, or, if the one dollar thrown away in New York upon the S. Commissioner be added, twenty-one dollars and three days of time, to say nothing of sundry risks of accidents by land and water travel.

"Well, if it will lead to a thorough separation of all interests and all intercourse with F. O. J., I shall not consider the time and money lost, yet, in conversation with Mr. Curtis, I have little hope of a change in Judge Curtis's views of the point in which he decides that Smith has an inchoate right, and our only chance of success is in the reversal of that decision by the Supreme Bench, and that after another year's suspense. . . .

"I wish there was some way of stopping this harassing, paralyzing litigation. I find my mind wholly unfit for the studies which the present state of the Telegraph requires from me, being distracted and irritated by the constant necessity for standing on the defensive. Smith will be Smith I know, and, therefore, as he is the appointed thorn to keep a proper ballast of humility in S. F. B. M. with his load of honors, why, be it so, if I can only have the proper strength and disposition to use the trial aright. . . . Write me some encouraging news if you can. How will the present calm in political affairs affect our California matters?"

The calm to which he referred was the apparent one which had settled down on the country after the election of Buchanan, and which, as everybody knows, was but the calm before the storm of our Civil War. He has this to say about the election in a letter to the Honorable John Y. Mason, our Minister to France: —

"I may congratulate you, my dear Sir, on the issue of the late election. My predictions have been verified. The country is quiet, and, as usual after the excitement of an election, has settled down into orderly acquiescence to the will of the majority, and into general good feeling. Europeans can hardly understand this truly anomalous phase of our American institutions; they do not understand that it is characteristic that 'we speak daggers but use none'; that we fight with ballots and not with bullets; that we have abundance of inkshed and little bloodshed, and that all that is explosive is blown off through newspaper safety-valves."

The events of the next few years were destined to shatter the peaceful visions of this lover of his country, for many daggers were drawn, the bullets flew thick and fast, and the bloodshed was appalling.

It is difficult to follow the history of the telegraph, in its relation to its inventor, through all the intricacies involved in the conflicting interests of various companies and men in this its formative period.

Morse himself was often at a loss to determine on the course which he should pursue, a course which would at the same time inure to his financial benefit and be in accordance with his high sense of right. Absolutely straightforward and honest himself, it was difficult for him to believe that others who spoke him fair were not equally sincere, and he was often imposed upon, and was frequently forced, in the exigencies of business, to be intimately associated with those whose ideas of right and wrong were far different from his own. The one person in whose absolute integrity he had faith was Amos Kendall, and yet he must sometimes have thought that his friend was too severe in his judgment of others, for I find in a letter of Mr. Kendall's of January 4, 1857, the following warning: —

"I earnestly beseech you to give up all idea of going out again on the cable-laying expedition. Your true friends do not comprehend how it is that you give your time, your labor, and your fame to build up an interest deliberately and unscrupulously hostile to all their interests and your own. . . . I believe that Peter Cooper is the only man among them who is sincerely your friend. As to Field, I have as little faith in him as I have in F. O. J. Smith. If you could get Cooper to take a stand in favor of the faithful observance of the contract for connection with the N. E. Union Line at Boston, he can put an end to all trouble, if, at the same time, he will refuse to concur in a further extension of their lines South."

In spite of this warning, or, perhaps, because Peter Cooper succeeded in overcoming Mr. Kendall's objections, Morse did go out on the next cable-laying expedition, and yet he found in the end that Mr. Kendall's suspicions were by no means unjustified. But of this in its proper place.

The United States Government had placed the steam frigate Niagara at the disposal of the cable company, and on her Morse, as the electrician of the American Company, sailed from New York on April 21, 1857. Arriving in London, he was again honored by many attentions and entertainments, including a dinner at the Lord Mayor's. The loading of the cable on board the ships designated for that purpose consumed, necessarily, some time, and Morse took advantage of this delay to visit Paris, at the suggestion of our Minister, Mr. Mason, in order to confer with the Premier, Count Walewski, with regard to the pecuniary indemnity which all agreed was due to him from the nations using his invention. This conference bore fruit, as we shall see later on.

In a letter to his wife from Paris he makes this amusing comment on the fashions of the day, after remarking on the dearth of female beauty in France: —

"You must consider me now as speaking of features only, for as to form, alas, that is under such a total crinoline eclipse that this season of total darkness in fashion's firmament forbids any speculation on that subject. The reign of crinoline amplitude is not only not removed, but is more dominant than ever. Who could have predicted that, because an heir to the French throne was in expectancy, all womankind, old and young, would so far sympathize with the amiable consort of Napoleon III as to be, in appearance at least, likely to flood the earth with heirs; that grave parliaments would be in solemn debate upon the pressing necessity of enlarging the entrances of royal palaces in order to meet the exigencies of enlarged crinolines; that the new carriages were all of increased dimensions to accommodate the crinoline? But so it is; it is the age of crinoline. . . . Talk no longer of chairs, they are no longer visible. Talk no longer of tête-à-têtes; two crinolines might get in sight of each other, at least by the use of the lorgnette, but as for conversation, that is out of the question except by speaking trumpets, by signs, and who knows but in this age of telegraphs crinoline may not follow the world's fashion and be a patroness of the Morse system."

All the preparations for the great enterprise of the laying of the cable proceeded slowly, and it was not until the latter part of July that the little fleet sailed from Liverpool on its way to the Cove of Cork and then to Valencia, on the west coast of Ireland, which was chosen as the European terminus of the cable. Morse wrote many pages of minute details to his wife, and from them I shall select the most important and interesting: —

"*July 28.* Here we are steaming our way towards Cork harbor, with most beautiful weather, along the Irish coast, which is in full view, and expecting to be in the Cove of Cork in the morning of to-morrow. . . . We left Liverpool yesterday morning, as I wrote you we should, and as we passed the ships of war in the harbor we were cheered from the rigging by the tars of the various vessels, and the flags of others were dipped as a salute, all of which were returned by us in kind. The landing stage and quays of Liverpool were densely crowded with people who waved their handkerchiefs as we slowly sailed by them.

"Two steamers accompanied us down to the bar filled with people, and then, after mutual cheering and firing of cannon from one of the steamers, they returned to port. . . . We shall be in Cork the remainder of the week, possibly sailing on Saturday, go round to Valencia and be ready to commence on Monday. Then, if all things are prosperous, we hope to reach Newfoundland in twenty days, and dear home again the first week in September. And yet there may be delays in this great work, for it is a vast and new one, so don't be impatient if I do not return quite so soon. The work must be thoroughly and well done before we leave it. . . .

"*Evening, ten o'clock.* We have had a beautiful day and have been going slowly along and expect to be in the Cove of Cork by daylight in the morning. The deck of our ship presents a curious appearance just now. Between the main and mizzen masts is an immense coil of one hundred and thirty miles of the cable, the rest is in larger coils below decks. Abaft the mizzen mast is a ponderous mass of machinery for regulating the paying out of the cable, a steam-engine and boiler complete, and they have just been testing it to see if all is right, and it is found right. We have the prospect of a fine moon for our expedition.

"I send you the copy of a prayer that has been read in the churches. I am rejoiced at the manner in which the Christian community views our enterprise. It is calculated to inspire my confidence of success. What the first message will be I cannot say, but if I send it it shall be, 'Glory to God in the highest, on earth peace and good will to men.' 'Not unto us, not unto us, but to Thy name be all the glory.'"

"*July 29, four o'clock afternoon.* On awaking this morning at five o'clock with the noise of coming to anchor, I found myself safely ensconced in one of the most beautiful harbors in the world, with Queenstown picturesquely rising upon the green hills from the foot of the bay. . . ."

"*August 1.* When I wrote the finishing sentence of my last letter I was suffering a little from a slight accident to my leg. We were laying out the cable from the two ships, the Agamemnon and Niagara, to connect the two halves of the cable together to experiment through the whole length of twenty-five hundred miles for the first time. In going down the side of the Agamemnon I had to cross over several small boats to reach the outer one, which was to take me on board the tug which had the connecting cable on board. In stepping from one to the other of the small boats, the water being very rough and the boats having a good deal of motion, I made a misstep, my right leg being on board the outer boat, and my left leg went down between the two boats scraping the skin from the upper part of the leg near the knee for some two or three inches. It pained me a little, but not much, still I knew from experience that, however slight and comparatively painless at the time,

I should be laid up the next day and possibly for several days.

"My warm-hearted, generous friend, Sir William O'Shaughnessy, was on board, and, being a surgeon, he at once took it in hand and dressed it, tell Susan, in good hydropathic style with cold water. I felt so little inconvenience from it at the time that I assisted throughout the day in laying the cable, and operating through it after it was joined, and had the satisfaction of witnessing the successful result of passing the electricity through twenty-five hundred miles at the rate of one signal in one and a quarter second. Since then Dr. Whitehouse has succeeded in telegraphing a message through it at the rate of a single signal in three quarters of a second. If the cable, therefore, is successfully laid so as to preserve continuity throughout, there is no doubt of our being able to telegraph through, and at a good commercial speed.

"I have been on my back for two days and am still confined to the ship. To-morrow I hope to be well enough to hobble on board the Agamemnon and assist in some experiments."

The accident to his leg was more serious than he at first imagined, and conditions were not improved by his using his leg more than was prudent.

"*August 3, eleven o'clock* A.M. I am still confined, most of the time on my back in my berth, quite to my annoyance in one respect, to wit, that I am unable to be on board the Agamemnon with Dr. Whitehouse to assist at the experiments. Yet I have so much to be thankful for that gratitude is the prevailing feeling.

"*Seven o'clock.* All the ships are under way from the Cove of Cork. The Leopard left first, then the Agamemnon, then the Susquehanna and the Niagara last; and at this moment we are off the Head of Kinsale in the following order: Niagara, Leopard, Agamemnon, Susquehanna. The Cyclops and another vessel, the Advice, left for Valencia on Saturday evening, and, with a beautiful night before us, we hope to be there also by noon to-morrow.

"This day three hundred and sixty-five years ago Columbus sailed on his first voyage of discovery and discovered America."

"*August 4.* Off the Skelligs light, of which I send you a sketch. A beautiful morning with head wind and heavy sea, making many seasick. We are about fifteen miles from our point of destination. Our companion ships are out of sight astern, except the Susquehanna, which is behind us only about a mile. In a few hours we hope to reach our expectant friends in Valencia and to commence the great work in earnest.

"Our ship is crowded with engineers, and operators, and delegates from the Governments of Russia and France, and the deck is a bewildering mass of machinery, steam-engines, cog-wheels, breaks, boilers, ropes of hemp and ropes of wire, buoys and boys, pulleys and sheaves of wood and iron, cylinders of wood and cylinders of iron, meters of all kinds, — anemometers, thermometers, barometers, electrometers, — steam-gauges, ships' logs — from the common log to Massey's log and Friend's log, to our friend Whitehouse's electro-magnetic log, which I think will prove to be the best of all, with a modification I have suggested. Thus freighted we expect to disgorge most of our solid cargo before reaching mid-ocean.

"I am keeping ready to close this at a moment's warning, so give all manner of love to all friends, kisses to whom kisses are due. I am getting almost impatient at the delays we necessarily encounter, but our great work must not be neglected. I have seen enough to know now that the Atlantic Telegraph is sure to be established, *for it is practicable*."

Was it a foreboding of what was to happen that caused him to add: —

"*We may not succeed in our first attempt;* some little neglect or accident may foil our present efforts, but the present enterprise will result in gathering stores of experience which will make the next effort certain. Not that I do not expect success now, but accidental failure now will not be the evidence of its impracticability.

"Our principal electrical difficulty is the slowness with which we must manipulate in order to be intelligible; twenty words in sixteen minutes is now the rate. I am confident we can get more after awhile, but the Atlantic Telegraph has its own rate of talking and cannot be urged to speak faster, any more than any other orator, without danger of becoming unintelligible.

"*Three o'clock* P.M. We are in Valencia Harbor. We shall soon come to anchor. A pilot who has just come to show us our anchorage ground says: 'There are a power of people ashore.'"

"*August 8.* Yesterday, at half past six P.M., all being right, we commenced again paying out the heavy shore-end, of which we had about eight miles to be left on the rocky bottom of the coast, to bear the attrition of the waves and to prevent injury to the delicate nerve which it incloses in its iron mail, and which is the living principle of the whole work. A critical time was approaching, it was when the end of the massive cable should pass overboard at the point where it joins the main and smaller cable. I was in my berth, by order of the surgeon, lest my injured limb, which was somewhat inflamed by the excitement of the day and too much walking about, should become worse.

"Above my head the heavy rumbling of the great wheels, over which the cable was passing and was being regulated, every now and then giving a tremendous thump like the discharge of artillery, kept me from sleep, and I knew they were approaching the critical point. Presently it came. The machinery stopped, and soon amid the voices I heard the unwelcome intelligence — 'The cable is broke.' Sure enough the smaller cable at this point had parted, but, owing to the prudent precautions of those superintending, the end of the great cable had been buoyed and the hawsers which had been attached secured it. The sea was moderate, the moonlight gave a clear sight of all, and in half an hour the joyous sound of 'All right' was heard, the machinery commenced a low and regular rumbling, like the purring of a great cat, which has continued from that moment (midnight) till the present moment uninterrupted.

"The coil on deck is most beautifully uncoiling at the rate of three nautical miles an hour. The day is magnificent, the land has almost disappeared and our companion ships are leisurely sailing with us at equal pace, and we are all, of course, in fine spirits. I sent you a telegraph dispatch this morning, thirty miles out, which you will duly receive with others that I shall send if all continues to go on without interruption. If you do receive any, preserve them with the greatest care, for they will be great curiosities."

"*August 10.* Thus far we have had most delightful weather, and everything goes on regularly and satisfactorily. You are aware we cannot stop night nor day in paying out. On Saturday we made our calculations that the first great coil, which is upon the main deck, would be completely paid out, and one of our critical movements, to wit, the change from this coil to the next, which is far forward, would be made by seven or eight o'clock yesterday morning (Sunday). So we were up and watching the last flake of the first coil gradually diminishing. Everything had been well prepared; the men were at their posts; it was an anxious moment lest a kink might occur. But, as the last round came up, the motion of the ship was slightly slackened, the men handled the slack cable handsomely, and in two minutes the change was made with perfect order, and the paying out from the second coil was as regularly commenced and at this moment continues, and at an increased rate to-day of five miles per hour.

"Last night, however, was another critical moment. On examining our chart of soundings we found the depth of the ocean gradually increasing up to about four hundred fathoms, and then the chart showed a sudden and great increase to seventeen hundred fathoms, and then a further increase to two thousand and fifty, nearly the greatest depth with which we should meet in the whole distance. We had, therefore, to watch the effect of this additional depth upon the straining of the cable. At two in the morning the effect showed itself in a greater strain and a more rapid tendency to run fast. We could check its speed, but it is a dangerous process. *Too sudden a check would inevitably snap the cable.* Too slack a rein would allow of its egress at such a wasting rate and at such a violent speed that we should lose too great a portion of the cable, and its future stopping within controllable limits be almost impossible. Hence our anxiety. All were on the alert; our expert engineers applied the brakes most judiciously, and at the moment I write — latitude 52° 28′ — the cable is being laid at the depth of two miles in its ocean bed as regularly and with as much facility as it was in the depth of a few fathoms. . . .

"*Six* P.M. We have just had a fearful alarm. 'Stop her! Stop her!' was reiterated from many voices on deck. On going up I perceived the cable had got out of its sheaves and was running out at great speed. All was confusion for a few moments. Mr. Canning, our friend, who was the engineer of the Newfoundland cable, showed great presence of mind, and to his coolness and skill, I think, is due the remedying of the evil. By rope stoppers the cable was at length brought to a standstill, and it strained most ominously, perspiring at every part great tar drops. But it held together long enough to put the cable on the sheaves again."

"*Tuesday, August 11.* Abruptly indeed am I stopped in my letter. This morning at 3.45 the cable parted, and we shall soon be on our way back to England."

Thus ended the first attempt to unite the Old World with the New by means of an electric nerve. Authorities differ as to who was responsible for the disaster, but the cause was proved to be what Morse had foreseen when he wrote: "Too sudden a check would inevitably snap the cable."

While, of course, disappointed, he was not discouraged, for under date of August 13, he writes: —

"Our accident will delay the enterprise but will not defeat it. I consider it a settled fact, from all I have seen, that it is perfectly practicable. It will surely be accomplished. There is no insurmountable difficulty that has for a moment appeared, none that has shaken my faith in it in the slightest degree. My report to the company as co-electrician will show everything right in that department. We got an electric current through till the moment of parting, so that electric connection was perfect, and yet the farther we paid out the feebler were the currents, indicating a difficulty which, however, I do not consider serious, while it is of a nature to require attentive investigation."

"*Plymouth, August 17.* Here I am still held by the leg and lying in my berth from which I have not moved for six days. I suffer but little pain unless I attempt to sit up, and the healing process is going on most favorably but slowly. . . . I have been here three days and have not yet had a glimpse of the beautiful country that surrounds us, and if we should be ordered to another port before I can be out I shall have as good an idea of Plymouth as I should have at home looking at a map."

While the wounded leg healed slowly, the plans of the company moved more deliberately still. A movement was on foot for the East India Company to purchase what remained of the cable for use in the Red Sea or the

Persian Gulf, so that the Atlantic Company could start afresh with an entirely new cable, and Morse hoped that this plan might be consummated at an early date so that he could return to America in the Niagara; but the negotiations halted from day to day and week to week. The burden of his letters to his wife is always that a decision is promised by "to-morrow," and finally he says in desperation: "To-day was to-morrow yesterday, but to-day has to-day another to-morrow, on which day, as usual, we are to know something. But as to-day has not yet gone, I wait with some anxiety to learn what it is to bring forth."

His letters are filled with affectionate longing to be at home again and with loving messages to all his dear ones, and at last he is able to say that his wound has completely healed, and that he has decided to leave the Niagara and sail from Liverpool on the Arabia, on September 19, and in due time he arrived at his beloved home on the Hudson.

While still intensely interested in the great cable enterprise, he begins to question the advisability of continuing his connection with the men against whom Mr. Kendall had warned him, for in a letter to his brother Richard, of October 15, 1857, he says: "I intend to withdraw altogether from the Atlantic Telegraph enterprise, as they who are prominent on this side of the water in its interests are using it with all their efforts and influence against my invention, and my interests, and those of my assignees, to whom I feel bound in honor to attach myself, even if some of them have been deceived into coalition with the hostile party."

It was, however, a great disappointment to him that

he was not connected with future attempts to lay the cable. His withdrawal was not altogether voluntary in spite of what he said in the letter from which I have just quoted. While he had been made an Honorary Director of the company in 1857, although not a stockholder, a law was subsequently passed declaring that only stockholders could be directors, even honorary directors. He had not felt financially able to purchase stock, but it was a source of astonishment to him and to others that a few shares, at least, had not been allotted to him for his valuable services in connection with the enterprise. He had, nevertheless, cheerfully given of his time and talents in the first attempt, although cautioned by Mr. Kendall.

He goes fully into the whole matter in a very long letter to Mr. John W. Brett, of December 27, 1858, in which he details his connection with the cable company, his regret and surprise at being excluded on the ground of his not being a stockholder, especially as, on a subsequent visit to Europe, he found that two other men had been made honorary directors, although they were not stockholders. He says that he learned also that "Mr. Field had represented to the Directors that I was hostile to the company, and was using my exertions to defeat the measures for aid from the United States Government to the enterprise, and that it was in consequence of these misrepresentations that I was not elected."

He says farther on: "I sincerely rejoiced in the consummation of the great enterprise, although prevented in the way I have shown from being present. I ought to have been with the cable squadron last summer. It was no fault of mine that I was not there. I hope Mr. Field can exculpate himself in the eyes of the Board, before the world, and before his own conscience, in the course he has taken."

On the margin of the letter-press copy of a letter written to Mr. Kendall on December 22, 1859, is a note in pencil written, evidently, at a later date: "Mr. Field has since manifested by his conduct a different temper. I have long since forgiven what, after all, may have been error of ignorance on his part."

The fact remains, however, that his connection with the cable company was severed, and that his relations with Messrs. Field, Cooper, etc., were decidedly strained. It is more than possible that, had he continued as electrician of the company, the second attempt might have been successful, for he foresaw the difficulty which resulted in failure, and, had he been the guiding mind, it would, naturally, have been avoided. The proof of this is in the following incident, which was related by a friend of his, Mr. Jacob S. Jewett, to Mr. Prime: —

"I thought it might interest you to know when and how Professor Morse received the first tidings of the success of the Atlantic Cable. I accompanied him to Europe on the steamer Fulton, which sailed from New York July 24, 1858. We were nearing Southampton when a sail boat was noticed approaching, and soon our vessel was boarded by a young man who sought an interview with Professor Morse, and announced to him that a message from America had just been received, the first that had passed along the wire lying upon the bed of the ocean.

"Professor Morse was, of course, greatly delighted, but, turning to me, said: '*This is very gratifying, but it is doubtful whether many more messages will be received*';

and gave as his reason that — 'the cable had been so long stored in an improper place that much of the coating had been destroyed, and the cable was in other respects injured.' His prediction proved to be true."

And Mr. Prime adds: "Had he been in the board of direction, had his judgment and experience as electrician been employed, that great calamity, which cost millions of money and eight years of delay in the use of the ocean telegraph, would, in all human probability, have been averted."

But it is idle to speculate on what might have been. His letters show that the action of the directors amazed and hurt him, and that it was with deep regret that he ceased to take an active part in the great enterprise the success of which he had been the first to prophesy.

Many other matters claimed his attention at this time, for, as usual upon returning from a prolonged absence, he found his affairs in more or less confusion, and his time for some months after his return was spent mainly in straightening them out. The winter was spent in New York with his family, but business calling him to Washington, he gives utterance, in a letter to his wife of December 16, to sentiments which will appeal to all who have had to do with the powers that be in the Government service: —

"As yet I have not had the least success in getting a proper position for Charles. A more thankless, repulsive business than asking for a situation under Government I cannot conceive. I would myself starve rather than ask such a favor if I were alone concerned. The modes of obtaining even a hearing are such as to drive a man of any sensitiveness to wish himself in the depths of the forest away from the vicinity of men, rather than encounter the airs of those on their temporary thrones of power. I cannot say what I feel. I shall do all I can, but anticipate no success. . . . I called to see Secretary Toucey for the purpose of asking him to put me in the way of finding some place for Charles, but, after sending in my card and waiting in the anteroom for half to three fourths of an hour, he took no notice of my card, just left his room, passed by deliberately the open door of the anteroom without speaking to me, and left the building. This may be all explained and I will charitably hope there was no intention of rudeness to me, but, unexplained, a ruder slight could not well be conceived."

The affection of the three Morse brothers for each other was unusually strong, and it is from the unreserved correspondence between Finley and Sidney that some of the most interesting material for this work has been gathered. Both of these brothers possessed a keen sense of humor and delighted in playful banter. The following is written in pencil on an odd scrap of paper and has no date: —

"When my brother and I were children my father one day took us each on his knee and said: 'Now I am going to tell you the character of each of you.' He then told us the fable of the Hare and the Tortoise. 'Now,' said he, 'Finley' (that is me), 'you are the Hare and Sidney, your brother, is the Tortoise. See if I am not correct in prophesying your future careers.' So ever since it has been a topic of banter between Sidney and me. Sometimes Sidney seemed to be more prosperous than I; then he would say, 'The old tortoise is ahead.' Then I would take a vigorous run and cry out to him, 'The hare is ahead.' For I am naturally quick and impulsive, and he sluggish and phlegmatic. So I am now going to give him the Hare riding the Tortoise as a piece of fun. Sidney will say: 'Ah! you see the Hare is obliged to ride on the Tortoise in order to get to the goal.' But I shall say: 'Yes, but the Tortoise could not get there unless the Hare spurred him up and guided him.'"

Both of these brothers achieved success, but, unfortunately for the moral of the old fable, the hare quite outdistanced the tortoise, without, however, kindling any spark of jealousy in that faithful heart.

While Sidney was still in Europe his brother writes to him on December 29, 1857: —

"I don't know what you must think of me for not having written to you since my return. It has not been for want of will but truly from the impossibility of withdrawing myself from an unprecedented pressure of more important duties, on which to *write* so that you could form any clear idea of them would be impossible. These duties arise from the state of my affairs thrown into confusion by the conduct of parties intent on controlling all my property. But, I am happy to state, my affairs are in a way of adjustment through the active exertions of my faithful agent and friend, Mr. Kendall, so far as his declining strength permits. . . . I wish you were near me so that we could exchange views on many subjects, particularly on the one which so largely occupies public attention everywhere. I have been collecting works pro and con on the Slavery question with a view of writing upon it. We are in perfect accord, I think, on that subject. I believe that you and I would be considered in New England as rank heretics, for, I confess, the more I study the subject the more I feel compelled to declare myself on the Southern side of the question.

"I care not for the judgment of men, however; I feel on sure ground while standing on Bible doctrine, and I have arrived at the conclusion that a fearful hallucination, not less absurd than that which beclouded some of the most pious and otherwise intelligent minds of the days of Salem witchcraft, has for a time darkened the moral atmosphere of the North."

The event has seemed to prove that it was the Southern sympathizers at the North, those "most pious and otherwise intelligent minds," whose moral atmosphere was darkened by a "fearful hallucination," for no one now claims that slavery is a divine institution because the Bible says, "Slaves, obey your masters."

I have stated that one of the purposes of Morse's visit to Europe in 1856 was to seek to persuade the various Governments which were using his telegraph to grant him some pecuniary remuneration. The idea was received favorably at the different courts, and resulted in a concerted movement initiated by the Count Walewski, representing France, and participated in by ten of the European nations. The sittings of this convention, or congress, were held in Paris from April, 1858, to the latter part of August, and the result is announced in a letter of Count Walewski to Morse of September 1:—

SIR, — It is with lively satisfaction that I have the honor to announce to you that a sum of four hundred thousand francs will be remitted to you, in four annuities, in the name of France, of Austria, of Belgium, of the Netherlands, of Piedmont, of Russia, of the

Holy See, of Sweden, of Tuscany and of Turkey, as an honorary gratuity, and as a reward, altogether personal, of your useful labors. Nothing can better mark than this collective act of reward the sentiment of public gratitude which your invention has so justly excited.

The Emperor has already given you a testimonial of his high esteem when he conferred upon you, more than a year ago, the decoration of a Chevalier of his order of the Legion of Honor. You will find a new mark of it in the initiative which his Majesty wished that his government should take in this conjuncture; and the decision that I charge myself to bring to your knowledge is a brilliant proof of the eager and sympathetic adhesion that his proposition has met with from the states I have just enumerated.

I pray you to accept on this occasion, sir, my personal congratulations, as well as the assurance of my sentiments of the most distinguished consideration.

While this letter is dated September 1, the amount of the gratuity agreed upon seems to have been made known soon after the first meeting of the convention, for on April 29, the following letter was written to Morse by M. van den Broek, his agent in all the preliminaries leading up to the convention, and who, by the way, was to receive as his commission one third of the amount of the award, whatever it might be: "I have this morning seen the secretary of the Minister, and from him learned that the sum definitely fixed is 400,000 francs, payable in four years. This does not by any means answer our expectations, and I am afraid you will be much disappointed, yet I used every exertion in my power, but without avail, to procure a grant of a larger sum."

It certainly was a pitiful return for the millions of dollars which Morse's invention had saved or earned for those nations which used it as a government monopoly, and while I find no note of complaint in his own letters, his friends were more outspoken. Mr. Kendall, in a letter of May 18, exclaims: "I know not how to express my contempt of the meanness of the European Governments in the award they propose to make you as *the* inventor of the Telegraph. I had set the sum at half a million dollars as the least that they could feel to be at all compatible with their dignity. I hope you will acknowledge it more as a tribute to the merits of your invention than as an adequate reward for it."

And in a letter of June 5, answering one of Morse's which must have contained some expressions of gratitude, Mr. Kendall says further: "In reference to the second subject of your letter, I have to say that it is only as a tribute to the superiority of your invention that the European grant can, in my opinion, be considered either 'generous' or 'magnanimous.' As an indemnity it is niggardly and mean."

It will be in place to record here the testimonials of the different nations of Europe to the Inventor of the Telegraph, manifested in various forms: —

France. A contributor to the honorary gratuity, and the decoration of the Legion of Honor.

Prussia. The Scientific Gold Medal of Prussia set in the lid of a gold snuff-box.

Austria. A contributor to the honorary gratuity, and the Scientific Gold Medal of Austria.

Russia. A contributor to the honorary gratuity.

Spain. The cross of Knight Commander de Numero of the order of Isabella the Catholic.

Portugal. The cross of a Knight of the Tower and Sword.

Italy. A contributor to the honorary gratuity, and the cross of a Knight of Saints Lazaro and Mauritio.

Württemberg. The Scientific Gold Medal of Württemberg.

Turkey. A contributor to the honorary gratuity, and the decoration in diamonds of the Nishan Iftichar, or Order of Glory.

Denmark. The cross of Knight Commander of the Dannebrog.

Holy See. A contributor to the honorary gratuity.

Belgium. A contributor to the honorary gratuity.

Holland. A contributor to the honorary gratuity.

Sweden. A contributor to the honorary gratuity.

Great Britain. Nationally nothing.

Switzerland. Nationally nothing.

Saxony. Nationally nothing.

The decorations and medals enumerated above, with the exception of the Danish cross, which had to be returned at the death of the recipient, and one of the medals, which mysteriously disappeared many years ago, are now in the Morse case at the National Museum in Washington, having been presented to that institution by the children and grandchildren of the inventor. It should be added that, in addition to the honors bestowed on him by foreign governments, he was made a member of the Royal Academy of Sciences of Sweden, a member of the Institute of France and of the principal scientific societies of the United States. It has been already noted in these pages that his *alma mater*, Yale, conferred on him the degree of LL.D.

I have said that I find no note of complaint in Morse's letters. Whatever his feelings of disappointment may have been, he felt it his duty to send the following letter to Count Walewski on September 15, 1858. Perhaps a slight note of irony may be read into the sentence accepting the gratuity, but, if intended, I fear it was too feeble to have reached its mark, and the letter is, as a whole and under the circumstances, almost too fulsome, conforming, however, to the stilted style of the time: —

On my return to Paris from Switzerland I have this day received, from the Minister of the United States, the most gratifying information which Your Excellency did me the honor to send to me through him, respecting the decision of the congress of the distinguished diplomatic representatives of ten of the August governments of Europe, held in special reference to myself.

You have had the considerate kindness to communicate to me a proceeding which reflects the highest honor upon the Imperial Government and its noble associates, and I am at a loss for language adequately to express to them my feelings of profound gratitude.

But especially, Your Excellency, do I want words to express towards the august head of the Imperial Government, and to Your Excellency, the thankful sentiments of my heart for the part so prominently taken by His Imperial Majesty, and by Your Excellency, in so generously initiating this measure for my honor in inviting the governments of Europe to a conference on the subject, and for so zealously and warmly advocating and perseveringly conducting to a successful termination, the measure in which the Imperial Government so magnanimously took the initiative.

I accept the gratuity thus tendered, on the basis of an honorary testimonial and a personal reward, with tenfold more gratification than could have been produced by a sum of money, however large, offered on the basis of a commercial negotiation.

I beg Your Excellency to receive my thanks, however inadequately expressed, and to believe that I appreciate Your Excellency's kind and generous services performed in the midst of your high official duties, consummating a proceeding so unique, and in a manner so graceful, that personal kindness has been beautifully blended with official dignity.

I will address respectively to the honorable ministers who were Your Excellency's colleagues a letter of thanks for their participation in this act of high honor to me.

I beg Your Excellency to accept the assurances of my lasting gratitude and highest consideration in subscribing myself

Your Excellency's most obedient humble servant,

SAMUEL F. B. MORSE.

CHAPTER XXXVII

SEPTEMBER 3, 1858 — SEPTEMBER 21, 1863

Visits Europe again with a large family party. — Regrets this. — Sails for Porto Rico with wife and two children. — First impressions of the tropics. — Hospitalities. — His son-in-law's plantation. — Death of Alfred Vail. — Smithsonian exonerates Henry. — European honors to Morse. — First line of telegraph in Porto Rico. — Banquet. — Returns home. — Reception at Poughkeepsie. — Refuses to become candidate for the Presidency. — Purchases New York house. — F. O. J. Smith claims part of European gratuity. — Succeeds through legal technicality. — Visit of Prince of Wales. — Duke of Newcastle. — War clouds. — Letters on slavery, etc. — Matthew Vassar. — Efforts as peacemaker. — Foresees Northern victory. — Gloomy forebodings. — Monument to his father. — Divides part of European gratuity with widow of Vail. — Continued efforts in behalf of peace. — Bible arguments in favor of slavery.

MANY letters of this period, including a whole letter-press copy-book, are missing, many of the letters in other copy-books are quite illegible through the fading of the ink, and others have been torn out (by whom I do not know) and have entirely disappeared. It will, therefore, be necessary to summarize the events of the remainder of the year 1858, and of some of the following years.

We find that, on July 24, 1858, Morse sailed with his family, including his three young boys, his mother-in-law and other relatives, a party of fifteen all told, for Havre on the steamer Fulton; that he was tendered a banquet by his fellow-countrymen in Paris, and that he was received with honor wherever he went. Travelling with a large family was a different proposition from the independence which he had enjoyed on his previous visits to Europe, when he was either alone or accompanied only by his wife and niece, and he pathetically

MORSE AND HIS YOUNGEST SON

remarks to his brother Sidney, in a letter of September 3, written from Interlaken: "It was a great mistake I committed in bringing my family. I have scarcely had one moment's pleasure, and am almost worn out with anxieties and cares. If I get back safe with them to Paris I hope, after arranging my affairs there, to go as direct as possible to Southampton, and settle them there till I sail in November. I am tired of travelling and long for the repose of Locust Grove, if it shall please our Heavenly Father to permit us to meet there again."

Before returning to the quiet of his home on the Hudson, however, he paid a visit which he had long had in contemplation. On November 17, 1858, he and his wife and their two younger sons sailed from Southampton for Porto Rico, where his elder daughter, Mrs. Edward Lind, had for many years lived, and where his younger daughter had been visiting while he was in Europe. He describes his first impressions of a tropical country in a letter to his mother-in-law, Mrs. Griswold, who had decided to spend the winter in Geneva to superintend the education of his son Arthur, a lad of nine: —

"In St. Thomas we received every possible attention. The Governor called on us and invited Edward and myself to breakfast (at 10.30 o'clock) the day we left. He lives in a fine mansion on one of the lesser hills that enclose the harbor, having directly beneath him on the slope, and only separated by a wall, the residence of Santa Anna. He was invited to be present, but he was ill (so he said) and excused himself. I presume his illness was occasioned by the thought of meeting an American from the States, for he holds the citizens of the States in perfect hatred, so much so as to refuse to receive United States money in change from his servants on their return from market.

"A few days in change of latitude make wonderful changes in feelings and clothing. When we left England the air was wintry, and thick woolen clothing and fires were necessary. The first night at sea blankets were in great demand. With two extra and my great-coat over all I was comfortably warm. In twenty-four hours the great-coat was dispensed with, then one blanket, then another, until a sheet alone began to be enough, and the last two or three nights on board this slight covering was too much. When we got into the harbor of St. Thomas the temperature was oppressive; our slightest summer clothing was in demand. Surrounded by pomegranate trees, magnificent oleanders, cocoa-nut trees with their large fruit some thirty feet from the ground, the aloe and innumerable, and to me strange, tropical plants, I could scarcely believe it was December. . . .

"We arrived on Thursday morning and remained until Monday morning, Edward having engaged a Long Island schooner, which happened to be in port, to take us to Arroyo. At four o'clock the Governor sent his official barge, under the charge of the captain of the port, a most excellent, intelligent, scientific gentleman, who had breakfasted with us at the Governor's in the morning, and in a few minutes we were rowed alongside of the schooner Estelle, and before dark were under way and out of the harbor. Our quarters were very small and close, but not so uncomfortable.

"At daylight in the morning of Tuesday we were sailing along the shores of Porto Rico, and at sunrise we found we were in sight of Guyama and Arroyo, and with our glasses we saw at a distance the buildings on Edward's estate. Susan had been advised of our coming and a flag was flying on the house in answer to the signal we made from the vessel. In two or three hours we got to the shore, as near as was safe for the vessel, and then in the doctor's boat, which had paid us an official visit to see that we did not bring yellow fever or other infectious disease, the kind doctor, an Irishman educated in America, took us ashore at a little temporary landing-place to avoid the surf. On the shore there were some handkerchiefs shaking, and in a crowd we saw Susan and Leila, and Charlie [his grandson] who were waiting for us in carriages, and in a few moments we embraced them all. The sun was hot upon us, but, after a ride of two or three miles, we came to the Henrietta, my dear Edward and Susan's residence, and were soon under the roof of a spacious, elegant and most commodious mansion. And here we are with midsummer temperature and vegetation, but a tropical vegetation, all around us.

"Well, we always knew that Edward was a prince of a man, but we did not know, or rather appreciate, that he has a princely estate and in as fine order as any in the island. When I say 'fine order,' I do not mean that it is laid out like the Bois de Boulogne, nor is there quite as much picturesqueness in a level plain of sugar canes as in the trees and shrubbery of the gardens of Versailles; but it is a rich and well-cultivated estate of some fourteen hundred acres, gradually rising for two or three miles from the sea-shore to the mountains, including some of them, and stretching into the valleys between them."

His visit to Porto Rico was a most delightful one to him in many ways, and I shall have more to say of it further on, but I digress for a moment to speak of two events which occurred just at this time, and which showed him that, even in this land of *dolce far niente*, he could not escape the griefs and cares which are common to all mankind.

Mr. Kendall, in a letter of February 20, announces the death of one of his early associates: "I presume you will have heard before this reaches you of the death of Alfred Vail. He had sold most of his telegraph stocks and told me when I last saw him that it was with difficulty he could procure the means of comfort for his family."

Morse had heard of this melancholy event, for, in a letter to Mr. Shaffner of February 22, he says: "Poor Vail! alas, he is gone. I only heard of the event on Saturday last. This death, and the death of many friends besides, has made me feel sad. Vail ought to have a proper notice. He was an upright man, and, although some ways of his made him unpopular with those with whom he came in contact, yet I believe his intentions were good, and his faults were the result more of ill-health, a dyspeptic habit, than of his heart."

He refers to this also in a letter to his brother Sidney of February 23: "Poor Vail is gone. He was the innocent cause of the original difficulty with the sensitive Henry, he all the time earnestly desirous of doing him honor."

And on March 30, he answers Mr. Kendall's letter:

"I regret to learn that poor Vail was so straitened in his circumstances at his death. I intend paying a visit to his father and family on my return. I may be able to relieve them in some degree."

This intention he fulfilled, as we shall see later on, and I wish to call special attention to the tone of these letters because, as I have said before, Morse has been accused of gross ingratitude and injustice towards Alfred Vail, whereas a careful and impartial study of all the circumstances of their connection proves quite the contrary. Vail's advocates, in loudly claiming for him much more than the evidence shows he was entitled to, have not hesitated to employ gross personal abuse of Morse in their newspaper articles, letters, etc., even down to the present day. This has made my task rather difficult, for, while earnestly desirous of giving every possible credit to Vail, I have been compelled to introduce much evidence, which I should have preferred to omit, to show the essential weakness of his character; he seems to have been foredoomed to failure. He undoubtedly was of great assistance in the early stages of the invention, and for this Morse always cheerfully gave him full credit, but I have proved that he did not invent the dot-and-dash alphabet, which has been so insistently claimed for him, and that his services as a mechanician were soon dispensed with in favor of more skilful men. I have also shown that he practically left Morse to his fate in the darkest years of the struggle to bring the telegraph into public use, and that, by his morbid suspicions, he hampered the efforts of Mr. Kendall to harmonize conflicting interests. For all this Morse never bore him any ill-will, but endeavored in every way to foster and safeguard his interests. That he did not succeed was no fault of his.

Another reminder that he was but human, and that he could not expect to sail serenely along on the calm seas of popular favor without an occasional squall, was given to him just at this time. Professor Joseph Henry had requested the Regents of the Smithsonian Institute to enquire into the rights and wrongs of the controversy between himself and Morse, which had its origin in Henry's testimony in the telegraph suits, tinged as this testimony was with bitterness on account of the omissions in Vail's book, and which was fanned into a flame by Morse's "Defense." The latter resented the fact that all these proceedings had taken place while he was out of the country, and without giving him an opportunity to present his side of the case. However, he shows his willingness to do what is right in the letter to Colonel Shaffner of February 22, from which I have already quoted: —

"Well, it has taken him four years to fire off his gun, and perhaps I am killed. When I return I shall examine my wounds and see if they are mortal, and, if so, shall endeavor to die becomingly. Seriously, however, if there are any new facts which go to exculpate Henry for his attack upon me before the courts at a moment when I was struggling against those who, from whatever motive, wished to deprive me of my rights, and even of my character, I shall be most happy to learn them, and, if I have unwittingly done him injustice, shall also be most happy to make proper amends. But as all this is for the future, as I know of no facts which alter the case, and as I am wholly unconscious of having done him any injustice, I must wait to see what he has put forth."

In a letter to his brother Sidney, of February 23, he philosophizes as follows: —

"I cannot avoid noticing a singular coincidence of events in my experience of life, especially in that part of it devoted to the invention of the Telegraph, to wit, that, when any special and marked honor has been conferred upon me, there has immediately succeeded some event of the envious or sordid character seemingly as a set-off, the tendency of which has been invariably to prevent any excess of exultation on my part. Can this be accident? Is it not rather the wise ordering of events by infinite wisdom and goodness to draw me away from repose in earthly honor to the more substantial and enduring honor that comes only from God? . . . I pray for wisdom to direct in such trials, and in any answer I may find it necessary to give to Henry or others, I desire most of all to be mindful of that charity which 'suffereth long, which vaunteth not itself, is not puffed up, hopeth all things, thinketh no evil.'"

This check to self-laudation came at an appropriate moment, as he said, for just at this time honors were being plentifully showered upon him. It was then that he was first notified of the bestowal of the Spanish decoration, and of the probability of Portugal's following suit. Perhaps even more gratifying still was his election as a member of the Royal Academy of Sciences of Sweden, for this was a recognition of his merits as a scientist, and not as a mere promoter, as he had been contemptuously called. On the Island of Porto Rico too he was being honored and fêted. On March 2, he writes: —

"I have just completed with success the construction and organization of the short telegraph line, the first on this island, initiating the great enterprise of the Southern Telegraph route to Europe from our shores, so far as to interest the Porto Ricans in the value of the invention.

"Yesterday was a day of great excitement here for this small place. The principal inhabitants of this place and Guayama determined to celebrate the completion of this little line, in which they take a great pride as being the first in the island, and so they complimented me with a public breakfast which was presided over by the lieutenant-colonel commandant of Guayama.

"The commandant and alcalde, the collector and captain of the port, with all the officials of the place, and the clergy of Guayama and Arroyo, and gentlemen planters and merchants of the two towns, numbering in all about forty, were present. We sat down at one o'clock to a very handsome breakfast, and the greatest enthusiasm and kind and generous feeling were manifested. My portrait was behind me upon the wall draped with the Spanish and American flags. I gave them a short address of thanks, and took the opportunity to interest them in the great Telegraph line which will give them communication with the whole world. I presume accounts will be published in the United States from the Porto Rico papers. Thus step by step (shall I not rather say *stride* by *stride?*) the Telegraph is compassing the world.

"My accounts from Madrid assure me that the government will soon have all the papers prepared for granting the concession to Mr. Perry, our former secretary of legation at Madrid, in connection with Sir James Carmichael, Mr. John W. Brett, the New York, Newfoundland and London Telegraph Company, and others. The recent consolidation plan in the United States has removed the only hesitation I had in sustaining this new enterprise, for I feared that I might unwittingly injure, by a counter plan, those it was my duty to support. Being now in harmony with the American Company and the Newfoundland Company, I presume all my other companies will derive benefit rather than injury from the success of this new and grand enterprise. At any rate I feel impelled to support all plans that manifestly tend to the complete circumvention of the globe, and the bringing into telegraphic connection all the nations of the earth, and this when I am not fully assured that present personal interests may not temporarily suffer. I am glad to know that harmonious arrangements are made between the various companies in the United States, although I have been so ill-used. I will have no litigation if I can avoid it. Even Henry may have the field in quiet, unless he has presented a case too flagrantly unjust to leave unanswered."

The short line of telegraph was from his son-in-law's house to his place of business on the bay, about two miles, and the building of it gave rise to the legend on the island that Morse conducted some of his first electrical experiments in Porto Rico, which, of course, is not true.

There is much correspondence concerning the proposed cable from Spain or Portugal by various routes to the West Indies and thence to the United States, but nothing came of it.

The rest of their stay in Porto Rico was greatly enjoyed by all in spite of certain drawbacks incidental to the tropics, to one of which he alludes in a letter to his sister-in-law, Mrs. Goodrich, who was then in Europe. Speaking of his wife he says: "She is dreadfully troubled with a plague which, if you have been in Italy, I am sure you are no stranger to. '*Pulci, pulci.*' If you have not had a colony of them settled upon you, and quartered, and giving you no quarter, you have been an exception to travellers in Italy. Well, I will pit any two *pulci* of Porto Rico against any ten you can bring from Italy, and I should be sure to see them bite the dust before the bites of our Porto Rico breed."

His letters are filled with apothegms and reflections on life in general and his own in particular, and they alone would almost fill a book. In a letter to Mr. Kendall, of March 30, we find the following: —

"I had hoped to return from honors abroad to enjoy a little rest from litigation at home, but, if I must take up arms, I hope to be able to use them efficiently in self-defense, and in a chivalrous manner as becometh a '*Knight.*' I have no reason to complain of my position abroad, but I suppose, as I am not yet under the ground, honors to a living inventor must have their offset in the attacks of envy and avarice.

"'Wrath is cruel, but who can stand before envy?' says the wise man. The contest with the envious is indeed an annoyance, but, if one's spirit is under the right guidance and revenge does not actuate the strife, victory is very certain. My position is now such before the world that I shall use it rather to correct my own temper than to make it a means of arrogant exultation."

He and his family left the island in the middle of April, 1859, and in due time reached their Poughkeepsie home. The "Daily Press" of that city gave the following account of the homecoming: —

"For some time previous to the hour at which the train was to arrive hundreds of people were seen flocking from all directions to the railroad depot, both in carriages and on foot, and when the train did arrive, and the familiar and loved form of Professor Morse was recognized on the platform of the car, the air was rent with the cheers of the assembled multitude. As soon as the cheers subsided Professor Morse was approached by the committee of reception and welcomed to the country of his birth and to the home of his adoption.

"A great procession was then formed composed of the carriages of citizens. The sidewalks were crowded with people on foot, the children of the public schools, which had been dismissed for the occasion, being quite conspicuous among them. Amid the ringing of bells, the waving of flags, and the gratulations of the people, the procession proceeded through a few of the principal streets, and then drove to the beautiful residence of Professor Morse, the band playing, as they entered the grounds, 'Sweet Home' and then 'Auld Lang Syne.'

"The gateways at the entrance had been arched with evergreens and wreathed with flowers. As the carriage containing their loved proprietor drove along the gravelled roads we noticed that several of the domestics, unable to restrain their welcomes, ran to his carriage and gave and received salutations. After a free interchange of salutations and a general 'shake-hands,' the people withdrew and left their honored guest to the retirement of his own beautiful home.

"So the world reverences its great men, and so it ought. In Professor Morse we find those simple elements of greatness which elevate him infinitely above the hero of any of the world's sanguinary conflicts, or any of the most successful aspirants after political power. He has benefited not only America and the world, but has dignified and benefited the whole race."

His friends and neighbors desired to honor him still further by a public reception, but this he felt obliged to decline, and in his letter of regret he expresses the following sentiments: "If, during my late absence abroad, I have received unprecedented honors from European nations, convened in special congress for the purpose, and have also received marks of honor from individual Sovereigns and from Scientific bodies, all which have gratified me quite as much for the honor reflected by them upon my country as upon myself, there are none of these testimonials, be assured, which have so strongly touched my heart as this your beautiful tribute of kindly feeling from esteemed neighbors and fellow-citizens."

Among the letters which had accumulated during his absence, Morse found one, written some time previously, from a Mr. Reibart, who had published his name as a candidate for the Presidency of the United States. In courteously declining this honor Morse drily adds: "There are hundreds, nay thousands, more able (not to say millions more willing) to take any office they can obtain, and perform its functions more faithfully and with more benefit to the country. While this is the case I do not feel that the country will suffer should one like myself, wearied with the struggles and litigations of half a century, desire to be excused from encountering the annoyances and misapprehensions inseparable from political life."

Thanks to the successful efforts of his good friend, Mr. Kendall, he was now financially independent, so much so that he felt justified in purchasing, in the fall of the year 1859, the property at 5 West Twenty-second Street, New York, where the winters of the remaining years of his life were passed, except when he was abroad. This house has now been replaced by a commercial structure, but a bronze tablet marks the spot where once stood the old-fashioned brown stone mansion.

While his mind was comparatively at rest regarding money matters, he was not yet free from vexatious litigation, and his opinion of lawyers is tersely expressed in a letter to Mr. Kendall of December 27, 1859: "I have not lost my respect for law but I have for its administrators; not so much for any premeditated dishonesty as for their stupidity and want of just insight into a case."

It was not long before he had a practical proof of the truth of this aphorism, for his "thorn in the flesh" never ceased from rankling, and now gave a new instance of the depths to which an unscrupulous man could descend. On June 9, 1860, Morse writes to his legal adviser, Mr. George Ticknor Curtis, of Boston: "You may remember that Smith, just before I sailed for Europe in 1858, intimated that he should demand of me a portion of the Honorary Gratuity voted to me by the congress of ten powers at Paris. I procured your opinion, as you know, and I had hoped that he would not insist on so preposterous a claim. I am, however, disappointed; he has recently renewed it. I have had some correspondence with him on the subject utterly denying any claim on his part. He proposes a reference, but I have not yet encouraged him to think I would assent. I wish your advice before I answer him."

It is difficult to conceive of a meaner case of extortion than this. As Morse says in a letter to Mr. Kendall, of August 3, 1860, after he had consented to a reference of the matter to three persons: "I have no apprehensions of the result except that I may be entrapped by some legal technicalities. Look at the case in an equitable point of view and, it appears to me, no intelligent, just men could give a judgment against me or in his favor. Smith's purchase into the telegraph, the consideration he gave, was his efforts to obtain a property in the invention abroad by letters patent or otherwise. In *such* property he was to share. No such property was created there. What can he then claim? The monies that he hazarded (taking his own estimate) were to the amount of some seven thousand dollars; and this was an advance, virtually a loan, to be paid back to him if he had created the property abroad. But his efforts being fruitless for that purpose, and of no value whatever to me, yet procured him one fourth patent interest in the United States, for which we know he has obtained at least $300,000. Is he not paid amply without claiming a portion of honorary gifts to me? Well, we shall see how legal men look at the matter."

One legal man of great brilliance gave his opinion

HOUSE AND LIBRARY AT 5 WEST 22D ST., NEW YORK

without hesitation, as we learn from a letter of Morse's to Mr. Curtis, of July 14: "I had, a day or two since, my cousin Judge Breese, late Senator of the United States from Illinois, on a visit to me. I made him acquainted with the points, after which he scouted the idea that any court of legal character could for a moment sustain Smith's claim. He thought my argument unanswerable, and playfully said: 'I will insure you against any claim from Smith for a bottle of champagne.'"

It is a pity that Morse did not close with the offer of the learned judge, for, in spite of his opinion, in spite of the opinion of most men of intelligence, in defiance of the perfectly obvious and proven fact that Smith had utterly failed in fulfilling his part of the contract, and that the award had been made to Morse "as a reward altogether personal" (*toute personelle*), the referees decided in Smith's favor. And on what did they base this remarkable decision? On the ground that in the contract of 1838 with Smith the word "otherwise" occurs. Property in Europe was to be obtained by "letters patent" or "otherwise." Of course no actual property had been obtained, and Smith had had no hand in securing the honorary gratuity, and it is difficult to follow the reasoning of these sapient referees. They were, on Smith's part, Judge Upham of New Hampshire; on Morse's, Mr. Hilliard, of Boston; and Judge Sprague, of the Circuit Court, Boston, chairman.

However, the decision was made, and Morse, with characteristic large-heartedness, submitted gracefully. On October 15, he writes to Mr. Curtis: "I ought, perhaps, with my experience to learn for the first time that

Law and *Justice* are not synonyms, but, with all deference to the opinion of the excellent referees, for each of whom I have the highest personal respect, I still think that they have not given a decision in strict conformity with Law. . . . I submit, however, to law with kindly feelings to all, and now bend my attention to repair my losses as best I may."

As remarked before, earlier in this volume, Morse, in his correspondence with Smith, always wrote in that courteous manner which becomes a gentleman, and he expresses his dissent from the verdict in this manner in a letter of November 20, in answer to one of Smith's, quibbling over the allowance to Morse by the referees of certain expenses: "Throwing aside as of no avail any discussion in regard to the equity of the decision of the referees, especially in the view of a conscientious and high-minded man, I now deal with the decision as it has been made, since, according to the technicalities of the law, it has been pronounced by honorable and honest men in accordance with their construction of the language of the deed in your favor. But 'He that 's convinced against his will is of the same opinion still,' and in regard to the intrinsic injustice of being compelled, by the strict construction of a general word, to pay over to you any portion of that which was expressly given to me as a personal and honorary *gratuity* by the European governments, my opinion is always as it has been, an opinion sustained by the sympathy of every intelligent and honorable man who has studied the merits of the case."

He was hard hit for a time by this unjust decision, and his correspondence shows that he regretted it most because it prevented him from bestowing as much in good works as he desired. He was obliged to refuse many requests which strongly appealed to him. His daily mail contained numerous requests for assistance in sums "from twenty thousand dollars to fifty cents," and it was always with great reluctance that he refused anybody anything.

However, as is usual in this life, the gay was mingled with the grave, and we find that he was one of the committee of prominent men to arrange for the entertainment of the Prince of Wales, afterward Edward VII, on his visit to this country. I have already referred to one incident of this visit when Morse, in an address to the Prince at the University of the City of New York, referred to the kindness shown him in London by the Earl of Lincoln, who was now the Duke of Newcastle and was in the suite of the Prince. Morse had hoped that he might have the privilege of entertaining H.R.H. at his country place on the Hudson, but the Duke of Newcastle, in a letter of October 8, 1860, regrets that this cannot be managed: —

I assure you I have not forgotten the circumstances which gave me the pleasure of your acquaintance in 1839, and I am very desirous of seeing you again during my short visit to this continent. I fear however that a visit by the Prince of Wales to your home, however I might wish it, is quite impracticable, although on our journey up the Hudson we shall pass so near you. Every hour of our time is fully engaged.

Is there any chance of seeing you in New York, or, if not, is there any better hope in Boston? If you should be in either during our stay, I hope you will be kind enough to call upon me. Pray let me have a line on Thursday at New York. I have lately been much interested in some electro-telegraphic inventions of yours which are new to me.

I am

Yours very truly,

NEWCASTLE.

Referring to another function in honor of the Prince, Morse says, in a letter to Mr. Kendall: "I did not see you after the so-styled Ball in New York, which was not a *ball* but a *levee* and a great jam. I hope you and yours suffered no inconvenience from it."

The war clouds in his beloved country were now lowering most ominously, and, true to his convictions, he exclaims in a letter to a friend of January 12, 1861: —

"Our politicians are playing with edged tools. It is easy to raise a storm by those who cannot control it. If I trusted at all in them I should despair of the country, but an Almighty arm makes the wrath of man to praise him, and he will restrain the rest. There is something so unnatural and abhorrent in this outcry of *arms* in one great family that I cannot believe it will come to a decision by the sword. Such counsels of force are in the court of passion, not of reason. Imagine such a conflict, imagine a victory, no matter by which side. Can the victors rejoice in the blood of brethren shed in a family brawl? Whose heart will thrill with pride at such success? No, no. I should as soon think of rejoicing that one of my sons had killed the other in a brawl.

"But I have not time to add. I hope for the best, and even can see beyond the clouds of the hour a brighter day. God bless the whole family, North, South, East and West. I will never divide them in my heart however they may be politically or geographically divided."

His hopes of a peaceful solution of the questions at issue between the North and the South were, of course, destined to be cruelly dashed, and he suffered much during the next few years, both in his feelings and in his purse, on account of the war. I have already shown that he, with many other pious men, believed that slavery was a divine institution and that, therefore, the abolitionists were entirely in the wrong; but that, at the same time, he was unalterably opposed to secession. Holding these views, he was misjudged in both sections of the country. Those at the North accused him of being a secessionist because he was not an abolitionist, and many at the South held that he must be an abolitionist because he lived at the North and did not believe in the doctrine of secession. Many pages of his letter-books are filled with vehement arguments upholding his point of view, and he, together with many other eminent men at the North, strove without success to avert the war. His former pastor at Poughkeepsie, the Reverend H. G. Ludlow, in long letters, with many Bible quotations, called upon him to repent him of his sins and join the cause of righteousness. He, in still longer letters, indignantly repelled the accusation of error, and quoted chapter and verse in support of his views. He was made the president of The American Society for promoting National Unity, and in one of his letters to Mr. Ludlow he uses forceful language: —

"The tone of your letter calls for extraordinary drafts on Christian charity. Your criticism upon and denunciation of a society planned in the interests of peace and good will to all, inaugurated by such men as Bishops McIlvaine and Hopkins, Drs. Krebs and Hutton, and Winslow, and Bliss, and Van Dyke, and Hawks, and Seabury, and Lord and Adams of Boston, and Wilson the missionary, and Styles and Boorman, and Professor Owen, and President Woods, and Dr. Parker, and my brothers, and many others as warm-hearted, praying, conscientious Christians as ever assembled to devise means for promoting peace — denunciations of these and such as these cannot but be painful in the highest degree. . . . I lay no stress upon these names other than to show that conscience in this matter has moved some Christians quite as strongly to view *Abolitionism* as a sin of the deepest dye, as it has other Christian minds to view Slavery as a sin, and so to condemn slaveholders to excommunication, and simply for being slaveholders.

"Who is to decide in a conflict of consciences? If the Bible be the umpire, as I hold it to be, then it is the Abolitionist that is denounced as worthy of excommunication; it is the Abolitionist from whom we are commanded to withdraw ourselves, while not a syllable of reproof do I find in the sacred volume administered to those who maintain, in the spirit of the gospel, the relation of *Masters and Slaves.* If you have been more successful, please point out chapter and verse. . . . I have no justification to offer for Southern *secession;* I have always considered it a remedy for nothing. It is, indeed, an expression of a sense of wrong, but, in turn, is itself a wrong, and two wrongs do not make a right."

I have quoted thus at some length from one of his

many polemics to show the absolute and fearless sincerity of the man, mistaken though he may have been in his major premise.

I shall quote from other letters on this subject as they appear in chronological order, but as no person of any mental caliber thinks and acts continuously along one line of endeavor, so will it be necessary in a truthful biography to change from one subject of activity to another, and then back again, in order to portray in their proper sequence the thoughts and actions of a man which go to make up his personality. For instance, while the outspoken views which Morse held on the subjects of slavery and secession made him many enemies, he was still held in high esteem, for it was in the year 1861 that the members of the National Academy of Design urged him so strongly to become their president again that he yielded, but on condition that it should be for one year only. And the following letter to Matthew Vassar, of Poughkeepsie, dated February 1, 1861, shows that he was actively interested in the foundation of the first college for women in this country: "Your favor of the 24th ulto. is received, and so far as I can further your magnificent and most generous enterprise, I will do so. I will endeavor to attend the meeting at the Gregory House on the 26th of the present month. May you long live to see your noble design in successful operation."

In spite of his deep anxiety for the welfare of his country, and in spite of the other cares which weighed him down, he could not resist the temptation to indulge in humor when the occasion offered. This humor is tinged with sarcasm in a letter of July 13, 1861, to

Mr. A. B. Griswold, his wife's brother, a prominent citizen of New Orleans. After assuring him of his undiminished affection, he adds: —

"And now see what a risk I have run by saying thus much, for, according to modern application of the definition of *treason*, it would not be difficult to prove me a traitor, and therefore amenable to the halter.

"For instance — treason is giving aid and comfort to the enemy; everybody south of a certain geographical line is an enemy; you live south of that line, ergo you are an enemy; I send you my love, you being an enemy; this gives you *comfort;* ergo, I have given comfort to the enemy; ergo, I am a traitor; ergo, I must be hanged."

As the war progressed he continued to express himself in forcible language against what he called the "twin heresies" — abolitionism and secession. He had done his best to avert the war. He describes his efforts in a letter of April 2, 1862, to Mr. George L. Douglas, of Louisville, Kentucky, who at that time was prominently connected with the Southern lines of the telegraph, and who had loyally done all in his power to safeguard Morse's interests in those lines: —

"You are correct in saying, in your answer as garnishee, that I have been an active and decided friend of Peace. In the early stages of the troubles, when the Southern Commissioners were in Washington, I devoted my time and influence and property, subscribing and paying in the outset five hundred dollars, to set on foot measures for preserving peace honorable to all parties. The attack on Fort Sumter struck down all these efforts (so far as my associates were concerned), but I was not personally discouraged, and I again addressed myself to the work of the Peacemaker, determining to visit *personally* both sections of the country, the Government at Washington, and the Government of the Confederates at Richmond, to ascertain if there were, by possibility, any means of averting war. And when, from physical inability and age, I was unable to undertake the duty personally, I defrayed from my own pocket the expenses of a friend in his performance of the same duties for me, who actually visited both Washington and Richmond and conferred with the Presidents and chiefs of each section on the subject. True his efforts were unsuccessful, and so nothing remained for me but to retire to the quiet of my own study and watch the vicissitudes of the awful storm which I was powerless to avert, and descry the first signs of any clearing up, ready to take advantage of the earliest glimmerings of light through the clouds."

He had no doubts as to the ultimate issue of the conflict, for, in a letter to his wife's sister, Mrs. Goodrich, of May 2, 1862, he reduces it to mathematics: —

"Sober men could calculate, and did calculate, the *military* issue, for it was a problem of mathematics and not at all of individual or comparative courage. A force of equal quality is to be divided and the two parts to be set in opposition to each other. If equally divided, they will be at rest; if one part equals 3 and the other 9, it does not require much knowledge of mathematics to decide which part will overcome the force of the other.

"Now this is the case here just now. Two thirds of the physical and material force of the country are at the North, and on this account *military* success, other things being equal, must be on the side of the North. Courage, justness of the cause, right, have nothing to do with it. War in our days is a game of chess. Two players being equal, if one begins the game with dispensing with a third of his best pieces, the other wins as a matter of course."

He was firmly of the opinion that England and other European nations had fomented, if they had not originated, the bad feeling between the North and the South, and at times he gave way to the most gloomy forebodings, as in a letter of July 23, 1862, to Mr. Kendall, who shared his views on the main questions at issue: —

"I am much depressed. There is no light in the political skies. Rabid abolitionism, with its intense, infernal hate, intensified by the same hate from secession quarters, is fast gaining the ascendancy. Our country is dead. God only can resuscitate it from its tomb. I see no hope of union. We are two countries, and, what is most deplorable, two hostile countries. Oh! how the nations, with England at their head, crow over us. It is the hour of her triumph; she has conquered by her arts that which she failed to do by her arms. If there was a corner of the world where I could hide myself, and I could consult the welfare of my family, I would sacrifice all my interests here and go at once. May God save us with his salvation. I have no heart to write or to do anything. Without a country! Without a country!"

He went even further, in one respect, in a letter to Mr. Walker, of Utica, of October 27, but his ordinarily keen prophetic vision was at fault: "Have you made up your mind to be under a future monarch, English or

French, or some scion of a European stock of kings? I shall not live to see it, I hope, but you may and your children will. I leave you this prophecy in black and white."

In spite of his occasional fits of pessimism he still strove with all his might, by letters and published pamphlets, to rescue his beloved country from what he believed were the machinations of foreign enemies. At the same time he did not neglect his more immediate concerns, and his letter-books are filled with loving admonitions to his children, instructions to his farmer, answers to inventors seeking his advice, or to those asking for money for various causes, etc.

He and his two brothers had united in causing a monument to be erected to the memory of their father and mother in the cemetery at New Haven, and he insisted on bearing the lion's share of the expense, as we learn from a letter written to his nephew, Sidney E. Morse, Jr., on October 10, 1862: —

"Above you have my check on Broadway Bank, New York, for five hundred dollars towards Mr. Ritter's bill.

"Tell your dear father and Uncle Sidney that this is the portion of the bill for the monument which I choose to assume. Tell them I have still a good memory of past years, when I was poor and received from them the kind attentions of affectionate brothers. I am now, through the loving kindness and bounty of our Heavenly Father, in such circumstances that I can afford this small testimonial to their former fraternal kindness, and I know no better occasion to manifest the long pent-up feelings of my heart towards them than by lightening, under the embarrassments of the times, the pecuniary burden of our united testimonial to the best of fathers and mothers."

This monument, a tall column surmounted by a terrestrial globe, symbolical of the fact that the elder Morse was the first American geographer, is still to be seen in the New Haven cemetery.

Another instance of the inventor's desire to show his gratitude towards those who had befriended him in his days of poverty and struggle is shown in a letter of November 17, 1862, to the widow of Alfred Vail: —

"You are aware that a sum of money was voted me by a special Congress, convened at Paris for the purpose, as a personal, honorary gratuity as the Inventor of the Telegraph. . . . Notwithstanding, however, that the Congress had put the sum voted me on the ground of a personal, honorary gratuity, I made up my mind in the very outset that I would divide to your good husband just that proportion of what I might receive (after due allowance and deduction of my heavy expenses in carrying through the transaction) as would have been his if the money so voted by the Congress had been the purchase money of patent rights. This design I early intimated to Mr. Vail, and I am happy in having already fulfilled in part my promise to him, when I had received the gratuity only in part. It was only the last spring that the whole sum, promised in four annual instalments (after the various deductions in Europe) has been remitted to me. . . . I wrote to Mr. Cobb [one of Alfred Vail's executors] some months ago, while he was in Washington, requesting an early interview to pay over the balance for you, but have never received an answer. . . . Could you not come to town this week, either with or without Mr. Cobb, as is most agreeable to you, prepared to settle this matter in full? If so, please drop me a line stating the day and hour you will come, and I will make it a point to be at home at the time."

In this connection I shall quote from a letter to Mr. George Vail, written much earlier in the year, on May 19: —

"It will give me much pleasure to aid you in your project of disposing of the '*original wire*' of the Telegraph, and if my certificate to its genuineness will be of service, you shall cheerfully have it. I am not at this moment aware that there is any quantity of this wire anywhere else, except it may be in the helices of the big magnets which I have at Poughkeepsie. These shall not interfere with your design.

"I make only one modification of your proposal, and that is, if any profits are realized, please substitute for my name the name of your brother Alfred's amiable widow."

Although the malign animosity of F. O. J. Smith followed him to his grave, and even afterwards, he was, in this year of 1862, relieved from one source of annoyance from him, as we learn from a letter of May 19 to Mr. Kendall: "I have had a settlement with Smith in full on the award of the Referees in regard to the 'Honorary Gratuity,' and with less difficulty than I expected."

Morse had now passed the Scriptural age allotted to man; he was seventy-one years old, and, in a letter of August 22, he remarks rather sorrowfully: "I feel that I am no longer young, that my career, whether for good or evil, is near its end, but I wish to give the energy and influence that remain to me to my country, to save it, if possible, to those who come after me."

All through the year 1863 he labored to this end, with alternations of hope and despair. On February 9, 1863, he writes to his cousin, Judge Sidney Breese: "A movement is commenced in the formation of a society here which promises good. It is for the purpose of Diffusing Useful Political Knowledge. It is backed up by millionaires, so far as funds go, who have assured us that funds shall not be wanting for this object. They have made me its president."

Through the agency of this society he worked to bring about "Peace with Honor," but, as one of their cardinal principles was the abandonment of abolitionism, he worked in vain. He bitterly denounced the Emancipation Proclamation, and President Lincoln came in for many hard words from his pen, being considered by him weak and vacillating. Mistaken though I think his attitude was in this, his opinions were shared by many prominent men of the day, and we must admit that for those who believed in a literal interpretation of the Bible there was much excuse. For instance, in a letter of September 21, 1863, to Martin Hauser, Esq., of Newbern, Indiana, he goes rather deeply into the subject: —

"Your letter of the 23d of last month I have just received, and I was gratified to see the evidences of an upright, honest dependence upon the only standard of right to which man can appeal pervading your whole letter. There is no other standard than the Bible, but our translation, though so excellent, is defective sometimes in giving the true meaning of the original languages in which the two Testaments are written; the Old Testament in Hebrew, the New Testament in Greek. Therefore it is that in words in the English translation about which there is a variety of opinion, it is necessary to examine the original Hebrew or Greek to know what was the meaning attached to these words by the writers of the original Bible. . . . I make these observations to introduce a remark of yours that the Bible does not contain anything like slavery in it because the words 'slave' and 'slavery' are not used in it (except the former twice) but that the word 'servant' is used.

"Now the words translated 'servant' in hundreds of instances are, in the original, 'slave,' and the very passage you quote, Noah's words — 'Cursed be Canaan, a servant of servants shall he be unto his brethren' — in the original Hebrew means exactly this — 'Cursed be Canaan, a *slave* of *slaves* shall he be.' The Hebrew word is '*ebed*,' which means a bond slave, and the words '*ebed ebadim*' translated 'slave of slaves,' means strictly *the most abject of slaves.*

"In the New Testament too the word translated 'servant' from the Greek is '*doulos*,' which is the same as '*ebed*' in the Hebrew, and always means a bond slave. Our word 'servant' formerly meant the same, but time and custom have changed its meaning with us, but the Bible word '*doulos*' remains the same, 'a slave.'"

It seems strange that a man of such a gentle, kindly disposition should have upheld the outworn institution of slavery, but he honestly believed, not only that it was ordained of God, but that it was calculated to benefit the enslaved race. To Professor Christy, of Cincinnati, he gives, on September 12, his reasons for this belief: —

"You have exposed in a masterly manner the fallacies of Abolitionism. There is a complete coincidence of views between us. My 'Argument,' which is nearly ready for the press, supports the same view of the necessity of slavery to the christianization and civilization of a barbarous race. My argument for the benevolence of the relation of master and slave, drawn from the four relations ordained of God for the organization of the social system (the fourth being the servile relation, or the relation of master and slave) leads conclusively to the recognition of some great benevolent design in its establishment.

"But you have demonstrated in an unanswerable manner by your statistics this benevolent design, bringing out clearly, from the workings of his Providence, the absolute necessity of this relation in accomplishing his gracious designs towards even the lowest type of humanity."

CHAPTER XXXVIII

FEBRUARY 26, 1864 — NOVEMBER 8, 1867

Sanitary Commission. — Letter to Dr. Bellows. — Letter on "loyalty." — His brother Richard upholds Lincoln. — Letters of brotherly reproof. — Introduces McClellan at preëlection parade. — Lincoln reëlected. — Anxiety as to future of country. — Unsuccessful effort to take up art again. — Letter to his sons. — Gratification at rapid progress of telegraph. — Letter to George Wood on two great mysteries of life. — Presents portrait of Allston to the National Academy of Design. — Endows lectureship in Union Theological Seminary. — Refuses to attend fifty-fifth reunion of his class. — Statue to him proposed. — Ezra Cornell's benefaction. — American Asiatic Society. — Amalgamation of telegraph companies. — Protest against stock manipulations. — Approves of President Andrew Johnson. — Sails with family for Europe. — Paris Exposition of 1867. — Descriptions of festivities. — Cyrus W. Field. — Incident in early life of Napoleon III. — Made Honorary Commissioner to Exposition. — Attempt on life of Czar. — Ball at Hôtel de Ville. — Isle of Wight. — England and Scotland. — The "Sounder." — Returns to Paris.

ALL the differences of those terrible years of fratricidal strife, all the heart-burnings, the bitter animosities, the family divisions, have been smoothed over by the soothing hand of time. I have neither the wish nor the ability to enter into a discussion of the rights and the wrongs of the causes underlying that now historic conflict, nor is it germane to such a work as this. While Morse took a prominent part in the political movements of the time, while he was fearless and outspoken in his views, his name is not now associated historically with those epoch-making events. It has seemed necessary, however, to make some mention of his convictions in order to make the portrait a true one. He continued to oppose the measures of the Administration; he did all in his power to hasten the coming of peace; he worked and voted for the election of McClellan to the Presidency, and when he and the other eminent men who believed as he did were outvoted, he bowed to the will of the majority with many misgivings as to the future. Although he was opposed to the war his heart bled for the wounded on both sides, and he took a prominent part in the National Sanitary Commission. He expresses himself warmly in a letter of February 26, 1864, to its president, Rev. Dr. Bellows: —

"There are some who are sufferers, great sufferers, whom we can reach and relieve without endangering political or military plans, and in the spirit of Him who ignored the petty political distinctions of Jew and Samaritan, and regarded both as entitled to His sympathy and relief, I cannot but think it is within the scope and interest of the great Sanitary Commission to extend a portion of their Christian regard to the unfortunate sufferers from this dreadful war, the prisoners in our fortresses, and to those who dwell upon the borders of the contending sections."

In a letter of March 23, to William L. Ransom, Esq., of Litchfield, Connecticut, he, perhaps unconsciously, enunciates one of the fundamental beliefs of that great president whom he so bitterly opposed: —

"I hardly know how to comply with your request to have a 'short, pithy, Democratic sentiment.' In glancing at the thousand mystifications which have befogged so many in our presumed intelligent community, I note one in relation to the new-fangled application of a common foreign word imported from the monarchies of Europe. I mean the word '*loyalty*,' upon which the changes are daily and hourly sung *ad nauseam*.

"I have no objection, however, to the word if it be

rightly applied. It signifies 'fidelity to a prince or sovereign.' Now if *loyalty* is required of us, it should be to the *Sovereign*. Where is this Sovereign? He is not the President, nor his Cabinet, nor Congress, nor the Judiciary, nor any nor all of the Administration together. Our Sovereign is on a throne above all these. He is the *People*, or *Peoples* of the States. He has issued his decree, not to private individuals only, but to his servant the President and to all his subordinate servants, and this sovereign decree is the Constitution. He who adheres faithfully to this written will of the Sovereign is *loyal*. He who violates the Constitution, this embodiment of the will of the Sovereign, is *disloyal*, whether he be a President, a Secretary, a member of Congress or of the Judiciary, or a simple citizen."

As a firm believer in the Democratic doctrine of States' Rights Morse, with many others, held that Lincoln had overridden the Constitution in his Emancipation Proclamation.

It was a source of grief to him just at this time that his brother Richard had changed his political faith, and had announced his intention of voting for the reëlection of President Lincoln. In a long letter of September 24, 1864, gently chiding him for thus going over to the Abolitionists, the elder brother again states his reasons for remaining firm in his faith: —

"I supposed, dear brother, that on that subject you were on the same platform with Sidney and myself. Have there been any new lights, any new aspects of it, which have rendered it less odious, less the 'child of Satan' than when you and Sidney edited the New York Observer before Lincoln was President? I have seen no reason to change my views respecting abolition. You well know I have ever considered it the logical progeny of Unitarianism and Infidelity. It is characterized by subtlety, hypocrisy and pharisaism, and one of the most melancholy marks of its speciousness is its influence in benumbing the gracious sensibilities of many Christian hearts, and blinding their eyes to their sad defection from the truths of the Bible.

"I know, indeed, the influences by which you are surrounded, but they are neither stronger nor more artful than those which our brave father manfully withstood in combating the monster in the cradle. I hope there is enough of father's firmness and courage in battling with error, however specious, to keep you, through God's grace, from falling into the embrace of the body-and-soul-destroying heresy of Abolitionism."

In another long letter to his brother Richard, of November 5, he firmly but gently upholds his view that the Constitution has been violated by Lincoln's action, and that the manner of amending the Constitution was provided for in that instrument itself, and that: "If that change is made in accordance with its provisions, no one will complain"; and then he adds: —

"But it is too late to give you the reasons of the political faith that I hold. When the excitement of the election is over, let it result as it may, I may be able to show you that my opinions are formed from deep study and observation. Now I can only announce them comparatively unsustained by the reasons for forming them.

"I am interrupted by a call from the committee requesting me to conduct General McClellan to the balcony of the Fifth Avenue Hotel this evening, to review the McClellan Legion and the procession. After my return I will continue my letter.

"*12 o'clock, midnight.* I have just returned, and never have I witnessed in any gathering of the people, either in Europe or in this country, such a magnificent and enthusiastic display. I conducted the General to the front of the balcony and presented him to the assemblage (a dense mass of heads as far as the eye could reach in every direction), and such a shout, which continued for many minutes, I never heard before, except it may have been at the reception in London of Blücher and Platoff after the battle of Waterloo. I leave the papers to give you the details. The procession was passing from nine o'clock to a quarter to twelve midnight, and such was the denseness of the crowd within the hotel, every entry and passageway jammed with people, that we were near being crushed. Three policemen before me could scarcely open a way for the General, who held my arm, to pass only a few yards to our room.

"After taking my leave I succeeded with difficulty in pressing my way through the crowd within and without the hotel, and have just got into my quiet library and must now retire, for I am too fatigued to do anything but sleep. Good-night."

A short time after this the election was held, and this enthusiastic advocate of what he considered the right learned the bitter lesson that crowds, and shouting, and surface enthusiasm do not carry an election. The voice of that Sovereign to whom he had sworn loyalty spoke in no uncertain tones, and Lincoln was overwhelmingly chosen by the votes of the People.

Morse was outvoted but not convinced, and I shall make but one quotation from a letter of November 9, to his brother Richard, who had also remained firm in spite of his brother's pleading: "My consolation is in looking up, and I pray you may be so enlightened that you may be delivered from the delusions which have ensnared you, and from the judgments which I cannot but feel are in store for this section of the country. When I can believe that my Bible reads '*cursed*' instead of '*blessed*' are the 'peacemakers,' I also shall cease to be a peace man. But while they remain, as they do, in the category of those that are blessed, I cannot be frightened at the names of 'copperhead' and 'traitor' so lavishly bestowed, with threats of hanging etc., by those whom you have assisted into power."

In a letter of Mr. George Wood's, of June 26, 1865, I find the following sentences: "I have to acknowledge your very carefully written letter on the divine origin of Slavery. . . . I hope you have kept a copy of this letter, for the time will come when you will have a biography written, and the defense you have made of your position, taken in your pamphlet, is unquestionably far better than he (your biographer) will make for you."

The letter to which Mr. Wood refers was begun on March 5, 1865, but finished some time afterwards. It is very long, too long to be included here, but in justice to myself, that future biographer, I wish to state that I have already given the main arguments brought forward in that letter, in quotations from previous letters, and that I have attempted no defense further than to emphasize the fact that, right or wrong, Morse was intensely sincere, and that he had the courage of his opinions.

Returning to an earlier date, and turning from matters political to the gentler arts of peace, we find that the one-time artist had always hoped that some day he could resume his brush, which the labors incident to the invention of the telegraph had compelled him to drop. But it seems that his hand, through long disuse, had lost its cunning. He bewails the fact in a letter of January 20, 1864, to N. Jocelyn, Esq.: —

"I have many yearnings towards painting and sculpture, but that rigid faculty called reason, so opposed often to imagination, reads me a lecture to which I am compelled to bow. To explain: I made the attempt to draw a short time ago; everything in the drawing seemed properly proportioned, but, upon putting it in another light, I perceived that every perpendicular line was awry. In other words I found that I could place no confidence in my eyes.

"No, I have made the sacrifice of my profession to establish an invention which is doing mankind a great service. I pursued it long enough to found an institution which, I trust, is to flourish long after I am gone, and be the means of educating a noble class of men in Art, to be an honor and praise to our beloved country when peace shall once more bless us throughout all our borders in one grand brotherhood of States."

The many letters to his children are models of patient exhortation and cheerful optimism, when sometimes the temptation to indulge in pessimism was strong. I shall give, as an example, one written on

May 9, 1864, to two of his sons who had returned to school at Newport: —

"Now we hope to have good reports of your progress in your studies. In spring, you know, the farmers sow their seed which is to give them their harvest at the close of the summer. If they were not careful to put the seed in the ground, thinking it would do just as well about August or September, or if they put in very little seed, you can see that they cannot expect to reap a good or abundant crop.

"Now it is just so in regard to your life. You are in the springtime of life. It is seed time. You must sow now or you will reap nothing by-and-by, or, if anything, only weeds. Your teachers are giving you the seed in your various studies. You cannot at present understand the use of them, but you must take them on trust; you must believe that your parents and teachers have had experience, and they know what will be for your good hereafter, what studies will be most useful to you in after life. Therefore buckle down to your studies diligently and very soon you will get to love your studies, and then it will be a pleasure and not a task to learn your lessons.

"We miss your *noise*, but, although agreeable quiet has come in place of it, we should be willing to have the noise if we could have our dear boys near us. You are, indeed, troublesome pleasures, but, after all, pleasant troubles. When you are settled in life and have a family around you, you will better understand what I mean."

In spite of the disorganization of business caused by the war, the value of telegraphic property was rapidly increasing, and new lines were being constantly built or proposed. Morse refers to this in a letter of June 25, 1864, to his old friend George Wood: —

"To you, as well as to myself, the rapid progress of the Telegraph throughout the world must seem wonderful, and with me you will, doubtless, often recur to our friend Annie's inspired message — 'What hath God wrought.' It is, indeed, his marvellous work, and to Him be the glory.

"Early in the history of the invention, in forecasting its future, I was accustomed to predict with confidence, 'It is destined to go round the world,' but I confess I did not expect to live to see the prediction fulfilled. It is quite as wonderful to me also that, with the thousand attempts to improve my system, with the mechanical skill of the world concentrated upon improving the mechanism, the result has been beautiful complications and great ingenuity, but no improvement. I have the gratification of knowing that my system, everywhere known as the 'Morse system,' is universally adopted throughout the world, because of its simplicity and its adaptedness to universality."

This remains true to the present day, and is one of the remarkable features of this great invention. The germ of the "Morse system," as jotted down in the 1832 sketch-book, is the basic principle of the universal telegraph of to-day.

In another letter to Mr. Wood, of September 11, 1864, referring to the sad death of the son of a mutual friend, he touches on two of the great enigmas of life which have puzzled many other minds: —

"It is one of those mysteries of Providence, one of those deep things of God to be unfolded in eternity, with the perfect vindication of God's wisdom and justice, that children of pious parents, children of daily anxiety and prayer, dedicated to God from their birth and trained to all human appearance 'in the way they should go,' should yet seem to falsify the promise that 'they should not depart from it.' It is a subject too deep to fathom.

"... It is my daily, I may say hourly, thought, certainly my constant wakeful thought at night, how to resolve the question: 'Why has God seen fit so abundantly to shower his earthly blessings upon me in my latter days, to bless me with every desirable comfort, while so many so much more deserving (in human eyes at least) are deprived of all comfort and have heaped upon them sufferings and troubles in every shape?'"

The memory of his student days in London was always dear to him, and on January 4, 1865, he writes to William Cullen Bryant: —

"I have this moment received a printed circular respecting the proposed purchase of the portrait of Allston by Leslie to be presented to the National Academy of Design.

"There are associations in my mind with those two eminent and beloved names which appeal too strongly to me to be resisted. Now I have a favor to ask which I hope will not be denied. It is that I may be allowed to present to the Academy that portrait in my own name. You can appreciate the arguments which have influenced my wishes in this respect. Allston was more than any other person my master in art. Leslie was my life-long cherished friend and fellow pupil, whom I loved as a brother. We all lived together for years in the closest intimacy and in the same house. Is there not then a fitness that the portrait of the master by one distinguished pupil should be presented by the surviving pupil to the Academy over which he presided in its infancy, as well as assisted in its birth, and, although divorced from Art, cannot so easily be divorced from the memories of an intercourse with these distinguished friends, an intercourse which never for one moment suffered interruption, even from a shadow of estrangement?"

It is needless to say that this generous offer was accepted, and Morse at the same time presented to the Academy the brush which Allston was using when stricken with his fatal illness.

As his means permitted he made generous donations to charities and to educational institutions, and on May 20, 1865, he endowed by the gift of $10,000 a lectureship in the Union Theological Seminary, making the following request in the letter which accompanied it: —

"If it be thought advisable that the name of the lectureship, as was suggested, should be the Morse Lectureship, I wish it to be distinctly understood that it is so named in honor of my venerated and distinguished father, whose zealous labors in the cause of theological education, and in various benevolent enterprises, as well as of geographical science, entitle his memory to preservation in connection with the efforts to diffuse the knowledge of our Lord and Saviour, Jesus Christ, and his gospel throughout the world."

Curiously enough I find no reference in the letters of the year 1865 to the assassination of President Lincoln, but I well remember being taken, a boy of eight, to our stable on the corner of Fifth Avenue and Twenty-first Street, from the second-floor windows of which we watched the imposing funeral cortège pass up the avenue.

The fifty-fifth reunion of his class of 1810 took place in this year, and Morse reluctantly decided to absent himself. The reasons why he felt that he could not go are given in a long letter of August 11 to his cousin, Professor E. S. Salisbury, and it is such a clear statement of his convictions that I am tempted to give it almost in its entirety: —

"I should have been most happy on many personal accounts to have been at the periodical meeting of my surviving classmates of 1810, and also to have renewed my social intercourse with many esteemed friends and relations in New Haven. But as I could not conscientiously take part in the proposed martial sectional glorification of those of the family who fell in the late lamentable family strife, and could not in any brief way or time explain the discriminations that were necessary between that which I approve and that which I most unqualifiedly condemn, without the risk of misapprehension, I preferred the only alternative left me, to absent myself altogether.

"You well know I never approved of the late war. I have ever believed, and still believe, if the warnings of far-seeing statesmen (Washington, Clay, and Webster among them) had been heeded, if, during the last thirty years of persistent stirring up of strife by angry words, the calm and Christian counsels of intelligent patriots had been followed at the North, and a strict observance of the letter and spirit of the Constitution had been sustained as the supreme law, instead of the insidious violations of its provisions, especially by New England, we should have had no war.

"As I contributed nothing to the war, so now I see no reason specially to exult in the display of brave qualities in an isolated portion of the family, qualities which no true American ever doubted were possessed by both sections of our country in an equal degree. Why then discriminate between alumni from the North and alumni from the South at a gathering in which alumni from both sections are expected to meet? . . . No, my dear cousin, the whole era of the war is one I wish not to remember. I would have no other memorial than a black cross, like those over the graves of murdered travellers, to cause a shudder whenever it is seen. It would be well if History could blot from its pages all record of the past four years. There is no glory in them for victors or vanquished. The only event in which I rejoice is the restoration of Peace, which never should have been interrupted. . . .

"I have no doubt that they who originated the recent demonstration honestly believed it to be *patriotic*, for every movement nowadays must take that shape to satisfy the morbid appetite of the popular mind. I cannot think it either in good taste or in conformity with sound policy for our collegiate institutions to foster this depraved appetite. Surely there is enough of this in the political harangues of the day for those who require such aids to patriotism without its being administered to by our colleges. That patriotism is of rather a suspicious character which needs such props. I love to see my children well clad and taking a proper pride in their attire, but I should not think them well instructed if I found them everywhere boasting of their fine clothes. A true nobleman is not forever boasting of his nobility for fear that his rank may not be recognized. The loudest boasts of patriotism do not come from the true possessors of the genuine spirit. Patriotism is not sectional nor local, it comprehends in its grasp the whole country. . . .

"I have said the demonstration at Commencement was in bad taste. Why? you will say. Because Commencement day brings together the alumni of the college from all parts of the Union, from the South as well as the North. They are to meet on some common ground, and that common ground is the love that all are supposed to bear to the old Alma Mater, cherished by memories of past friendships in their college associations. The late Commencement was one of peculiar note. It was the first after the return of peace. The country had been sundered; the ties of friendship and of kindred had been broken; the bonds of college affection were weakened if not destroyed. What an opportunity for inaugurating the healing process! What an occasion for the display of magnanimity, of mollifying the pain of humiliation, of throwing a veil of oblivion over the past, of watering the perishing roots of fraternal affection and fostering the spirit of genuine union! But no. The Southern alumnus may come, but he comes to be humiliated still further. Can he join in the plaudits of those by whom he has been humbled? You may applaud, but do not ask him to join in your acclamations. He may be mourning the death of father, brother, yes, of mother and sister, by the very hands of those you are glorifying. Do not aggravate his sorrow by requiring him to join you in such a demonstration.

"No, my dear cousin, it was in *bad taste* to say the least of it, and it was equally *impolitic* to intercalate such a demonstration into the usual and appropriate exercises of the week. You expect, I presume, to have pupils from the South as heretofore; will such a sectional display be likely to attract them or to repel them? If they can go elsewhere they will not come to you. They will not be attracted by a perpetual memento before their eyes of your triumph over them. It was not politic. It is no improvement for Christian America to show less humanity than heathen Rome. The Romans never made demonstrations of triumph over the defeat of their countrymen in a civil war. It is no proof of superior civilization that we refuse to follow Roman example in such cases.

"My dear cousin, I have written you very frankly, but I trust you will not misunderstand me as having any personal reproaches to make for the part you have taken in the matter. We undoubtedly view the field from different standpoints. I concede to you conscientious motives in what you do. You are sustained by those around you, men of intellect, men of character. I respect them while I differ from them. I appeal, however, to a higher law, and that, I think, sustains me."

His strong and outspoken stand for what he believed to be the right made him many enemies, and he was called hard names by the majority of those by whom he was surrounded at the North; and yet the very fearlessness with which he advocated an unpopular point of view undoubtedly compelled increased respect for him. A proof of this is given in a letter to his daughter, Mrs. Lind, of December 28, 1865: —

"I also send you some clippings from the papers giving you an account of some of the doings respecting a statue proposed to me by the Common Council. The Mayor, who is a personal friend of mine, you see has vetoed the resolutions, not from a disapproval of their character, but because he did not like the locality proposed. He proposes the Central Park, and in this opinion all my friends concur.

"I doubt if they will carry the project through while I am alive, and it would really seem most proper to wait until I was gone before they put up my monument. I have nothing, however, to say on the subject. I am gratified, of course, to see the manifestation of kindly feeling, but, as the tinder of vainglory is in every human heart, I rather shrink from such a proposed demonstration lest a spark of flattery should kindle that tinder to an unseemly and destructive flame. I am not blind to the popularity, world-wide, of the Telegraph, and a sober forecast of the future foreshadows such a statue in some place. If ever erected I hope the prominent mottoes upon the pedestal will be: '*Not unto us, not unto us, but to God be the glory*,' and the first message or telegram: '*What hath* God *wrought*.'"

He says very much the same thing in a letter to his friend George Wood, of January 15, 1866, and he also says in this letter, referring to some instance of benevolent generosity by Mr. Kendall: —

"Is it not a noticeable fact that the wealth acquired by the Telegraph has in so many conspicuous instances been devoted to benevolent purposes? Mr. Kendall is prominent in his expenditures for great Christian enterprises, and think of Cornell, always esteemed by me as an ingenious and shrewd man, when employed by me to set the posts and put up the wire for the first line of Telegraphs between Washington and Baltimore, yet thought to be rather close and narrow-minded by those around him. But see, when his wealth had increased by his acquisition of Telegraph stock to millions (it is said), what enlarged and noble plans of public benefit were conceived and brought forth by him. I have viewed his course with great gratification as the evidence of God's blessing on *what He hath wrought.*"

It has been made plain, I think, that Morse was essentially a leader in every movement in which he took an interest, whether it was artistic, scientific, religious, or political. This is emphasized by the number of requests made to him to assume the presidency of all sorts of organizations, and these requests multiplied as he advanced in years. Most of them he felt compelled to decline, for, as he says in a letter of March 13, 1866, declining the presidency of the Geographical and Statistical Society: "I am at an age when I find it necessary rather to be relieved from the cares and responsibilities already resting upon me, than to take upon me additional ones."

In many other cases he allowed his name to be used as vice-president or member, when he considered the object of the organization a worthy one, and his benefactions were only limited by his means.

He did, however, accept the presidency of one association just at this time, the American Asiatic Society, in which were interested such men as Gorham Abbott,

Dr. Forsyth, E. H. Champlin, Thomas Harrison, and Morse's brother-in-law, William M. Goodrich. The aims of this society were rather vast, including an International Congress to be called by the Emperor Napoleon III, for the purpose of opening up and controlling the great highways from the East to the West through the Isthmus of Suez and that of Panama; also the colonization of Palestine by the Jews, and other commercial and philanthropic schemes. I cannot find that anything of lasting importance was accomplished by this society, so I shall make no further mention of it, although there is much correspondence about it.

The following, from a letter to Mr. Kendall of March 19, 1866, explains itself: "If I understand the position of our Telegraph interests, they are now very much as you and I wished them to be in the outset, not cut up in O'Reilly fashion into irresponsible parts, but making one grand whole like the Post-Office system. It is becoming, doubtless, a *monopoly*, but no more so than the Post-Office system, and its unity is in reality a public advantage if properly and uprightly managed, and this, of course, will depend on the character of the managers. Confidence must be reposed somewhere, and why not in upright and responsible men who are impelled as well by their own interest to have their matters conducted with fairness and with liberality."

As a curious commentary on his misplaced faith in the integrity of others, I shall quote from a letter of January 4, 1867, to E. S. Sanford, Esq., which also shows his abhorrence of anything like crooked dealing in financial matters: —

"I wish when you again write me you would give me,

in confidence, the names of those in the Board of the Western Union who are acting in so dishonorable and tricky a manner. I think I ought to know them in order

Jesse Hoyt. Esqr. Superintendent.
Halifax.
Nova Scotia
All my patents expire on the eleventh (11th) day of April eighteen hundred and Sixty Seven. (1867)
Saml. F. B. Morse.

TELEGRAM SHOWING MORSE'S CHARACTERISTIC DEADHEAD, WHICH HE ALWAYS USED TO FRANK HIS MESSAGES

to avoid them, and resist them in the public interest. It is a shame that an enterprise which, honestly conducted, is more than usually profitable, should be conducted on the principles of sharpers and tricksters.

"So far as the Russian Extension is concerned, I should judge from your representation that, as a stockholder in that enterprise to the amount of $30,000, the plan would conduce to my immediate pecuniary benefit. But so would the *robbery of the safe of a bank*. If wealth can be obtained only by such swindles, I prefer poverty. You have my proxy and I have the utmost confidence in your management. Do by me as you would do for yourself, and I shall be satisfied. . . . In regard to any honorable propositions made in the Board be conciliatory and compromising, but any scheme to oppress the smaller stockholders for the benefit of the larger resist to the death. I prefer to sacrifice all my stock rather than have such a stigma on my character as such mean, and I will add villainous, conduct would be sure to bring upon all who engaged in it."

In this connection I shall also quote from another letter to Mr. Sanford, of February 15, 1867: "If Government thinks seriously of purchasing the Telegraph, and at this late day adopting my early suggestion that it ought to belong to the Post-Office Department, be it so if they will now pay for it. They must now pay millions for that which I offered to them for one hundred thousand dollars, and gave them a year for consideration ere they adopted it."

There are but few references to politics in the letters of this period, but I find the following in a letter of March 20, 1866, to a cousin: "You ask my opinion of our President. I did not vote for him, but I am agreeably surprised at his masterly statesmanship, and hope, by his firmness in resisting the extreme radicals, he will preserve the Union against now the greatest enemies we have to contend against. I mean those who call themselves Abolitionists. . . . President Johnson deserves the support of all true patriots, and he will have it against all the 'traitors' in the country, by whatever soft names of loyalty they endeavor to shield themselves."

Appeals of all kinds kept pouring in on him, and, in courteously refusing one, on April 17, he uses the following language: "I am unable to aid you. I cannot, indeed, answer a fiftieth part of the hundreds of applications made to me from every section of the country *daily* — I might say *hourly* — for yours is the third this morning and it is not yet 12 o'clock."

After settling his affairs at home in his usual methodical manner, Morse sailed with his wife and his four young children, and Colonel John R. Leslie their tutor, for Europe on the 23d of June, 1866, prepared for an extended stay. He wished to give his children the advantages of travel and study in Europe, and he was very desirous of being in Paris during the Universal Exposition of 1867.

There is a gap in the letter-books until October, 1866, but from the few letters to members of the family which have been preserved, and from my own recollections, we know that the summer of 1866 was most delightfully spent in journeying through France, Germany, and Switzerland. The children were now old enough not to be the nuisances they seem to have been in 1858, for we find no note of complaint on that account.

In September he returned with his wife, his daughter, and his youngest son to Paris, leaving his two older sons with their tutor in Geneva. As he wished to make Paris his headquarters for nearly a year, he sought and found a furnished apartment at No. 10 Avenue du Roi de Rome (now the Avenue du Trocadero), and he writes to his mother-in-law on September 22: "We are fortunate in having apartments in a new building, or rather one newly and completely repaired throughout. All the apartments are newly furnished with elegant furniture, we having the first use of it. We have ample rooms, not large, but promising more comfort for winter residence than if they were larger. The situation is on a wide avenue and central for many purposes; close to the

Champs Elysées, near also to the Bois de Boulogne, and within a few minutes walk of the Champ de Mars, so that we shall be most eligibly situated to visit the great Exposition when it opens in April."

His wife's sister, Mrs. Goodrich, with her husband and daughters, occupied an apartment in the same building; his grandson Charles Lind was also in Paris studying painting, and before the summer of the next year other members of his family came to Paris, so that at one time eighteen of those related to him by blood or marriage were around him. To a man of Morse's affectionate nature and loyalty to family this was a source of peculiar joy, and those Parisian days were some of the happiest of his life. The rest of the autumn and early winter were spent in sight-seeing and in settling his children in their various studies.

The brilliance of the court of Napoleon III just before the *débâcle* of 1870 is a matter of history, and it reached its high-water mark during the Exposition year of 1867, when emperors, kings, and princes journeyed to Paris to do homage to the man of the hour. Court balls, receptions, gala performances at opera and theatre, and military reviews followed each other in bewildering but well-ordered confusion, and Morse, as a man of world-wide celebrity, took part in all of them. He and his wife and his young daughter, a girl of sixteen, were presented at court, and were fêted everywhere. In a letter to his mother-in-law he gives a description of his court costume on the occasion of his first presentation, when he was accompanied only by his brother-in-law, Mr. Goodrich:—

"We received our cards inviting us to the soirée and to pass the evening with their majesties on the

16th of January (Wednesday evening). '*En uniforme*' was stamped upon the card, so we had to procure court dresses. Mr. Goodrich, as is the custom in most cases, hired his; I had a full suit made for me. A *chapeau bras*, with gold lace loop, a blue coat, with standing collar, single breasted, richly embroidered with gold lace, the American eagle button, white silk lining, vest light cashmere with gilt buttons, pantaloons with a broad stripe of gold lace on the outside seams, a small sword, and patent-leather shoes or boots completed the dress of ordinary mortals like Brother Goodrich, but for *extra*ordinary mortals, like my humble republican self, I was bedizened with all my orders, seven decorations, covering my left breast. If thus accoutred I should be seen on Broadway, I should undoubtedly have a numerous escort of a character not the most agreeable, but, as it was, I found myself in very good and numerous company, none of whom could consistently laugh at his neighbors."

After describing the ceremony of presentation he continues: —

"Occasionally both the emperor and empress said a few words to particular individuals. When my name was mentioned the emperor said to me, 'Your name, sir, is well known here,' for which I thanked him; and the empress afterwards said to me, when my name was mentioned, 'We are greatly indebted to you, sir, for the Telegraph,' or to that effect. Afterwards Mr. Bennett, the winner of the yacht race, engaged for a moment their particular regards. . . . [I wonder if the modest inventor appreciated the irony of this juxtaposition.] After the dancers were fully engaged, the refreshment-room, the Salon of Diana, was opened, and, as in our less aristocratic country, the tables attracted a great crowd, so that the doors were guarded so as to admit the company by instalments. I had in vain for some time endeavored to gain admittance, and was waiting patiently quite at a distance from the door, which was thronged with ladies and high dignitaries, when a gentleman who guarded the door, and who had his breast covered with orders, addressed me by name, asking me if I was not Professor Morse. Upon replying in the affirmative, quite to my surprise, he made way for me to the door and, opening it, admitted me before all the rest. I cannot yet divine why this special favor was shown to me.

"The tables were richly furnished. I looked for bonbons to carry home to the children, but when I saw some tempting looking almonds and candies and mottoes, to my surprise I found they were all composed of fish put up in this form, and the mottoes were of salad."

It is good to know that Morse, ever willing to forgive and forget, was again on terms of friendly intercourse with Cyrus W. Field, who was then in London, as the following letter to him, dated March 1, 1867, will show: —

"Singular as it may seem, I was in the midst of your speech before the Chamber of Commerce reception to you in New York, perusing it with deep interest, when my valet handed me your letter of the 27th ulto.

"I regret exceedingly that I shall not have the great pleasure I had anticipated, with other friends here, who were prepared to receive you in Paris with the welcome

MORSE IN OLD AGE

you so richly deserve. You invite me to London. I have the matter under consideration. March winds and that boisterous channel have some weight in my decision, but I so long to take you by the hand and to get posted upon Telegraph matters at home, that I feel disposed to make the attempt. But without positively saying 'yes,' I will see if in a few days I can so arrange my affairs as to have a few hours with you before you sail on the 20th.

"I send you by book post the proceedings of the banquet given to our late Minister, Bigelow, in which you will see my remarks on the great enterprise with which your name will forever be so honorably associated and justly immortalized."

It will be remembered that the Atlantic cable was finally successfully laid on July 27, 1866, and that to Cyrus Field, more than to any other man, was this wonderful achievement due.

In a letter of March 4, 1867, to John S. C. Abbott, Esq., Morse gives the following interesting incident in the life of Napoleon III: —

"In 1837, I was one of a club of gentlemen in New York who were associated for social and informal intellectual converse, which held weekly meetings at each other's houses in rotation. Most of these distinguished men are now deceased. The club consisted of such men as Chancellor Kent, Albert Gallatin, Peter Augustus Jay, Reporter Johnson, Dr. (afterwards Bishop) Wainwright, the President and Professors of Columbia College, the Chancellor and Professors of the New York City University, Dr. Augustus Smith, Messrs. Goodhue and De Rham of the mercantile class, and John C.

Hamilton, Esq. and ex-Governor W. B. Lawrence from the literary ranks.

"Among the rules of the club was one permitting any member to introduce to the meetings distinguished strangers visiting the city. At one of the reunions of the club the place of meeting was at Chancellor Kent's. On assembling the chancellor introduced to us Louis Napoleon, a son of the ex-King of Holland, a young man pale and contemplative, somewhat reserved. This reserve we generally attributed to a supposed imperfect acquaintance with our language. At supper he sat on the right of the Chancellor at the head of the table. Mr. Gallatin was opposite the Chancellor at the foot of the table, and I was on his right.

"In the course of the evening, while the conversation was general, I drew the attention of Mr. Gallatin to the stranger, observing that I did not trace any resemblance in his features to his world-renowned uncle, yet that his forehead indicated great intellect. 'Yes,' replied Mr. Gallatin, 'there is a great deal in that head of his, but he has a strange fancy. Can you believe it, he has the impression that he will one day be the Emperor of the French; can you conceive of anything more ridiculous?'

"Certainly at that period, even to the sagacious eye of Mr. Gallatin, such an idea would naturally seem too improbable to be entertained for a moment, but, in the light of later events, and the actual state of things at present, does not the fact show that, even in his darkest hours, there was in this extraordinary man that unabated faith in his future which was a harbinger of success; a faith which pierced the dark clouds which surrounded him, and realized to him in marvellous prophetic vision that which we see at this day and hour fully accomplished?"

Morse must have penned these words with peculiar satisfaction, for they epitomized his own sublime faith in his future. In 1837 he also was passing through some of his darkest hours, but he too had had faith, and now, thirty years afterwards, his dreams of glory had been triumphantly realized, he was an honored guest of that other man of destiny, and his name was forever immortalized.

The spring and early summer of 1867 were enjoyed to the full by the now venerable inventor and his family. The Exposition was a source of never-ending joy to him, and he says of it in a letter to his son-in-law, Edward Lind: —

"You will hear all sorts of stories about the Exposition. The English papers (some of them), in John Bull style, call it a humbug. Let me tell you that, imperfect as it is in its present condition, going on rapidly to completion, it may without exaggeration be pronounced the eighth wonder of the world. It is the world in epitome. I came over with my children to give them the advantage of thus studying the world in anticipation of what I now see, and I can say that the two days only in which I have been able to glance through parts of its vast extent, have amply repaid me for my voyage here. I believe my children will learn more of the condition of the arts, agriculture, customs, manufactures and mineral and vegetable products of the world in five weeks than they could by books at home in five years, and as many years' travel."

He was made an Honorary Commissioner of the United States to the Exposition, and he prepared an elaborate and careful report on the electrical department, for which he received a bronze medal from the French Government. Writing of this report to his brother Sidney, he says: "This keeps me so busy that I have no time to write, and I have so many irons in the fire that I fear some must burn. But father's motto was — 'Better wear out than rust out,' — so I keep at work."

In a letter to his friend, the Honorable John Thompson, of Poughkeepsie, he describes one of his dissipations: —

"Paris now is the great centre of the world. Such an assemblage of sovereigns was never before gathered, and I and mine are in the midst of the great scenes and fêtes. We were honored, a few evenings ago, with cards to a very select fête given by the emperor and empress at the Tuilleries to the King and Queen of the Belgians, the Prince of Wales and Prince Alfred, to the Queen of Portugal, the Grand Duchess Marie of Russia, sister of the late Emperor Nicholas, a noble looking woman, the Princess Metternich of Austria, and many others.

"The display was gorgeous, and as the number of guests was limited (only one thousand!) there was more space for locomotion than at the former gatherings at the Palace, where we were wedged in with some four thousand. There was dancing and my daughter was solicited by one of the gentlemen for a set in which Prince Alfred and the Turkish Ambassador danced, the latter with an American belle, one of the Miss Beckwiths. I allowed her to dance in this set once. The Empress is truly a beautiful woman and of unaffected manners."

In a long letter to his brother Sidney, of June 8, he describes some of their doings. At the Grand Review of sixty thousand troops he and his wife and eldest son were given seats in the Imperial Tribune, a little way behind the emperor and the King of Prussia, who were so soon to wage a deadly war with each other. On the way back from the review the following incident occurred: —

"After the review was over we took our carriage to return home. The carriages and cortège of the imperial personages took the right of the Cascade (which you know is in full view from the hippodrome of Longchamps). We took the left side and were attracted by the report of firearms on our left, which proceeded from persons shooting at pigeons from a trap. Soon after we heard a loud report on our right from a pistol, which attracted no further attention from us than the remark which I made that I did not know that persons were allowed to use firearms in the Bois. We passed on to our home, and in the evening were informed of the atrocious attempt upon the Emperor of Russia's life. The pistol report which I heard was that of the pistol of the assassin."

Farther on in this letter he describes the grand fête given by the City of Paris to the visiting sovereigns at the Hotel de Ville. There were thirty-five thousand applications for tickets, but only eight thousand could be granted. Of these Morse was gratified to receive three: —

"Well, the great fête of Saturday the 8th is over. I

despair of any attempt properly to describe its magnificence. I send you the papers. . . . Such a blaze of splendor cannot be conceived or described but in the descriptions of the Arabian Nights. We did not see half the display, for the immense series of gorgeous halls, lighted by seventy thousand candles, with fountains and flowers at every turn, made one giddy to see even for a moment. We had a good opportunity to scan the features of the emperors, the King of Prussia and the renowned Bismarck, with those of the beautiful empress and the princesses and princes and other distinguished persons of their suite.

"I must tell you (for family use only) that the Emperor Napoleon made to me a marked recognition as he passed along. Sarah and I were standing upon two chairs overlooking the front rank of those ranged on each side. The emperor gave his usual bow on each side, but, as he came near us, he gave an unusual and special bow to me, which I returned, and he then, with a smile, gave me a second bow so marked as to draw the attention of those around, who at once turned to see to whom this courtesy was shown. I should not mention this but that Sarah and others observed it as an unusual mark of courtesy."

Feeling the need of rest after all the gayety and excitement of Paris, Morse and part of his family retired to Shanklin, on the Isle of Wight, where in a neat little furnished cottage — Florence Villa — they spent part of two happy months. Then with his wife and daughter and youngest son he journeyed in leisurely fashion through England and Scotland, returning to Paris in October. Here he spent some time in working on his report to the United States Government as Commissioner to the Exposition.

Among his notes I find the following, which seems to me worthy of record: —

"*The Sounder.* Mr. Prescott, I perceive, is quoted as an authority. He is not reliable on many points and his work should be used with caution. His work was originally written in the interest of those opposing my patents, and his statements are, many of them, grossly unjust and strongly colored with prejudice. Were he now to reprint his work I am convinced he would find it necessary, for the sake of his reputation, to expunge a great deal, and to correct much that he has misstated and misapprehended.

"He manifests the most unpardonable ignorance or wilful prejudice in regard to the *Sounder*, now so-called. The possibility of reading by sound was among the earliest modes noticed in the first instrument of 1835, and it was in consequence of observing this fact that, in my first patent specifications drawn up in 1837–1838, I distinctly specify these *sounds* of the signs, and they were secured in my letters patent. Yet Mr. Prescott makes it an accidental discovery, and in 1860 (the date of his publication) he wholly ignores my agency in this mode. The sounder is but the pen-lever deprived of the pen. In everything else it is the same. The sound of the letter is given with and without the pen."

On November 8, 1867, he writes from Paris to his friend, the Honorable John Thompson: —

"I am still held in Paris for the completion of my labors, but hope in a few days to be relieved so that we may leave for Dresden, where my boys are pursuing their studies in the German language. . . . I am yet doubtful how long a sojourn we may make in Dresden, and whether I shall winter there or in Paris, but I am inclined to the latter. We wish to visit Italy, but I am not satisfied that it will be pleasant or even safe to be there just now. The Garibaldian inroad upon the Pontifical States is, indeed, for the moment suppressed, but the end is not yet.

"Alas for poor Italy! How hard to rid herself of evils that have become chronic. Why cannot statesmen of the Old World learn the great truth that most of their perplexities in settling the questions of international peace arise from the unnatural union of Church and State? He who said 'My kingdom is not of this world' uttered a truth pregnant with consequences. The attempt to rule the State by the Church or the Church by the State is equally at war with his teachings, and until these are made the rule of conduct, whether for political bodies or religious bodies, there will be the sword and not peace.

"I see by the papers that the reaction I have long expected and hoped for has commenced in our country. It is hailed here by intelligent and cool-headed citizens as a good omen for the future. The Radicals have had their way, and the people, disgusted, have at length given their command — 'Thus far and no farther.'"

CHAPTER XXXIX

NOVEMBER 28, 1867 — JUNE 10, 1871

Goes to Dresden. — Trials financial and personal. — Humorous letter to E. S. Sanford. — Berlin. — The telegraph in the war of 1866. — Paris. — Returns to America. — Death of his brother Richard. — Banquet in New York. — Addresses of Chief Justice Chase, Morse, and Daniel Huntington. — Report as Commissioner finished. — Professor W. P. Blake's letter urging recognition of Professor Henry. — Morse complies. — Henry refuses to be reconciled. — Reading by sound. — Morse breaks his leg. — Deaths of Amos Kendall and George Wood. — Statue in Central Park. — Addresses of Governor Hoffman and William Cullen Bryant. — Ceremonies at Academy of Music. — Morse bids farewell to his children of the telegraph.

It will not be necessary to record in detail the happenings of the remainder of this last visit to Europe. Three months were spent in Dresden, with his children and his sister-in-law's family around him. The same honors were paid to him here as elsewhere on the continent. He was received in special audience by the King and Queen of Saxony, and men of note in the scientific world eagerly sought his counsel and advice. But, apart from so much that was gratifying to him, he was just then called upon to bear many trials and afflictions of various kinds and degrees, and it is marvellous, in reading his letters, to note with what great serenity and Christian fortitude, yet withal, with what solicitude, he endeavored to bear his cross and solve his problems. As he advanced in years an increasing number of those near and dear to him were taken from him by death, and his letters of Christian sympathy fill many pages of the letter books. There were trials of a domestic nature, too intimate to be revealed, which caused him deep sorrow, but which he bravely and optimistically strove to meet. Clouds, too, obscured his financial horizon; investments in certain mining ventures, entered into with high hopes, turned out a dead loss; the repayment of loans, cheerfully made to friends and relatives, was either delayed or entirely defaulted; and, to cap the climax, the Western Union Telegraph Company, in which most of his fortune was invested, passed one dividend and threatened to pass another. He had provided for this contingency by a deposit of surplus funds before his departure for Europe, but he was fearful of the future.

In spite of all this he could not refrain from treating the matter lightly and humorously in a letter to Mr. E. S. Sanford of November 28, 1867, written from Dresden: "Your letter gave me both pleasure and pain. I was glad to hear some particulars of the condition of my '*basket*,' but was pained to learn that the *hens*' eggs instead of swelling to *goose* eggs, and even to *ostrich* eggs (as some that laid them so enthusiastically anticipated when they were so closely packed), have shrunk to *pigeons*' eggs, if not to the diminutive *sparrows*'. To keep up the figure, I am thankful there are any left not addled."

He was all the time absorbed in the preparation of his report as Commissioner to the Paris Exposition, and it was, of course, a source of great gratification to him to learn from the answers to his questions sent to the telegraph officers of the whole world, that the Morse system was practically the only one in general use. As one of his correspondents put it — "The cry is, 'Give us the Morse.'"

The necessity for the completion of this work, and

his desire to give his children every advantage of study, kept him longer in Europe than he had expected, and he writes to his brother Sidney on December 1, 1867: "I long to return, for age creeps on apace, and I wish to put my house in order for a longer and better journey to a better home."

In the early part of February, 1868, he and his wife and daughter and youngest son left Dresden for Paris, stopping, however, a few days in Berlin. Mr. George Bancroft was our minister at the Prussian court, and he did all that courtesy could suggest to make the stay of his distinguished countryman a pleasant one. He urged him to stay longer, so that he might have the pleasure of presenting him at court, but this honor Morse felt obliged to decline. The inventor did, however, find time to visit the government telegraph office, of which Colonel (afterwards General) von Chauvin was the head, and here he received an ovation from all the operators, several hundred in number, who were seated at their instruments in what was then the largest operating-room in the world.

Another incident of his visit to Berlin I shall give in the words of Mr. Prime: —

"Not to recount the many tributes of esteem and respect paid him by Dr. Siemens, and other gentlemen eminent in the specialty of telegraphy, one other unexpected compliment may be mentioned. The Professor was presented to the accomplished General Director of the Posts of the North German Bund, Privy Councillor von Phillipsborn, in whose department the telegraph had been comprised before Prussia became so great and the centre of a powerful confederation.

"At the time of their visit the Director was so engaged, and that, too, in another part of the Post-Amt, that the porter said it was useless to trouble him with the cards. The names had not been long sent up, however, before the Director himself came hurriedly down the corridor into the antechamber, and, scarcely waiting for the hastiest of introductions, enthusiastically grasped both the Professor's hands in his own, asking whether he had 'the honor of speaking to Dr. Morse,' or, as he pronounced it 'Morzey.'

"When, after a brief conversation, Mr. Morse rose to go, the Director said that he had just left a conference over a new post and telegraph treaty in negotiation between Belgium and the Bund, and that it would afford him great pleasure to be permitted to present his guest to the assembled gentlemen, including the Belgian Envoy and the Belgian Postmaster-General. There followed, accordingly, a formal presentation with an introductory address by the Director, who, in excellent English, thanked Mr. Morse in the name of Prussia and of all Germany for his great services, and speeches by the principal persons present — the Belgian envoy, Baron de Nothomb, very felicitously complimenting the Professor in French.

"Succeeding the hand-shaking the Director spoke again, and, in reply, Mr. Morse gratefully acknowledged the courtesy shown to him, adding: 'It is very gratifying to me to hear you say that the Telegraph has been and is a means of promoting peace among men. Believe me, gentlemen, my remaining days shall be devoted to this great object.' . . .

"The Director then led his visitors into a small, cosily

furnished room, saying as they entered: 'Here I have so often thought of you, Mr. Morse, but I never thought I should have the honor of receiving you in my own private room.'

"After they were seated the host, tapping upon a small table, continued: 'Over this passed the important telegrams of the war of 1866.' Then, approaching a large telegraph map on the wall, he added: 'Upon this you can see how invaluable was the telegraph in the war. Here,' — pointing with the forefinger of his right hand, — 'here the Crown Prince came down through Silesia. This,' indicating with the other forefinger a passage through Bohemia, 'was the line of march of Prince Friedrich Carl. From this station the Crown Prince telegraphed Prince Friedrich Carl, always over Berlin, "Where are you?" The answer from this station reached him, also over Berlin. The Austrians were here,' placing the thumb on the map below and between the two fingers. 'The next day Prince Friedrich Carl comes here,' — the left forefinger joined the thumb, — 'and telegraphs the fact, always over Berlin, to the Crown Prince, who hurries forward here.' The forefinger of the right hand slipped quickly under the thumb as if to pinch something, and the narrator looked up significantly.

"Perhaps the patriotic Director thought of the July afternoon when, eagerly listening at the little mahogany-topped table, over which passed so many momentous messages, he learned that the royal cousins had effected a junction at Königgrätz, a junction that decided the fate of Germany and secured Prussia its present proud position, a junction which but for his modest visitor's invention, the telegraph, 'always over Berlin,' would have been impossible."

Returning to Paris with his family, he spent some months at the Hôtel de la Place du Palais Royal, principally in collecting all the data necessary to the completion of his report, which had been much delayed owing to the dilatoriness of those to whom he had applied for facts and statistics. On April 14, 1868, he says in a letter to the Honorable John Thompson: "Pleasant as has been our European visit, with its advantages in certain branches of education, our hearts yearn for our American home. We can appreciate, I hope, the good in European countries, be grateful for European hospitality, and yet be thorough Americans, as we all profess to be notwithstanding the display of so many defects which tend to disgrace us in the eyes of the world."

On May 18 he writes to Senator Michel Chevalier: "And now, my dear sir, farewell. I leave beautiful Paris the day after to-morrow for my home on the other side of the Atlantic, more deeply impressed than ever with the grandeur of France, and the liberality and hospitality of her courteous people, so kindly manifested to me and mine. I leave Paris with many regrets, for my age admonishes me that, in all probability, I shall never again visit Europe."

Sailing from Havre on the St. Laurent, on May 22, he and his family reached, without untoward incident, the home on the Hudson, and on June 21 he writes to his son Arthur, who had remained abroad with his tutor:—

"You see by the date where we all are. Once more

I am seated at my table in the half octagon study under the south verandah. Never did the Grove look more charming. Its general features the same, but the growth of the trees and shrubbery greatly increased. Faithful Thomas Devoy has proved himself to be a truly honest and efficient overseer. The whole farm is in fine condition. . . .

"On Thursday last I was much gratified with Mr. Leslie's letter from Copenhagen, with his account of your reception by the King of Denmark. How gratifying to me that the portrait of Thorwaldsen has given such pleasure to the king, and that he regards it as the best likeness of the great sculptor."

The story of Morse's presentation to the King of Denmark of the portrait, painted in Rome in 1831, has already been told in the first volume of this work. The King, as we learn from the above quotation, was greatly pleased with it, and in token of his gratification raised Morse to the rank of Knight Commander of the Dannebrog, the rank of Knight having been already conferred on the inventor by the King's predecessor on the throne.

In another letter to Colonel Leslie, of November 2, 1868, brief reference is made to matters political: —

"To-morrow is the important day for deciding our next four years' rulers. I am glad our Continental brethren cannot read our newspapers of the present day, otherwise they must infer that our choice of rulers is made from a class more fitted for the state's prison than the state thrones, and elevation to a scaffold were more suited to the characters of the individual candidates than elevation to office. But in a few days matters will calm down, and the business of the nation will assume its wonted aspect.

"I have not engaged in this warfare. As a citizen I have my own views, and give my vote on general principles, but am prepared to learn that my vote is on the defeated side. I presume that Grant will be the president, and I shall defer to the decision like a peaceable citizen. The day after to-morrow you will know as well as we shall the probable result. The Telegraph is telling upon the world, and its effect upon human affairs is yet but faintly appreciated."

In this letter he also speaks of the death of his youngest brother, Richard C. Morse, who died at Kissingen on September 22, 1868, and in a letter to his son Arthur, of October 11, he again refers to it, and adds: "It is a sad blow to all of us but particularly to the large circle of his children. Your two uncles and your father were a three-fold cord, strongly united in affection. It is now sundered. The youngest is taken first, and we that remain must soon follow him in the natural course of things."

Farther on in this letter he says: "I attended the funeral of Mr. L—— a few weeks ago. I am told that he died of a broken heart from the conduct of his graceless son Frank, and I can easily understand that the course he has pursued, and his drunken habits, may have killed his father with as much certainty as if he had shot him. Children have little conception of the effect of their conduct upon their parents. They never know fully these anxieties until they are parents themselves."

But his skies were not all grey, for in addition to his

satisfaction in being once more at home in his own beloved country, and in his quiet retreat on the Hudson, he was soon to be the recipient of a signal mark of respect and esteem by his own countrymen, which proved that this prophet was not without honor even in his own country.

NEW YORK, November 30th, 1868.

PROFESSOR S. F. B. MORSE, LL.D.

SIR, — Many of your countrymen and numerous personal friends desire to give definite expression to the fact that this country is in full accord with European nations in acknowledging your title to the position of father of Modern Telegraphy, and at the same time in a fitting manner to welcome you to your home.

They, therefore, request that you will name a day on which you will favor them with your company at a public banquet.

With great respect we remain,

Very truly your friends.

Here follow the names of practically every man of prominence in New York at that time.

Morse replied on December 4: —

To the Hon. Hamilton Fish, Hon. John T. Hoffman, Hon. Wm. Dennison, Hon. A. G. Curtin, Hon. Wm. E. Dodge, Peter Cooper, Esq., Daniel Huntington, Esq., Wm. Orton, Esq., A. A. Low, Esq., James Brown, Esq., Cyrus W. Field, Esq., John J. Cisco, Esq., and others.

GENTLEMEN, — I have received your flattering request of the 30th November, proposing the compliment of a public banquet to me, and asking me to appoint a day on which it would be convenient for me to meet you.

Did your proposal intend simply a personal compliment I should feel no hesitation in thanking you cordially for this evidence of your personal regard, while I declined your proffered honor; but I cannot fail to perceive that there is a paramount patriotic duty connected with your proposal which forbids me to decline your invitation.

In accepting it, therefore, I would name (in view of some personal arrangements) Wednesday the 30th inst. as the day which would be most agreeable to me.

Accept, Gentlemen, the assurance of the respect of

Your obedient servant,

Samuel F. B. Morse.

The banquet was given at Delmonico's, which was then on the corner of Fifth Avenue and Fourteenth Street, and was presided over by Chief Justice Salmon P. Chase, who had been the leading counsel *against* Morse in his first great lawsuit, but who now cheerfully acknowledged that to Morse and America the great invention of the telegraph was due. About two hundred men sat down at the tables, among them some of the most eminent in the country. Morse sat at the right of Chief Justice Chase, and Sir Edward Thornton, British Ambassador, on his left. When the time for speechmaking came, Cyrus Field read letters from President Andrew Johnson; from General Grant, President-elect; from Speaker Colfax, Admiral Farragut, and many others. He also read a telegram from Governor Alexander H.

Bullock of Massachusetts: "Massachusetts honors her two sons — Franklin and Morse. The one conducted the lightning safely from the sky; the other conducts it beneath the ocean from continent to continent. The one tamed the lightning; the other makes it minister to human wants and human progress."

From London came another message: —

"CYRUS W. FIELD, New York. The members of the joint committee of the Anglo-American and Atlantic Telegraph Companies hear with pleasure of the banquet to be given this evening to Professor Morse, and desire to greet that distinguished telegraphist, and wish him all the compliments of the season."

Mr. Field added: "This telegram was sent from London at four o'clock this afternoon, and was delivered into the hands of your committee at 12.50." This, naturally, elicited much applause and laughter.

Speeches then followed by other men prominent in various walks of life. Sir Edward Thornton said that he "had great satisfaction in being able to contribute his mite of that admiration and esteem for Professor Morse which must be felt by all for so great a benefactor of his fellow creatures and of posterity."

Chief Justice Chase introduced the guest of the evening in the following graceful words: —

"Many shining names will at once occur to any one at all familiar with the history of the Telegraph. Among them I can pause to mention only those of Volta, the Italian, to whose discoveries the battery is due; Oersted, the Dane, who first discovered the magnetic properties of the electric current; Ampere and Arago, the Frenchmen, who prosecuted still further and most successfully similar researches; then Sturgeon, the Englishman, who may be said to have made the first electro-magnet; next, and not least illustrious among these illustrious men, our countryman Henry, who first showed the practicability of producing electro-magnetic effects by means of the galvanic current at distances infinitely great; and finally Steinheil, the German, who, after the invention of the Telegraph in all its material parts was complete, taught, in 1837, the use of the ground as part of the circuit. These are some of those searchers for truth whose names will be long held in grateful memory, and not among the least of their titles to gratitude and remembrance will be the discoveries which contributed to the possibility of the modern Telegraph.

"But these discoveries only made the Telegraph possible. They offered the brilliant opportunity. There was needed a man to bring into being the new art and the new interest to which they pointed, and it is the providential distinction and splendid honor of the eminent American, who is our guest to-night, that, happily prepared by previous acquirements and pursuits, he was quick to seize the opportunity and give to the world the first recording Telegraph.

"Fortunate man! thus to link his name forever with the greatest wonder and the greatest benefit of the age! [great applause] . . . I give you 'Our guest, Professor S. F. B. Morse, the man of science who explored the laws of nature, wrested electricity from her embrace, and made it a missionary in the cause of human progress.'"

As the venerable inventor rose from his chair, overcome with profound emotion which was almost too great to be controlled, the whole assembly rose with him, and cheer after cheer resounded through the hall for many minutes. When at last quiet was restored, he addressed the company at length, giving a resumé of his struggles and paying tribute to those who had befriended and assisted him in his time of need — to Amos Kendall, who sat at the board with him and whose name called forth more cheers, to Alfred Vail, to Leonard Gale, and, in the largeness of his heart, to F. O. J. Smith. It will not be necessary to give his remarks in full, as the history of the invention has already been given in detail in the course of this work, but his concluding remarks are worthy of record: —

"In casting my eyes around I am most agreeably greeted by faces that carry me back in memory to the days of my art struggles in this city, the early days of the National Academy of Design.

"Brothers (for you are yet brothers), if I left your ranks you well know it cost me a pang. I did not leave you until I saw you well established and entering on that career of prosperity due to your own just appreciation of the important duties belonging to your profession. You have an institution which now holds and, if true to yourselves, will continue to hold a high position in the estimation of this appreciative community. If I have stepped aside from Art to tread what seems another path, there is a good precedent for it in the lives of artists. Science and Art are not opposed. Leonardo da Vinci could find congenial relaxation in scientific researches and invention, and our own Fulton was a painter whose scientific studies resulted in steam navigation. It may not be generally known that the important invention of the *percussion cap* is due to the scientific recreations of the English painter Shaw.

"But I must not detain you from more instructive speech. One word only in closing. I have claimed for America the origination of the modern Telegraph System of the world. Impartial history, I think, will support that claim. Do not misunderstand me as disparaging or disregarding the labors and ingenious modifications of others in various countries employed in the same field of invention. Gladly, did time permit, would I descant upon their great and varied merits. Yet in tracing the birth and pedigree of the modern Telegraph, 'American' is not the highest term of the series that connects the past with the present; there is at least one higher term, the highest of all, which cannot and must not be ignored. If not a sparrow falls to the ground without a definite purpose in the plans of infinite wisdom, can the creation of an instrumentality so vitally affecting the interests of the whole human race have an origin less humble than the Father of every good and perfect gift?

"I am sure I have the sympathy of such an assembly as is here gathered if, in all humility and in the sincerity of a grateful heart, I use the words of inspiration in ascribing honor and praise to Him to whom first of all and most of all it is preëminently due. 'Not unto us, not unto us, but to God be all the glory.' Not what hath man, but 'What hath God wrought?'"

More applause followed as Morse took his seat, and other speeches were made by such men as Professor Goldwin Smith, the Honorable William M. Evarts, A. A. Low, William Cullen Bryant, William Orton, David

Dudley Field, the Honorable William E. Dodge, Sir Hugh Allan, Daniel Huntington, and Governor Curtin of Pennsylvania.

While many of these speeches were most eloquent and appropriate, I shall quote from only one, giving as an excuse the words of James D. Reid in his excellent work "The Telegraph in America": "As Mr. Huntington's address contains some special thoughts showing the relationship of the painter to invention, and is, besides, a most affectionate and interesting tribute to his beloved master, Mr. Morse, it is deemed no discourtesy to the other distinguished speakers to give it nearly entire."

I shall, however, omit some portions which Mr. Reid included.

"In fact, however, every studio is more or less a laboratory. The painter is a chemist delving into the secrets of pigments, varnishes, mixtures of tints and mysterious preparations of grounds and overlaying of colors; occult arts by which the inward light is made to gleam from the canvas, and the warm flesh to glow and palpitate.

"The studio of my beloved master, in whose honor we have met to-night, was indeed a laboratory. Vigorous, life-like portraits, poetic and historic groups, occasionally grew upon his easel; but there were many hours — yes, days — when absorbed in study among galvanic batteries and mysterious lines of wires, he seemed to us like an alchemist of the middle ages in search of the philosopher's stone.

"I can never forget the occasion when he called his pupils together to witness one of the first, if not the first, successful experiment with the electric telegraph. It was in the winter of 1835–36. I can see now that rude instrument, constructed with an old stretching-frame, a wooden clock, a home-made battery and the wire stretched many times around the walls of the studio. With eager interest we gathered about it as our master explained its operation while, with a click, click, the pencil, by a succession of dots and lines, recorded the message in cypher. The idea was born. The words circled that upper chamber as they do now the globe.

"But we had little faith. To us it seemed the dream of enthusiasm. We grieved to see the sketch upon the canvas untouched. We longed to see him again calling into life events in our country's history. But it was not to be; God's purposes were being accomplished, and now the world is witness to his triumph. Yet the love of art still lives in some inner corner of his heart, and I know he can never enter the studio of a painter and see the artist silently bringing from the canvas forms of life and beauty, but he feels a tender twinge, as one who catches a glimpse of the beautiful girl he loved in his youth whom another has snatched away.

"Finally, my dear master and father in art, allow me in this moment of your triumph in the field of discovery, to greet you in the name of your brother artists with 'All hail.' As an artist you might have spent life worthily in turning God's blessed daylight into sweet hues of rainbow colors, and into breathing forms for the delight and consolation of men, but it has been His will that you should train the lightnings, the sharp arrows of his anger, into the swift yet gentle messengers of Peace and Love."

Morse's wife and his daughter and other ladies had been present during the speeches, but they began to take their leave after Mr. Huntington's address, although the toastmaster arose to announce the last toast, which was "The Ladies." So he said: "This is the most inspiring theme of all, but the theme itself seems to be vanishing from us. Indeed [after a pause], has already vanished. [After another pause and a glance around the room.] And the gentleman who was to have responded seems also to have vanished with his theme. I may assume, therefore, that the duties of the evening are performed, and its enjoyments are at an end."

The unsought honor of this public banquet, in his own country, organized by the most eminent men of the day, calling forth eulogies of him in the public press of the whole world, was justly esteemed by Morse as one of the crowning events of his long career; but an even greater honor was still in store for him, which will be described in due season.

The early months of 1869 were almost entirely devoted to his report as Commissioner, which was finally completed and sent to the Department of State in the latter part of March. In this work he received great assistance from Professor W. P. Blake, who was "In charge of publication," and who writes to him on March 29: "I have had only a short time to glance at it as it was delivered towards the close of the day, but I am most impressed by the amount of labor and care you have so evidently bestowed upon it."

Professor Blake wrote another letter on August 21, which I am tempted to give almost in its entirety: —

"I feel it to be my duty to write to you upon another

point regarding your report, upon which I know that you are sensitive, but, as I think you will see that my motives are good, and that I sincerely express them, I believe you will not be offended with me although my views and opinions may not coincide exactly with yours. I allude to the mention which you make of some of the eminent physicists who have contributed by their discoveries and experiments to our knowledge of the phenomena of electro-magnetism.

"On page 9 of the manuscript you observe: 'The application of the electro-magnet, the invention of Arago and Sturgeon (first combined and employed by Morse in the construction of the generic telegraph) to the purposes also of the semaphore, etc.'

"Frankly, I am pained not to see the name of Henry there associated with those of Arago and Sturgeon, for it is known and generally conceded among men of science that his researches and experiments and the results which he reached were of radical importance and value, and that they deservedly rank with those of Ampere, Arago and Sturgeon.

"I am aware that, by some unfortunate combination of circumstances, the personal relations of yourself and Professor Henry are not pleasant. I deplore this, and it would be an intense satisfaction to me if I could be the humble means of bringing about a harmonious and honorable adjustment of the differences which separate you. I write this without conference with Professor Henry or his friends. I do it impartially, first, in the line of my duty as editor (but not now officially); second, as a lover of science; third, with a patriotic desire to secure as much as justly can be for the scientific reputation of the country; and fourth, with a desire to promote harmony between all who are concerned in increasing and disseminating knowledge, and particularly between such sincere lovers of truth and justice as I believe both yourself and Professor Henry to be.

"I do not find that Professor Henry anywhere makes a claim which trenches upon your claim of first using the electro-magnet for writing or printing at a distance — the telegraph as distinguished from the semaphore. This he cannot claim, for he acknowledges it to be yours. You, on the other hand, do not claim the semaphoric use of electricity. I therefore do not see any obstacle to an honorable adjustment of the differences which separate you, and which, perhaps, make you disinclined to freely associate Professor Henry's name with those of other promoters of electrical science.

"Your report presents a fitting opportunity to effect this result. A magnanimous recognition by you of Professor Henry's important contributions to the science of electro-magnetism appears to me to be all that is necessary. They can be most appropriately and gracefully acknowledged in your report, and you will gain rather than lose by so doing. Such action on your part would do more than anything else could to secure for you the good will of all men of science, and to hasten a universal and generous accord of all the credit for your great gift to civilization that you can properly desire.

"Now, my dear sir, with this frank statement of my views on this point, I accept your invitation, and will go to see you at your house to talk with you upon this point and others, perhaps more agreeable, but if, after this expression of my inclinations, you will not deem me a welcome guest, telegraph me not to come — I will not take it unkindly."

To this Morse replied on August 23: "Your most acceptable letter, with the tone and spirit of which I am most gratified, is just received, for which accept my thanks. I shall be most happy to see you and freely to communicate with you on the subject mentioned, and with the sincere desire of a satisfactory result."

The visit was paid, but the details of the conversation have not been preserved. However, we find in Morse's report, on page 10, the following: "In 1825, Mr. Sturgeon, of England, made the first electro-magnet in the horseshoe form by loosely winding a piece of iron wire with a spiral of copper wire. In the United States, as early as 1831, the experimental researches of Professor Joseph Henry were of great importance in advancing the science of electro-magnetism. He may be said to have carried the electro-magnet, in its lifting powers, to its greatest perfection. Reflecting upon the principle of Professor Schweigger's galvanometer, he constructed magnets in which great power could be developed by a very small galvanic element. His published paper in 1831 shows that he experimented with wires of different lengths, and he noted the amount of magnetism which could be induced through them at various lengths by means of batteries composed of a single element, and also of many elements. He states that the magnetic action of 'a current from a trough composed of many pairs is at least not sensibly diminished by passing through a long wire,' and he incidentally noted the bearing of this fact upon the project of an electro-magnetic telegraph [semaphore?].

"In more recent papers, first published in 1857, it appears that Professor Henry demonstrated before his pupils the practicability of ringing a bell, by means of electro-magnetism, at a distance."

Whether Professor Blake was satisfied with this change from the original manuscript is not recorded. Morse evidently thought that he had made the *amende honorable*, but Henry, coldly proud man that he was, still held aloof from a reconciliation, for I have been informed that he even refused to be present at the memorial services held in Washington after the death of Morse.

In a letter of May 10, 1869, to Dr. Leonard Gale, some interesting facts concerning the reading by sound are given: —

"The fact that the lever action of the earliest instrument of 1835 by its click gave the sound of the numerals, as embodied in the original type, is well known, nor is there anything so remarkable in that result. . . . When you first saw the instrument in 1836 this was so obvious that it scarcely excited more than a passing remark, but, after the adaptation of the dot and space, with the addition of the line or dash, in forming the alphabetic signs (which, as well as I can remember, was about the same date, late in 1835 or early in 1836) then I noticed that the different letters had each their own individual sounds, and could also be distinguished from each other by the sound. The fact did not then appear to me to be of any great importance, seeming to be more curious than useful, yet, in reflecting upon it, it seemed desirable to secure this result by specifying it in my letters patent, lest it might be used as an *evasion* in indicating my novel alphabet without recording it. Hence the *sounds* as well as the imprinted signs were specified in my letters patent.

"As to the time when these sounds were *practically* used, I am unable to give a precise date. I have a distinct recollection of one case, and proximately the date of it. The time of the incident was soon after the line was extended from Philadelphia to Washington, having a way station at Wilmington, Delaware. The Washington office was in the old post-office, in the room above it. I was in the operating room. The instruments were for a moment silent. I was standing at some distance near the fire-place conversing with Mr. Washington, the operator, who was by my side. Presently one of the instruments commenced writing and Mr. Washington listened and smiled. I asked him why he smiled. 'Oh!' said he, 'that is Zantzinger of the Philadelphia office, but he is operating from Wilmington.' 'How do you know that?' 'Oh! I know his touch, but I must ask him why he is in Wilmington.' He then went to the instrument and telegraphed to Zantzinger at Wilmington, and the reply was that he had been sent from Philadelphia to regulate the relay magnet for the Wilmington operator, who was inexperienced in operating. . . .

"I give this instance, not because it was the *first*, but because it is one which I had specially treasured in my memory and frequently related as illustrative of the practicality of reading by *sound* as well as by the written record. This must have occurred about the year 1846."

A serious accident befell the aged inventor, now seventy-nine years old, in July, 1869. He slipped on the stairs of his country house and fell with all his weight on his left leg, which was broken in two places. This mishap confined him to his bed for three months, and many feared that, owing to his advanced age, it would be fatal. But, thanks to his vigorous constitution and his temperate life, he recovered completely. He bore this affliction with Christian fortitude. In a letter to his brother Sidney, of August 14, he says: "The healing process in my leg is very slow. The doctor, who has just left me, condemns me to a fortnight more of close confinement. I have other troubles, for they come not singly, but all is for the best."

Troubles, indeed, came not singly, for, in addition to sorrows of a domestic nature, his friends one by one were taken from him by death, and on November 12, 1869, he writes to William Stickney, Esq., son-in-law of Amos Kendall: —

"Although prepared by recent notices in the papers to expect the sad news, which a telegram this moment received announces to me, of the death of my excellent, long-tried friend Mr. Kendall, I confess that the intelligence has come with a shock which has quite unnerved me. I feel the loss as of a *father* rather than of a brother in age, for he was one in whom I confided as a father, so sure was I of affectionate and sound advice. . . .

"I need not tell you how deeply I feel this sad bereavement. I am truly and severely bereaved in the loss of such a friend, a friend, indeed, upon whose faithfulness and unswerving integrity I have ever reposed with perfect confidence, a confidence which has never been betrayed, and a friend to whose energy and skill, in the conduct of the agency which I had confided to him, I owe (under God) the comparative comfort which a kind

Providence has permitted me to enjoy in my advanced age."

In the following year he was called upon to mourn the death of still another of his good friends, for, on August 24, 1870, George Wood died very suddenly at Saratoga.

While much of sadness and sorrow clouded the evening of the life of this truly great man, the sun, ere it sank to rest, tinged the clouds with a glory seldom vouchsafed to a mortal, for he was to see a statue erected to him while he was yet living. Of many men it has been said that — "Wanting bread they receive only a stone, and not even that until long after they have been starved to death." It was Morse's good fortune not only to see the child of his brain grow to a sturdy manhood, but to be honored during his lifetime to a truly remarkable degree.

The project of a memorial of some sort to the Inventor of the Telegraph was first broached by Robert B. Hoover, manager of the Western Union Telegraph office, Allegheny City, Pennsylvania. The idea once started spread with the rapidity of the electric fluid itself, and, under the able management of James D. Reid, a fund was raised, partly by dollar subscriptions largely made by telegraph operators all over the country, including Canada, and it was decided that the testimonial should take the form of a bronze statue to be erected in Central Park, New York. Byron M. Pickett was chosen as the sculptor, and the Park Commission readily granted permission to place the statue in the park.

It was at first hoped that the unveiling might take place on the 27th of April, 1871, Morse's eightieth birthday; but unavoidable delays arose, and it was not until the 10th of June that everything was in readiness. It was a perfect June day and the hundreds of telegraphers from all parts of the country, with their families, spent the forenoon in a steamboat excursion around the city. In the afternoon crowds flocked to the park where, near what is now called the "Inventor's Gate," the statue stood in the angle between two platforms for the invited guests. Morse himself refused to attend the ceremonies of the unveiling of his counterfeit presentment, as being too great a strain on his innate modesty. Some persons and some papers said that he was present, but, as Mr. James D. Reid says in his "Telegraph in America," "Mr. Morse was incapable of such an indelicacy. . . . Men of refinement and modesty would justly have marvelled had they seen him in such a place."

At about four o'clock the Governor of New York, John T. Hoffman, delivered the opening address, saying, in the course of his speech: "In our day a new era has dawned. Again, for the second time in the history of the world, the power of language is increased by human agency. Thanks to Samuel F. B. Morse men speak to one another now, though separated by the width of the earth, with the lightning's speed and as if standing face to face. If the inventor of the alphabet be deserving of the highest honors, so is he whose great achievement marks this epoch in the history of language — the inventor of the Electric Telegraph. We intend, so far as in us lies, that the men who come after us shall be at no loss to discover his name for want of recorded testimony."

Governor Claflin, of Massachusetts, and William

Orton, president of the Western Union Telegraph Company, then drew aside the drapery amidst the cheers and applause of the multitude, while the Governor's Island band played the "Star-Spangled Banner."

William Cullen Bryant, who was an early friend of the inventor, then presented the statue to the city in an eloquent address, from which I shall quote the following words: —

"It may be said, I know, that the civilized world is already full of memorials which speak the merit of our friend and the grandeur and utility of his invention. Every telegraphic station is such a memorial. Every message sent from one of these stations to another may be counted among the honors paid to his name. Every telegraphic wire strung from post to post, as it hums in the wind, murmurs his eulogy. Every sheaf of wires laid down in the deep sea, occupying the bottom of soundless abysses to which human sight has never penetrated, and carrying the electric pulse, charged with the burden of human thought, from continent to continent, from the Old World to the New, is a testimonial to his greatness. . . . The Latin inscription in the church of St. Paul's in London, referring to Sir Christopher Wren, its architect, — 'If you would behold his monument, look around you,' — may be applied in a far more comprehensive sense to our friend, since the great globe itself has become his monument."

The Mayor of New York, A. Oakey Hall, accepted the statue in a short speech, and, after a prayer by the Reverend Stephen H. Tyng, D.D., the assembled multitude joined in singing the doxology, and the ceremonies at the park were ended.

But other honors still awaited the venerable inventor, for, on the evening of that day, the old Academy of Music on Fourteenth Street was packed with a dense throng gathered together to listen to eulogies on this benefactor of his race, and to hear him bid farewell to his children of the Telegraph. A table was placed in the centre of the stage on which was the original instrument used on the first line from Washington to Baltimore. This was connected with all the lines of telegraph extending to all parts of the world. The Honorable William Orton presided, and, after the Reverend Howard Crosby had opened the ceremonies with prayer, speeches were delivered by Mr. Orton, Dr. George B. Loring, of Salem, and the Reverend Dr. George W. Samson.

At nine o'clock Mr. Orton announced that all lines were clear for the farewell message of the inventor to his children; that this message would be flashed to thousands of waiting operators all over the world, and that answers would be received during the course of the evening. The pleasant task of sending the message had been delegated to Miss Sadie E. Cornwell, a skilful young operator of attractive personality, and Morse himself was to manipulate the key which sent his name, in the dots and dashes of his own alphabet, over the wires.

The vast audience was hushed into absolute silence as Miss Cornwell clicked off the message which Morse had composed for the occasion: "Greeting and thanks to the Telegraph fraternity throughout the world. Glory to God in the highest, on earth peace, good will to men."

As Mr. Orton escorted Morse to the table a tremendous burst of applause broke out, but was silenced by a gesture from the presiding officer, and again the great audience was still. Slowly the inventor spelled out the letters of his name, the click of the instrument being clearly heard in every part of the house, and as clearly understood by the hundreds of telegraphers present, so that without waiting for the final dot, which typified the letter e, the whole vast assembly rose amid deafening cheers and the waving of handkerchiefs.

It was an inspiring moment, and the venerable man was almost overcome by his emotions, and sat for some time with his head buried in his hands, striving to regain his self-control.

When the excitement had somewhat subsided, Mr. Orton said: "Thus the Father of the Telegraph bids farewell to his children."

The current was then switched to an instrument behind the scenes, and answers came pouring in, first from near-by towns and cities, and then from New Orleans, Quebec, San Francisco, Halifax, Havana, and finally from Hongkong, Bombay, and Singapore.

Mr. Reid has given a detailed account of these messages in his "Telegraph in America," but I shall not pause to reproduce them here; neither shall I quote from the eloquent speeches which followed, delivered by General N. P. Banks, the Reverend H. M. Gallagher, G. K. Walcott, and James D. Reid. After Miss Antoinette Sterling had sung "Auld Lang Syne," to the great delight of the audience, who recalled her several times, Chief Justice Charles P. Daly introduced Professor Morse in an appropriate address.

As the white-haired inventor, in whose honor this great demonstration had been organized, stepped forward to deliver his valedictory, he was greeted with another round of cheering and applause. At first almost overcome by emotion, he soon recovered his self-control, and he read his address in a clear, resonant voice which carried to every part of the house. The address was a long one, and as most of it is but a recapitulation of what has been already given, I shall only quote from it in part: —

"FRIENDS AND CHILDREN OF THE TELEGRAPH, — When I was solicited to be present this evening, in compliance with the wishes of those who, with such zeal and success, responded to the suggestion of one of your number that a commemorative statue should be erected in our unrivaled Park, and which has this day been placed in position and unveiled, I hesitated to comply. Not that I did not feel a wish in person to return to you my heartfelt thanks for this unique proof of your personal regard, but truly from a fear that I could use no terms which would adequately express my appreciation of your kindness. Whatever I say must fall short of expressing the grateful feelings or conflicting emotions which agitate me on an occasion so unexampled in the history of invention. Gladly would I have shrunk from this public demonstration were it not that my absence to-night, under the circumstances, might be construed into an apathy which I do not feel, and which your overpowering kindness would justly rebuke. . . .

"You have chosen to impersonate in my humble effigy an invention which, cradled upon the ocean, had its birth in an American ship. It was nursed and cherished not so much from personal as from patriotic motives. Forecasting its future, even at its birth, my most powerful stimulus to perseverance through all the perils and trials of its early days — and they were neither few nor insignificant — was the thought that it must inevitably be world-wide in its application, and, moreover, that it would everywhere be hailed as a grateful American gift to the nations. It is in this aspect of the present occasion that I look upon your proceedings as intended, not so much as homage to an individual, as to the invention, 'whose lines [from America] have gone out through all the earth, and their words to the end of the world.'

"In the carrying-out of any plan of improvement, however grand or feasible, no single individual could possibly accomplish it without the aid of others. We are none of us so powerful that we can dispense with the assistance, in various departments of the work, of those whose experience and knowledge must supply the needed aid of their expertness. It is not sufficient that a brilliant project be proposed, that its modes of accomplishment are foreseen and properly devised; there are, in every part of the enterprise, other minds and other agencies to be consulted for information and counsel to perfect the whole plan. The Chief Justice, in delivering the decision of the Supreme Court, says: 'It can make no difference whether he [the inventor] derives his information from books or from conversation with men skilled in the science.' And: 'The fact that Morse sought and obtained the necessary information and counsel from the best sources, and acted upon it, neither impairs his rights as an inventor nor detracts from his merits.'

"The inventor must seek and employ the skilled mechanician in his workshop to put the invention into practical form, and for this purpose some pecuniary means are required as well as mechanical skill. Both these were at hand. Alfred Vail, of Morristown, New Jersey, with his father and brother, came to the help of the unclothed infant, and with their funds and mechanical skill put it into a condition to appear before the Congress of the nation. To these New Jersey friends is due the first important aid in the progress of the invention. Aided also by the talent and scientific skill of Professor Gale, my esteemed colleague in the University, the Telegraph appeared in Washington in 1838, a suppliant for the means to demonstrate its power. To the Honorable F. O. J. Smith, then chairman of the House Committee of Commerce, belongs the credit of a just appreciation of the new invention, and of a zealous advocacy of an experimental essay, and the inditing of an admirably written report in its favor, signed by every member of the committee. . . . To Ezra Cornell, whose noble benefactions to his state and the country have placed his name by the side of Cooper and Peabody high on the roll of public benefactors, is due the credit of early and effective aid in the superintendence and erection of the first public line of telegraph ever established."

After paying tribute to the names of Amos Kendall, Cyrus Field, Volta, Oersted, Arago, Schweigger, Gauss and Weber, Steinheil, Daniell, Grove, Cooke, Dana, Henry, and others, he continued: —

"There is not a name I have mentioned, and many whom I have not mentioned, whose career in science or experience in mechanical and engineering and nautical tactics, or in financial practice, might not be the theme of volumes rather than of brief mention in an ephemeral address.

"To-night you have before you a sublime proof of the grand progress of the Telegraph in its march round the globe. It is but a few days since that our veritable antipodes became telegraphically united to us. We can speak to and receive an answer in a few seconds of time from Hongkong in China, where ten o'clock to-night here is ten o'clock in the day there, and it is, perhaps, a debatable question whether their ten o'clock is ten to-day or ten to-morrow. China and New York are in interlocutory communication. We know the fact, but can imagination realize the fact?

"But I must not further trespass on your patience at this late hour. I cannot close without the expression of my cordial thanks to my long-known, long-tried and honored friend Reid, whose unwearied labors early contributed so effectively to the establishment of telegraph lines, and who, in a special manner as chairman of your Memorial Fund, has so faithfully, and successfully, and admirably carried to completion your flattering design. To the eminent Governors of this state and the state of Massachusetts, who have given to this demonstration their honored presence; to my excellent friend the distinguished orator of the day; to the Mayor and city authorities of New York; to the Park Commissioners; to the officers and managers of the various, and even rival, telegraph companies, who have so cordially united on this occasion; to the numerous citizens, ladies and gentlemen; and, though last not least, to every one of my large and increasing family of telegraph children who have honored me with the proud title of Father, I tender my cordial thanks."

CHAPTER XL

JUNE 14, 1871 — APRIL 16, 1872

Nearing the end. — Estimate of the Reverend F. B. Wheeler. — Early poem. — Leaves "Locust Grove" for last time. — Death of his brother Sidney. — Letter to Cyrus Field on neutrality of telegraph. — Letter of F. O. J. Smith to H. J. Rogers. — Reply by Professor Gale. — Vicious attack by F. O. J. Smith. — Death prevents reply by Morse. — Unveils statue of Franklin in last public appearance. — Last hours. — Death. — Tributes of James D. Reid, New York "Evening Post," New York "Herald," and Louisville "Courier-Journal." — Funeral. — Monument in Greenwood Cemetery. — Memorial services in House of Representatives, Washington. — Address of James G. Blaine. — Other memorial services. — Mr. Prime's review of Morse's character. — Epilogue.

THE excitement caused by all these enthusiastic demonstrations in his honor told upon the inventor both physically and mentally, as we learn from a letter of June 14, 1871, to his daughter Mrs. Lind and her husband: —

"So fatigued that I can scarcely keep my eyes open, I nevertheless, before retiring to my bed, must drop you a line of enquiry to know what is your condition. We have only heard of your arrival and of your first unfavorable impressions. I hope these latter are removed, and that you are both benefiting by change of air and the waters of the Clifton Springs.

"You know how, in the last few days, we have all been overwhelmed with unusual cares. The grand ceremonies of the Park and the Academy of Music are over, but have left me in a good-for-nothing condition. Everything went off splendidly, indeed, as you will learn from the papers. . . . I find it more difficult to bear up with the overwhelming praise that is poured out without measure, than with the trials of my former life. There is something so remarkable in this universal laudation that the effect on me, strange as it may seem, is rather depressing than exhilarating.

"When I review my past life and see the way in which I have been led, I am so convinced of the faithfulness of God in answer to the prayers of faith, which I have been enabled in times of trial to offer to Him, that I find the temper of my mind is to constant praise. 'Bless the Lord, Oh my soul, and forget not all his benefits!' is ever recurring to me. It is doubtless this continued referring all to Him that prevents this universal demonstration of kindly feeling from puffing me up with the false notion that I am anything but the feeblest of instruments. I cannot give you any idea of the peculiar feelings which gratify and yet oppress me."

He had planned to cross the ocean once more, partly as a delegate to Russia from the Evangelical Alliance, and partly to see whether it would not be possible to induce Prussia and Switzerland and other European nations, from whom he had as yet received no pecuniary remuneration, to do him simple justice. But, for various reasons, this trip was abandoned, and from those nations he never received anything but medals and praise.

So the last summer of the aged inventor's life was spent at his beloved Locust Grove, not free from care and anxiety, as he so well deserved, but nevertheless, thanks to his Christian philosophy, in comparative serenity and happiness. His pastor in Poughkeepsie, the Reverend F. B. Wheeler, says of him in a letter to Mr. Prime: "In his whole character and in all his relations he was one of the most remarkable men of his age. He was one who drew all who came in contact with him to his heart, disarming all prejudices, silencing all cavil. In his family he was light, life, and love; with those in his employ he was ever considerate and kind, never exacting and harsh, but honorable and just, seeking the good of every dependent; in the community he was a pillar of strength and beauty, commanding the homage of universal respect; in the Church he walked with God and men."

That he was a man of great versatility has been shown in the recital of his activities as artist, inventor, and writer; that he had no mean ability as a poet is also on record. On January 6, 1872, he says in a letter to his cousin, Mrs. Thomas R. Walker: "Some years ago, when both of us were younger, I remember addressing to you a trifle entitled 'The Serenade,' which, on being shown to Mr. Verplanck, was requested for publication in the 'Talisman,' edited and conducted by him and Mr. Sands. I have not seen a copy of that work for many years, and have preserved no copy of 'The Serenade.' If you have a copy I should be pleased to have it."

He was delicately discreet in saying "some years ago," for this poem was written in 1827 as the result of a wager between Morse and his young cousin, he having asserted that he could write poetry as well as paint pictures, and requesting her to give him a theme. It seems that the young lady had been paid the compliment of a serenade a few nights previously, but she had, most unromantically, slept through it all, so she gave as her theme "The Serenade," and the next day Morse produced the following poem: —

THE SERENADE

Haste! 't is the stillest hour of night,
The Moon sheds down her palest light,
And sleep has chained the lake and hill,
The wood, the plain, the babbling rill;
And where yon ivied lattice shows
My fair one slumbers in repose.
Come, ye that know the lovely maid,
And help prepare the serenade.
Hither, before the night is flown,
Bring instruments of every tone.
But lest with noise ye wake, not lull,
Her dreaming fancy, ye must cull
Such only as shall soothe the mind
And leave the harshest all behind.
Bring not the thundering drum, nor yet
The harshly-shrieking clarionet,
Nor screaming hautboy, trumpet shrill,
Nor clanging cymbals; but, with skill,
Exclude each one that would disturb
The fairy architects, or curb
The wild creations of their mirth,
All that would wake the soul to earth.
Choose ye the softly-breathing flute,
The mellow horn, the loving lute;
The viol you must not forget,
And take the sprightly flageolet
And grave bassoon; choose too the fife,
Whose warblings in the tuneful strife,
Mingling in mystery with the words,
May seem like notes of blithest birds.

Are ye prepared? Now lightly tread
As if by elfin minstrels led,
And fling no sound upon the air
Shall rudely wake my slumbering fair.
Softly! Now breathe the symphony,
So gently breathe the tones may vie
In softness with the magic notes
In visions heard; music that floats
So buoyant that it well may seem,
With strains ethereal in her dream,
One song of such mysterious birth
She doubts it comes from heaven or earth.

Play on! My loved one slumbers still.
Play on! She wakes not with the thrill
Of joy produced by strains so mild,
But fancy moulds them gay and wild.
Now, as the music low declines,
'T is sighing of the forest pines;
Or 't is the fitful, varied war
Of distant falls or troubled shore.
Now, as the tone grows full or sharp,
'T is whispering of the Æolian harp.
The viol swells, now low, now loud,
'T is spirits chanting on a cloud
That passes by. It dies away;
So gently dies she scarce can say
'T is gone; listens; 't is lost she fears;
Listens, and thinks again she hears.
As dew drops mingling in a stream
To her 't is all one blissful dream,
A song of angels throned in light.
Softly! Away! Fair one, good-night.

In the autumn of 1871 Morse returned with his family to New York, and it is recorded that, with an apparent premonition that he should never see his beloved Locust Grove again, he ordered the carriage to stop as he drove out of the gate, and, standing up, looked long and lovingly at the familiar scene before telling the coachman to drive on. And as he passed the rural cemetery on the way to the station he exclaimed: "Beautiful! beautiful! but I shall not lie there. I have prepared a place elsewhere."

Not long after his return to the city death once more laid its heavy hand upon him in the loss of his sole surviving brother, Sidney. While this was a crushing blow, for these two brothers had been peculiarly attached to each other, he bore it with Christian resignation, confident that the separation would be for a short time only — "We must soon follow, I also am over eighty years, and am waiting till my change comes."

But his mind was active to the very end, and he never ceased to do all in his power for the welfare of mankind. One of the last letters written by him on a subject of public importance was sent on December 4, 1871, to Cyrus Field, who was then attending an important telegraphic convention in Rome: —

"Excuse my delay in writing you. The excitement occasioned by the visit of the Grand Duke Alexis has but just ceased, and I have been wholly engrossed by the various duties connected with his presence. I have wished for a few calm moments to put on paper some thoughts respecting the doings of the great Telegraphic Convention to which you are a delegate.

"The Telegraph has now assumed such a marvellous position in human affairs throughout the world, its influences are so great and important in all the varied concerns of nations, that its efficient protection from injury has become a necessity. It is a powerful advocate for universal peace. Not that of itself it can command a 'Peace, be still!' to the angry waves of human passions, but that, by its rapid interchange of thought and opinion, it gives the opportunity of explanations to acts and to laws which, in their ordinary wording, often create doubt and suspicion. Were there no means of quick explanation it is readily seen that doubt and suspicion, working on the susceptibilities of the public mind, would engender misconception, hatred and strife. How important then that, in the intercourse of nations, there should be the ready means at hand for prompt correction and explanation.

"Could there not be passed in the great International Convention some resolution to the effect that, in whatever condition, whether of Peace or War between the nations, the Telegraph should be deemed a sacred thing, to be by common consent effectually protected both on the land and beneath the waters?

"In the interest of human happiness, of that 'Peace on Earth' which, in announcing the advent of the Saviour, the angels proclaimed with 'good will to men,' I hope that the convention will not adjourn without adopting a resolution asking of the nations their united, effective protection to this great agent of civilization."

Richly as he deserved that his sun should set in an unclouded sky, this was not to be. Sorrows of a most intimate nature crowded upon him. He was also made the victim of a conscienceless swindler who fleeced him of many thousand dollars, and, to crown all, his old and indefatigable enemy, F. O. J. Smith, administered a cowardly thrust in the back when his weakening powers prevented him from defending himself with his old-time vigor. From a very long letter written by Smith on December 11, 1871, to Henry J. Rogers in Washington, I shall quote only the first sentences: —

Dear Sir, — In my absence your letter of the 11th ult. was received here, with the printed circular of the National Monumental Society, in reply to which I feel constrained to say if that highly laudable association resolves "to erect at the national capital of the United States a memorial monument" to symbolize in statuary of colossal proportions the "history of the electro-magnetic telegraph," before that history has been authentically written, it is my conviction that the statue most worthy to stand upon the pedestal of such monument would be that of the man of true science, who explored the laws of nature ahead of all other men, and was "the first to wrest electro-magnetism from Nature's embrace and make it a missionary to the cause of human progress," and that man is Professor Joseph Henry, of the Smithsonian Institution.

Professor Morse and his early coadjutors would more appropriately occupy, in groups of high relief, the sides of that pedestal, symbolizing, by their established merits and coöperative works, the grandeur of the researches and resulting discoveries of their leader and chief, who was the first to announce and to demonstrate to a despairing world, by actual mechanical agencies, the practicability of an electro-magnetic telegraph through any distances.

Much more of the same flatulent bombast follows which it will not be necessary to introduce here. While Morse himself naturally felt some delicacy in noticing such an attack as this, he found a willing and efficient champion in his old friend (and the friend of Henry as well) Professor Leonard D. Gale, who writes to him on January 22, 1872: —

"I have lately seen a mean, unfair, and villainous letter of F. O. J. Smith, addressed to H. J. Rogers (officer of the Morse Monumental Association), alleging that the place on the monument designed to be occupied by the statue of Morse, should be awarded to Henry; that Morse was not a scientific man, etc., etc. It was written in his own peculiar style. The allegations were so outrageous that I felt it my duty to reply to it without delay. As Smith's letter was to Rogers, as an officer of the Association, I sent my reply to the same person. I enclose a copy herewith.

"Mrs. Gale suggests an additional figure to the group on the monument — a serpent with the face of F. O. J. S., biting the heel of Morse, but with the fangs extracted."

Professor Gale's letter to Henry J. Rogers is worthy of being quoted in full: —

"I have just read a letter from F. O. J. Smith, dated December 11, 1871, addressed to you, and designed to throw discredit on Morse's invention of the Telegraph, the burden of which seems to be rebuke to the designer of the monument, for elevating Morse to the apex of the monument and claiming for Professor J. Henry, of the Smithsonian Institution, that high distinction.

"The first question of an impartial inquirer is: 'To which of these gentlemen is the honor due?' To ascertain this we will ask a second question: 'Was the subject of the invention a *machine*, or was it *a new fact in science?*' The answer is: 'It was a *machine*.' The first was Morse's, the latter was Henry's. Henry stated that electric currents might be sent through long distances applicable to telegraphic purposes. Morse took the facts as they then existed, invented a machine, harnessed the steed therein, and set the creature to work. There is honor due to Henry for his great discovery of the scientific principle; there is honor also due to Morse for his invention of the ingenious machine which accomplishes the work.

"Men of science regard the discovery of a new fact in science as a higher attainment than the application of it to useful purposes, while the world at large regards the *application* of the principle or fact in science to the useful arts as of paramount importance. All honor to the discoverer of a new fact in science; equal honor to him who utilizes that fact for the benefit of mankind.

"Has the world forgotten what Robert Fulton did for the navigation of the waters by steamboats? It was he who first applied steam to propel a vessel and navigated the Hudson for the first time with steam and paddle-wheels and vessel in 1807. Do not we honor him as the Father of steamboats? Yet Fulton did not invent steam, nor the steam-engine, nor paddle-wheels, nor the vessel. He merely adapted a steam-engine to a vessel armed with paddle-wheels. The combination was his invention.

"There is another example on record. Cyrus H. McCormick, the Father of the Reaping and Mowing Machine, took out the first successful patent in 1837, and is justly acknowledged the world over as the inventor of this great machine. Although one hundred and forty-six patents were granted in England previous to McCormick's time, they are but so many unsuccessful efforts to perfect a practical machine. The cutting apparatus, the device to raise and lower the cutters, the levers, the platform, the wheels, the framework, had all been used before McCormick's time. But McCormick was the first genius able to put these separate devices together in a practical, harmonious operation. The combination was his invention.

"Morse did more. He invented the form of the various parts of his machine as well as their combination; he was the first to put such a machine into practical operation; and for such a purpose who can question his title as the Inventor of the Electric Telegraph?"

To the letter of Professor Gale, Morse replied on January 25: —

"Thank you sincerely for your effective interference in my favor in the recent, but not unexpected, attack of F. O. J. S. I will, so soon as I can free myself from some very pressing matters, write you more fully on the subject. Yet I can add nothing to your perfectly clear exposition of the difference between a discovery of a principle in science and its application to a useful purpose. . . . As for Smith's suggestion of putting Henry on the top of the proposed monument, I can hardly suppose Professor H. would feel much gratification on learning the character of his zealous advocate. It is simply a matter of spite; carrying out his intense and smothered antipathy to me, and not for any particular regard for Professor H.

"As I have had nothing to do with the proposed monument, I have no feeling on the subject. If they who have the direction of that monument think the putting of Professor H. on the apex will meet the applause of the public, including the expressed opinion of the entire world, by all means put him there. I certainly shall make no complaint."

The monument was never erected, and this effort of Smith's to humiliate Morse proved abortive. But his spite did not end there, as we learn from the following letter written by Morse on February 26, 1872, to the Reverend Aspinwall Hodge, of Hartford, Connecticut, the husband of one of his nieces: —

"Some unknown person has sent me the advance sheets of a work (the pages between 1233 and 1249) publishing in Hartford, the title of which is not given, but I think is something like 'The Great Industries of the United States.' The pages sent me are entitled 'The American Magnetic Telegraph.' They contain the most atrocious and vile attack upon me which has ever appeared in print. I shall be glad to learn who are the publishers of this work, what are the characters of the publishers, and whether they will give me the name or names of the author or authors of this diatribe, and whether they vouch for the character of those who furnished the article for their work.

"I know well enough, indeed, who the libellers are and their motives, which arise from pure spite and revenge for having been legally defeated parties in cases relating to the Telegraph before the courts. To you I can say the concocters of this tirade are F. O. J. Smith, of bad notoriety, and Henry O'Reilly.

"Are the publishers responsible men, and are they aware of the character of those who have given them that article, particularly the moral character of Smith, notorious for his debaucheries and condemned in court for subornation of perjury, and one of the most revengeful men, who has artfully got up this tirade because my agent, the late Honorable Amos Kendall, was compelled to resist his unrighteous claim upon me for some $25,000 which, after repeated trials lasting some twelve years, was at length, by a decision of the Supreme Court of the United States, decided against him, and he was adjudged to owe me some $14,000?

"Mr. Kendall, previous to his decease, managed the case which has thus resulted. The necessity of seizing some property of his in the city of Williamsburg, through the course of the legal proceedings, has aroused his revengeful feelings, and he has openly threatened that he would be revenged upon me for it, and he has for two or three years past with O'Reilly been concocting this mode of revenge.

"If the publishers are respectable men, I think they will regret that they have been the dupes of these arch conspirators. If not too late to suppress that article I should be glad of an interview with them, in which I will satisfy them that they have been most egregiously imposed upon."

This was the last flash of that old fire which, when he was sufficiently aroused by righteous indignation at unjust attacks, had enabled him to strike out vigorously in self-defense, and had won him many a victory. He was now nearing the end of his physical resources. He had fought the good fight and he had no misgivings as to the verdict of posterity on his achievements. He could fight no more, willing and mentally able though he was to confound his enemies again. He must leave it to others to defend his fame and good name in the future. The last letter which was copied into his letter-press book was written on March 14, not three weeks before the last summons came to him, and it refers to his old enemy who thus pursued him even to the brink of the grave. It is addressed to F. J. Mead, Esq.: —

"Although forbidden to read or write by my physician, who finds me prostrate with a severe attack of neuralgia in the head, I yet must thank you for your kind letter of the 12th inst.

"I should be much gratified to know what part Professor Henry has taken, if any, in this atrocious and absurd attack of F. O. J. S. I have no fears of the result, but no desire either to suspect any agency on the part of Professor Henry. It is difficult for me to conceive that a man in his position should not see the true position of the matter."

This vicious attack had no effect upon his fame. Dying as soon as it was born, choked by its own venom, it was overwhelmed by the wave of sorrow and sympathy which swept over the earth at the announcement of the death of the great inventor.

His last public appearance was on January 17, 1872, when he, in company with Horace Greeley, unveiled the statue of Benjamin Franklin in Printing House Square, New York. It was a very cold day, but, against the advice of his physician and his family, he insisted on being present. As he drove up in his carriage and, escorted by the committee, ascended to the platform, he was loudly cheered by the multitude which had assembled. Standing uncovered in the biting air, he delivered the following short address: —

"MR. DE GROOT AND FELLOW-CITIZENS, — I esteem it one of my highest honors that I should have been designated to perform the office of unveiling this day the fine statue of our illustrious and immortal Franklin. When requested to accept this duty I was confined to my bed, but I could not refuse, and I said: 'Yes, if I have to be lifted to the spot!'

"Franklin needs no eulogy from me. No one has more reason to venerate his name than myself. May his illustrious example of devotion to the interest of universal humanity be the seed of further fruit for the good of the world."

Morse was to have been an honored guest at the

banquet in the evening, where in the speeches his name was coupled with that of Franklin as one of the great benefactors of mankind; but, yielding to the wishes of his family, he remained at home. He had all his life been a sufferer from severe headaches, and now these neuralgic pains increased in severity, no doubt aggravated by his exposure at the unveiling. When the paroxysms were upon him he walked the floor in agony, pressing his hands to his temples; but these seizures were, mercifully, not continuous, and he still wrote voluminous letters, and tried to solve the problems which were thrust upon him, even to the end.

One of the last acts of his life was to go down town with his youngest son, whose birthday was the 29th of March, to purchase for him his first gold watch, and that watch the son still carries, a precious memento of his father.

Gradually the pains in the head grew less severe, but great weakness followed, and he was compelled to keep to his bed, sinking into a peaceful, painless unconsciousness relieved by an occasional flash of his old vigor. To his pastor, Reverend Dr. William Adams, he expressed his gratitude for the goodness of God to him, but added: "The best is yet to come." He roused himself on the 29th of March, the birthday of his son, kissing him and gazing with pleasure on a drawing sent to the boy by his cousin, Mary Goodrich, pronouncing it excellent.

Shortly before the end pneumonia set in, and one of the attending physicians, tapping on his chest, said "This is the way we doctors telegraph"; and the dying man, with a momentary gleam of the old humor lighting up his fading eyes, whispered, "Very good." These were the last words spoken by him.

From a letter written by one who was present at his bedside to another member of the family I shall quote a few words: "He is fast passing away. It is touching to see him so still, so unconscious of all that is passing, waiting for death. He has suffered much with neuralgia of the head, increased of late by a miserable pamphlet by F. O. J. S. Poor dear man! Strange that they could not leave him in peace in his old age. But now all sorrow is forgotten. He lies like a quiet infant. Heaven is opening to him with its peace and perfect rest. The doctor calls his sickness 'exhaustion of the brain.' He looks very handsome; the light of Heaven seems shining on his beautiful eyes."

On April 1, consciousness returned for a few moments and he recognized his wife and those around him with a smile, but without being able to speak. Then he gradually sank to sleep and on the next day he gently breathed his last.

His faithful and loving friend, James D. Reid, in the Journal of the Telegraph, of which he was editor, paid tribute to his memory in the following touching words:—

"In the ripeness and mellow sunshine of the end of an honored and protracted life Professor Morse, the father of the American Telegraph system, our own beloved friend and father, has gone to his rest. The telegraph, the child of his own brain, has long since whispered to every home in all the civilized world that the great inventor has passed away. Men, as they pass each other on the street, say, with the subdued voice of personal sorrow, 'Morse is dead.' Yet to us he lives. If he is dead it is only to those who did not know him.

"It is not the habit of ardent affection to be garrulous in the excitement of such an occasion as this. It would fain gaze on the dead face in silence. The pen, conscious of its weakness, hesitates in its work of endeavoring to reveal that which the heart can alone interpret in a language sacred to itself, and by tears no eye may ever see. For such reason we, who have so much enjoyed the sweetness of the presence of this venerable man, now so calm in his last sacred sleep, to whom he often came, with his cheerful and gentle ways, as to a son, so confiding of his heart's tenderest thoughts, so free in the expression of his hopes of the life beyond, find difficulty in making the necessary record of his decease. We can only tell what the world has already known by the everywhere present wires, that, on the evening of Tuesday, April 2, Professor Morse, in the beautiful serenity of Christian hope, after a life extended beyond fourscore years, folded his hands upon his breast and bade the earth, and generation, and nation he had honored, farewell."

In the "Evening Post," probably from the pen of his old friend William Cullen Bryant, was the following: —

"The name of Morse will always stand in the foremost rank of the great inventors, each of whom has changed the face of society and given a new direction to the growth of civilization by the application to the arts of one great thought. It will always be read side by side with those of Gutenberg and Schoeffer, or Watt and Fulton. This eminence he fairly earned by one splendid invention. But none who knew the man will be satisfied to let this world-wide and forever growing monument be the sole record of his greatness.

"Had he never thought of the telegraph he would still receive, in death, the highest honors friendship and admiration can offer to distinguished and varied abilities, associated with a noble character. In early life he showed the genius of a truly great artist. In after years he exercised all the powers of a masterly scientific investigator. Throughout his career he was eminent for the loftiness of his aims, for his resolute faith in the strength of truth, for his capacity to endure and to wait, and for his fidelity alike to his convictions and to his friends.

"His intellectual eminence was limited to no one branch of human effort, but, in the judgment of men who knew him best, he had endowments which might have made him, had he not been the chief of inventors, the most powerful of advocates, the boldest and most effective of artists, the most discerning of scientific physicians, or an administrative officer worthy of the highest place and of the best days in American history."

The New York "Herald" said: —

"Morse was, perhaps, the most illustrious American of his age. Looking over the expanse of the ages, we think more earnestly and lovingly of Cadmus, who gave us the alphabet; of Archimedes, who invented the lever; of Euclid, with his demonstrations in geometry; of Faust, who taught us how to print; of Watt, with his development of steam, than of the resonant orators who inflamed the passions of mankind, and the gallant chieftains who led mankind to war. We decorate history with our Napoleons and Wellingtons, but it was better for the world that steam was demonstrated to be an active, manageable force, than that a French Emperor and his army should win the battle of Austerlitz. And when a Napoleon of peace, like the dead Morse, has passed away, and we come to sum up his life, we gladly see that the world is better, society more generous and enlarged, and mankind nearer the ultimate fulfillment of its earthly mission because he lived and did the work that was in him."

The Louisville "Courier-Journal" went even higher in its praise: —

"If it is legitimate to measure a man by the magnitude of his achievements, the greatest man of the nineteenth century is dead. Some days ago the electric current brought us the intelligence that S. F. B. Morse was smitten with paralysis. Since then it has brought us the bulletins of his condition as promptly as if we had been living in the same square, entertaining us with hopes which the mournful sequel has proven to be delusive, for the magic wires have just thrilled with the tidings to all nations that the father of telegraphy has passed to the eternal world. Almost as quietly as the all-seeing eye saw the soul depart from that venerable form, mortal men, thousands of miles distant, are apprised of the same fact by the swift messenger which he won from the unknown — speaking, as it goes around its world-wide circuit, in all the languages of earth.

"Professor Morse took no royal road to this discovery. Indeed it is never a characteristic of genius to seek such roads. He was dependent, necessarily, upon facts and principles brought to light by similar diligent, patient minds which had gone before him. Volta, Galvani,

Morcel, Grove, Faraday, Franklin, and a host of others had laid a basis of laws and theories upon which he humbly and reverently mounted and arranged his great problem for the hoped-for solution. But to him was reserved the sole, undivided glory of discovering the priceless gem, 'richer than all its tribe,' which lay just beneath the surface, and around which so many *savans* had blindly groped.

"He is dead, but his mission was fully completed. It has been no man's fortune to leave behind him a more magnificent legacy to earth, or a more absolute title to a glorious immortality. To the honor of being one of the most distinguished benefactors of the human race, he added the personal and social graces and virtues of a true gentleman and a Christian philosopher. The memory of his private worth will be kept green amid the immortelles of sorrowing friendship for a lifetime only, but his life monument will endure among men as long as the human race exists upon earth."

The funeral services were held on Friday, April 5, at the Madison Square Presbyterian Church. At eleven o'clock the long procession entered the church in the following order:—

Rev. Wm. Adams, D.D., Rev. F. B. Wheeler, D.D.

COFFIN.

PALL-BEARERS.

William Orton,	Cyrus W. Field,
Daniel Huntington,	Charles Butler,
Peter Cooper,	John A. Dix,
Cambridge Livingston,	Ezra Cornell.

The Family.

Governor Hoffman and Staff.
Members of the Legislature.

Directors of the New York, Newfoundland and London Telegraph Company.
Directors of the Western Union Telegraph Company and officers and operators.
Members of the National Academy of Design.
Members of the Evangelical Alliance.
Members of the Chamber of Commerce.
Members of the Association for the Advancement of Science and Art.
Members of the New York Stock Exchange.
Delegations from the Common Councils of New York, Brooklyn and Poughkeepsie and many of the Yale Alumni.
The Legislative Committee: Messrs. James W. Husted, L. Bradford Prince, Samuel J. Tilden, Severn D. Moulton and John Simpson.

The funeral address, delivered by Dr. Adams, was long and eloquent, and near the conclusion he said: —

"To-day we part forever with all that is mortal of that man who has done so much in the cause of Christian civilization. Less than one year ago his fellow-citizens, chiefly telegraphic operators, who loved him as children love a father, raised his statue in Central Park. To-day all we can give him is a grave. That venerable form, that face so saintly in its purity and refinement, we shall see no more. How much we shall miss him in our homes, our churches, in public gatherings, in the streets and in society which he adorned and blessed. But his life has been so useful, so happy and so complete that, for him, nothing remains to be wished. Congratulate the man who, leaving to his family, friends and country a name spotless, untarnished, beloved of nations, to be repeated in foreign tongues and by sparkling seas, has died in the bright and blessed hope of everlasting life.

"Farewell, beloved friend, honored citizen, public benefactor, good and faithful servant!"

The three Morse brothers were united in death as they had been in life. In Greenwood Cemetery a little hill had been purchased by the brothers and divided into three equal portions. On the summit of the hill there now stands a beautiful three-sided monument, and at its base reposes all that is mortal of these three upright men, each surrounded by those whom they had loved on earth, and who have now joined them in their last resting place.

Resolutions of sympathy came to the family from all over the world, and from bodies political, scientific, artistic, and mercantile, and letters of condolence from friends and from strangers.

In the House of Representatives, in Washington, the Honorable S. S. Cox offered a concurrent resolution, declaring that Congress has heard — "with profound regret of the death of Professor Morse, whose distinguished and varied abilities have contributed more than those of any other person to the development and progress of the practical arts, and that his purity of private life, his loftiness of scientific aims, and his resolute faith in truth, render it highly proper that the Representatives and Senators should solemnly testify to his worth and greatness."

This was unanimously agreed to. The Honorable Fernando Wood, after a brief history of the legislation which resulted in the grant of $30,000 to enable Morse to test his invention, added that he was proud to say that his name had been recorded in the affirmative on that historic occasion, and that he was then the only living member of either house who had so voted.

Similar resolutions were passed in the Senate, and a committee was appointed by both houses to arrange for a suitable memorial service, and, on April 9, the following letter was sent to Mrs. Morse by A. S. Solomons, Chairman of the Committee of Arrangements: —

Dear Madam, — Congress and the citizens of Washington purpose holding memorial services in honor of your late respected husband in the Hall of the House of Representatives, on Tuesday evening next, the 16th of April, and have directed me to request that yourself and family become the guests of the nation on that truly solemn occasion. If agreeable, be good enough to inform me when you will likely be here.

The widow was not able to accept this graceful invitation, but members of the family were present.

The Hall was crowded with a representative audience. James G. Blaine, Speaker of the House, presided, assisted by Vice-President Colfax. President Grant and his Cabinet, Judges of the Supreme Court, Governors of States, and other dignitaries were present in person or by proxy. In front of the main gallery an oil portrait of Morse had been placed, and around the frame was inscribed the historic first message: "What hath God wrought."

After the opening prayer by Dr. William Adams, Speaker Blaine said: —

"Less than thirty years ago a man of genius and learning was an earnest petitioner before Congress for a small pecuniary aid that enabled him to test certain occult theories of science which he had laboriously evolved. To-night the representatives of forty million people assemble in their legislative hall to do homage and honor to the name of 'Morse.' Great discoverers and inventors rarely live to witness the full development and perfection of their mighty conceptions, but to him whose death we now mourn, and whose fame we celebrate, it was, in God's good providence, vouchsafed otherwise. The little thread of wire, placed as a timid experiment between the national capital and a neighboring city, grew and lengthened and multiplied with almost the rapidity of the electric current that darted along its iron nerves, until, within his own lifetime, continent was bound unto continent, hemisphere answered through ocean's depths unto hemisphere, and an encircled globe flashed forth his eulogy in the unmatched elements of a grand achievement.

"Charged by the House of Representatives with the agreeable and honorable duty of presiding here, and of announcing the various participants in the exercises of the evening, I welcome to this hall those who join with us in this expressive tribute to the memory and to the merit of a great man."

After Mr. Blaine had concluded his remarks the exercises were conducted as follows: —

Resolutions by the Honorable C. C. Cox, M.D., of Washington, D.C.

Address by the Honorable J. W. Patterson, of New Hampshire.

Address by the Honorable Fernando Wood, of New York.

Vocal music by the Choral Society of Washington.

Address by the Honorable J. A. Garfield, of Ohio.

Address by the Honorable S. S. Cox, of New York.

Address by the Honorable N. P. Banks, of Massachusetts.

Vocal music by the Choral Society of Washington.

Benediction by the Reverend Dr. Wheeler of Poughkeepsie.

Once again the invention which made him famous paid marvellous tribute to the man of science. While less than a year before, joyous messages of congratulation had flashed over the wires from the four quarters of the globe, to greet the living inventor, now came words of sorrow and condolence from Europe, Asia, Africa, and America mourning that inventor dead, and again were they read to a wondering audience by that other man of indomitable perseverance, Cyrus W. Field.

On the same evening memorial services were held in Faneuil Hall, Boston, at which the mayor of the city presided, and addresses were made by Josiah Quincy, Professor E. N. Horsford, the Honorable Richard H. Dana, and others.

Other cities all over the country, and in foreign lands, held commemorative services, and every telegraph office in the country was draped in mourning, in sad remembrance of him whom all delighted to call "Father."

Mr. Prime, in his closing review of Morse's character, uses the following words: —

"It is not given to mortals to leave a perfect example for the admiration and imitation of posterity, but it is safe to say that the life and character of few men, whose history is left on record, afford less opportunity for criticism than is found in the conspicuous career of the Inventor of the Telegraph.

"Having followed him step by step from the birth to the grave, in public, social and private relations; in struggles with poverty, enemies and wrongs; in courts of law, the press and halls of science; having seen him tempted, assailed, defeated, and again in victory, honor and renown; having read thousands of his private letters, his essays and pamphlets, and volumes in which his claims are canvassed, his merits discussed and his character reviewed; having had access to his most private papers and confidential correspondence, in which all that is most secret and sacred in the life of man is hid — it is right to say that, in this mass of testimony by friends and foes, there is not a line that requires to be erased or changed to preserve the lustre of his name. . . .

"It was the device and purpose of those who sought to rob him of his honors and his rights to depreciate his intellectual ability and his scientific attainments. But among all the men of science and of learning in the law, there was not one who was a match for him when he gave his mind to a subject which required his perfect mastery. . . .

"He drew up the brief with his own hand for one of the distinguished counsel in a great lawsuit involving his patent rights, and his lawyer said it was the argument that carried conviction to every unprejudiced mind.

"Such was the versatility and variety of his mental endowments that he would have been great in any department of human pursuits. His wonderful rapidity of thought was associated with patient, plodding perseverance, a combination rare but mightily effective.

He leaped to a possible conclusion, and then slowly developed the successive steps by which the end was gained and the result made secure. He covered thousands of pages with his pencil notes, annotated large and numerous volumes, filled huge folios with valuable excerpts from newspapers, illustrated processes of thought with diagrams, and was thus fortified and enriched with stores of knowledge and masses of facts, so digested, combined and arranged, that he had them at his easy command to defend the past or to help him onward to fresh conquests in the fields of truth. Yet such was his modesty and reticence in regard to himself that none outside of his household were aware of his resources, and his attainments were only known when displayed in self-defense. Then they never failed to be ample for the occasion, as every opponent had reason to remember.

"Yet he was gentle as he was great. Many thought him weak because he was simple, childlike and unworldly. Often he suffered wrong rather than resist, and this disposition to yield was frequently his loss. The firmness, tenacity and perseverance with which he fought his foes were the fruits of his integrity, principle and profound convictions of right and duty. . . . His nature was a rare combination of solid intellect and delicate sensibility. Thoughtful, sober and quiet, he readily entered into the enjoyments of domestic and social life, indulging in sallies of humor, and readily appreciating and greatly enjoying the wit of others. Dignified in his intercourse with men, courteous and affable with the gentler sex, he was a good husband, a judicious father, a generous and faithful friend.

"He had the misfortune to incur the hostility of men

who would deprive him of his merit and the reward of his labors. But this is the common fate of great inventors. He lived until his rights were vindicated by every tribunal to which they could be referred, and acknowledged by all civilized nations, and he died leaving to his children a spotless and illustrious name, and to his country the honor of having given birth to the only Electro-Magnetic Recording Telegraph whose line is gone out through all the earth, and its words to the end of the world."

And now my pleasant task is ended. After the lapse of so many years it has been possible for me to introduce much more evidence of a personal nature, to reveal the character of those with whom Morse had to contend, than would have been discreet or judicious during the lifetime of some of the actors in the drama. Many attempts have been made since the death of the inventor to minimize his fame, and to exalt others at his expense, but, while these attempts have seemed to triumph for a time, while they may have influenced a few minds and caused erroneous attributions to be made in some publications, their effect is ephemeral, for "Truth is mighty and will prevail," and the more carefully and exhaustively this complicated subject is studied, the more apparent will it be that Morse never claimed more than was his due; that his upright, truth-loving character, as revealed in his intimate correspondence and in the testimony of his contemporaries, forbade his ever stooping to deceit or wilful appropriation of the ideas of others.

A summary, in as few words as possible, of what

Morse actually invented or discovered may be, at this point, appropriate.

In 1832, he conceived the idea of a true electric telegraph — a writing at a distance by means of the electro-magnet. The use of the electro-magnet for this purpose was original with him; it was entirely different from any form of telegraph devised by others, and he was not aware, at the time, that any other person had even combined the words "electric" and "telegraph."

The mechanism to produce the desired result, roughly drawn in the 1832 sketch-book, was elaborated and made by Morse alone, and produced actual results in 1835, 1836, and 1837. Still further perfected by him, with the legitimate assistance of others, it became the universal telegraph of to-day, holding its own and successfully contending with all other plans of telegraphs devised by others.

He devised and perfected the dot-and-dash alphabet.

In 1836, he discovered the principle of the relay.

In 1838, he received a French patent for a system of railway telegraph, which also embodies the principle of the police and fire-alarm telegraph. At the same time he suggested a practical form of military telegraph.

In 1842, he laid the first subaqueous cable.

In 1842, he discovered, with Dr. Fisher, the principle of duplex telegraphy, and he was also the first to experiment with wireless telegraphy.

In addition to his electrical inventions and discoveries he was the first to experiment with the Daguerreotype in America, and, with Professor Draper, was the first in the world to take portraits by this means, Daguerre himself not thinking it possible.

The verdict of the world, as pronounced at the time of his death, has been strengthened with the lapse of years. He was one of the first to be immortalized in the Hall of Fame. His name, like those of Volta, Galvani, Ampere, and others, has been incorporated into everyday speech, and is now used to symbolize the language of that simple but marvellous invention which brings the whole world into intimate touch.

THE END

INDEX

The Riverside Press
CAMBRIDGE . MASSACHUSETTS
U . S . A

Printed in the United States
By Bookmasters